Nanostructure Science and Technology

Series Editor: David J. Lockwood, FRSC
National Research Council of Canada
Ottawa, Ontario, Canada

For other titles published in this series, go to
www.springer.com/series/6331

Tetsuya Osaka · Madhav Datta
Yosi Shacham-Diamand
Editors

Electrochemical Nanotechnologies

Springer

Editors
Tetsuya Osaka
Faculty of Science and Engineering
Waseda University
Okubo
Shinjuku-ku, Tokyo
Japan
osakatets@waseda.jp

Madhav Datta
Cooligy Precision Cooling
Emerson Network Power
800 Maude Avenue
Mountain View, CA
USA
madhav.datta@emerson.com

Yosi Shacham-Diamand
Faculty of Engineering
Tel Aviv University
Israel
YosiSh@tauex.tau.ac.il

ISBN 978-1-4419-1423-1 e-ISBN 978-1-4419-1424-8
DOI 10.1007/978-1-4419-1424-8
Springer New York Dordrecht Heidelberg London

Library of Congress Control Number: 2009942264

Printed on acid-free paper

Springer is part of Springer Science+Business Media (www.springer.com)

Foreword to *Electrochemical Nanotechnology*

The advent of nanotechnology has prompted an unprecedented revolution in many fields of science and technology. Outstanding progress in medicine, biology, and energy, just to cite a few, became feasible only by passing from bulk to nano dimensions. Thus, we can safely state that nanotechnology has led to the achievement of a fundamental step forwards in the understanding of life, in the benefit of human kind and in the renewal of our energy scenario.

Books that critically review these advances are obviously of valuable importance to the scientific community. This particularly holds for the present monograph, which is focused on a special and key branch of nanotechnology, namely electrochemical nanotechnology. Indeed, electrochemistry has particularly benefited from nanotechnology. Reduction to the nano size has allowed researchers to achieve results otherwise impossible to imagine. By reducing the size of materials, e.g., battery materials, to the nanodimension, jumps in power density and in cycle life have been obtained, finally upgrading devices which were limited to modest uses to highly sophisticated applications. Nanotechnology has also helped to reach unheard of goals in energy conversion by improving the efficiency of solar cells, in medicine by allowing the development of drug dispensers, and in electronics, by opening the route to nanofabrication techniques.

Thus, there is no doubt that this book is timely and necessary; and one could not imagine better editors for its assemblage and completion than the three scientists involved. The great experience and authority of Professor Osaka, Professor Shacham-Diamand, and Dr. Datta is a guarantee of excellence for this book. They selected key topics in the field and called for outstanding co-authors to complete their description and evaluation. The reader will in fact be very pleased by the clear yet rigorous writing, and quite interested by the description and evaluation, as well as by the scientific and technological implications of the topics treated here.

The content of this book reflects the experience of the Editors who decided to focus on those areas that are expected to play a key role in the advancing science and technology. The interested reader will learn from this book how to use nanotechnology to fabricate and evaluate electrochemical energy conversion and storage devices, such as lithium batteries, fuel cells and supercapacitors; magnetic storage devices, i.e., soft and alloy films; bio-chips for sensing genetic changes and water toxicity; and MEMS/NEMS devices for a variety of important applications.

Really, a book that nobody involved in the field or interested in entering it should miss.

Bruno Scrosati
Full Professor of Electrochemistry
University of Rome
Sapienza, Italy

Preface

Nanotechnology is now the foundation for advances in many fields of science and technology. In this monograph, the term "Electrochemical Nanotechnology (ECNT)" is defined as *Nanoprocessing by Means of Electrochemical Techniques*. The use of ECNT is increasing in various electronics applications, and in some cases, it is extended for applications in other fields. Understanding processes for the fabrication of nano-sized films and structures is essential for the development of new precision nanofabrication techniques. This introductory book reviews selected topics of the application of ECNT with the aim of understanding the wider applicability of ECNT in evolving nanoindustries and thus facilitating the creation of new applications in the future.

The idea of writing this book evolved as a result of a series of symposia. The first symposium was on "Electrochemical Microsystem Technologies (EMT)" held in 1996 in Düsseldorf-Grevenbroich, Germany, followed by the second held in 1998 at Waseda University, Tokyo, the third held in 2000 in Garmisch-Partenkirchen, Germany, and the fourth symposium in Düsseldorf, Germany held in 2002. Because of the evolving importance of nanotechnology, the title of the fifth symposium at Waseda University, Tokyo in 2004 was changed to "International Symposium on Electrochemical Micro & Nanosystem Technologies (EMNT)," which was followed by the sixth symposium held in Bonn in 2006. The 2008 symposium was held in Israel. During the period of more than a decade, micro/nano technologies have grown not only in the electrochemical field but also in several interdisciplinary areas. These advances have impacted microelectronics, sensors, materials science, and corrosion, and new fields of research have been generated to promote interaction between biology, medicine, and microelectronics. Such interactions have led to novel approaches to miniaturization with an increased ability to fabricate structures with high lateral and vertical resolution.

The content of this book is focused mainly on research activities in nanotechnology at Waseda University. Accordingly, a majority of contributors were selected from the members of Waseda University, although experts from other organizations also contributed chapters on specific subjects of their expertise. As indicated earlier, this volume provides an overview of nanotechnology applications in selected high technology areas with particular emphasis on the near-term and future advances in these fields. The chapters in the book are classified under four different headings: Nanotechnology for energy devices, Nanotechnology for magnetic storage devices, Nanotechnology for bio-chip applications, and Nanotechnology for MEMS/Packaging.

We express our thanks to all authors, referees and advisors for their help and support in making the publication of this book possible.

Shinjuku-ku, Tokyo — Prof. Tetsuya Osaka
Mountain View, CA — Dr. Madhav Datta
Tel Aviv, Israel — Prof. Yosi Shacham-Diamand

Contents

Contributors

Dr. Takahiro Arakawa
School of Science and Engineering, Waseda University, 3-4-1 Okubo, Shinjuku-ku, Tokyo 169-8555, Japan

Dr. Madhav Datta
Cooligy Precision Cooling, Emerson Network Power, 800 Maude Avenue, Mountain View, CA 94043, USA
madhav.datta@emerson.com

Dr. Kaoru Dokko
Graduate School of Urban Environmental Science, Tokyo Metropolitan University, 1-1 Minami-Ohsawa, Hachioji, Tokyo 192-0397, Japan

Dr. Thomas George
ViaLogy Corporation. Altadena, CA 91001, USA

Mr. Takuma Hachisu
Faculty of Science and Engineering, Waseda University, 3-4-1 Okubo, Shinjuku-ku, Tokyo 169-8555, Japan

Dr. Carlos M. Hangarter
Department of Chemical and Environmental Engineering, University of California-Riverside, Riverside, CA 92521, USA

Prof. Naoki Honda
Akita research institute of advanced technology (AIT), 4-21 Sanuki, Araya, Akita 010-1623, Japan
Faulty of Engineering, Tohoku Institute of Technology, 35-1 Yagiyama, Kasumicho, Taihakuku, Sendai, Miyagi 982-8577, Japan
n_honda@ait.pre

Dr. Masanori Ishizuka
School of Science and Engineering, Waseda University, 3-4-1 Okubo, Shinjuku-ku, Tokyo 169-8555, Japan

Prof. Kiyoshi Kanamura
Graduate School of Urban Environmental Science, Tokyo Metropolitan University, 1-1 Minami-Ohsawa, Hachioji, Tokyo 192-0397, Japan
kanamura-kiyoshi@c.metro-u.ac.jp

Prof. Tadashi Matsunaga
Department of Biotechnology, Tokyo University of Agriculture and Technology, 2-24-16 Nakamachi, Koganei-city, Tokyo 184-8588, Japan
tmatsuna@cc.tuat.ac.jp

Prof. Jun Mizuno
School of Science and Engineering, Waseda University, 3-4-1 Okubo, Shinjuku-ku, Tokyo 169-8555, Japan

Prof. Toshiyuki Momma
Waseda Institute for Advanced Study, Waseda University, 3-4-1 Okubo, Shinjuku-ku, Tokyo 169-8555, Japan
momma@waseda.jp

Dr. Hitomi Mukaibo
Faculty of Science and Engineering, Waseda University, 3-4-1 Okubo, Shinjuku-ku, Tokyo 169-8555, Japan

Dr. Hirokazu Munakata
Graduate School of Urban Environmental Science, Tokyo Metropolitan University, 1-1 Minami-Ohsawa, Hachioji, Tokyo 192-0397, Japan

Prof. Nosang V. Myung
Department of Chemical and Environmental Engineering, University of California-Riverside, Riverside, CA 92521, USA
myung@engr.ucr.edu

Dr. Daisuke Niwa
Consolidated Research Institute for Advanced Science and Medical Care, Waseda University 513 Wasedatsurumaki-chi, Shinjuku-ku, Tokyo 162-0041, Japan

Prof. Tetsuya Osaka
Faculty of Science and Engineering, Waseda University, 3-4-1 Okubo, Shinjuku-ku, Tokyo 169-8555, Japan
osakatets@waseda.jp

Dr. Kazuhiro Ouchi
Akita research institute of advanced technology (AIT), 4-21 Sanuki, Araya, Akita 010-1623, Japan
ouchi@ait.pref.akita.jp

Dr. Jong-Eun Park
Faculty of Science and Engineering, Waseda University, 3-4-1 Okubo, Shinjuku-ku, Tokyo 169-8555, Japan

Dr. R. Popovtzer
School of Engineering, Bar Ilan Univerisity, Ramat Gan 59700, Israel
rachelap@eng.biu.ac.il

Prof. Y. Rishpon
Faculty of Life Sciences, Tel Aviv University, Tel Aviv 69978, Israel

Dr. Junji Sasano
Consolidated Research Institute for Advanced Science and Medical Care, Waseda University513 Wasedatsurumaki-chi, Shinjuku-ku, Tokyo 162-0041, Japan

Dr. Hironobu Sato
School of Science and Engineering, Waseda University, 3-4-1 Okubo, Shinjuku-ku, Tokyo 169-8555, Japan

Prof. Y. Shacham-Diamand
Faculty of Engineering, Tel Aviv University, Tel Aviv 69978, Israel
YosiSh@tauex.tau.ac.il

Dr. Takahiro Shimizu
Faculty of Science and Engineering, Waseda University, 3-4-1 Okubo, Shinjuku-ku, Tokyo 169-8555, Japan

Prof. Shuichi Shoji
School of Science and Engineering, Waseda University, 3-4-1 Okubo, Shinjuku-ku, Tokyo 169-8555, Japan

Dr. Atsushi Sugiyama
Waseda Institute for Advanced Study, 1-6-1 Nishiwaseda, Shinjuku-ku, 169-8050 Tokyo, Japan
sugiyama@waseda.jp

Prof. Tsuyoshi Tanaka
Department of Biotechnology, Tokyo University of Agriculture and Technology, 2-24-16 Nakamachi, Koganei city, Tokyo 184-8588, Japan
tsuyo@cc.tuat.ac.jp

Prof. Kiyoshi Yamakawa
Akita research institute of advanced technology (AIT), 4-21 Sanuki, Araya, Akita 010-1623, Japan
yamakawa@ait.pref.akita.jp

Dr. Tokihiko Yokoshima
High Density Interconnection Group, Nanoelectronics Research Institute (NeRI), National Institute of Advanced Industrial Science and Technology (AIST), Tsukuba Central 2, 1-1-1 Umezono, Tsukuba, Ibaraki 305-8568, Japan
t.yokoshima@aoni.waseda.jp

Dr. Masahiro Yoshino
Faculty of Science and Engineering, Waseda University, 3-4-1 Okubo, Shinjuku-ku, Tokyo 169-8555, Japan

Chapter 1
Introduction

Tetsuya Osaka

Electrochemical nanotechnology utilizes electrochemical processes and techniques. We have been publishing several books in series, on electrochemical nanotechnologies. [1–6] This book deals mainly with applications of electrochemical nanotechnology in the fields of magnetic recording, ULSI interconnection, energy devices, bio-analysis, and bio-electrochemistry.

Nanotechnologies, in general, are concerned not only with downsizing and miniaturization of products, but also with ideas for creating new systems and new materials involving the scale of nanometers. A typical example is the invention of GMR, for which Fert and Gruenberg received the 2007 Nobel Prize in Physics. The system is based exactly on the nano-order combination of magnetic materials with thin films.

Figure 1.1 shows the change in areal density of magnetic recording device with time, in which the areal density of hard disk drive (HDD) is plotted against the calendar year. The areal density of 1 terabit/in^2 was thought to be the achievable limit in view of the physical limitation of super paramagnetism. However, recent assessment assumes the possibility of achieving higher densities on new systems of DTR (discrete track recording), BPR (bit patterned recording), and TAMR (thermal-assisted magnetic recording). This example shows that nano-order arrangements and relevant ideas are becoming increasingly important in recent years.

Needless to say, highly functional magnetic film is a key component of future magnetic recording devices, and the development of new magnetic films will contribute to the fabrication of new magnetic devices, as has been demonstrated in the past. For example, the growth rate of the areal density in the past was controlled mainly by the development of head core material of soft magnetic film with a high-saturation magnetic flux density. To meet the demand for the GHz response, it was necessary for the soft magnetic film to possess properties of low magnetostriction, high electrical resistivity, high thermal stability, low film stress, and high corrosion resistance.

T. Osaka (✉)
Faculty of Science and Engineering, Waseda University, 3-4-1 Okubo, Shinjuku-ku, Tokyo, 169-8555, Japan
e-mail: osakatets@waseda.jp

T. Osaka et al. (eds.), *Electrochemical Nanotechnologies,*
Nanostructure Science and Technology, DOI 10.1007/978-1-4419-1424-8_1,

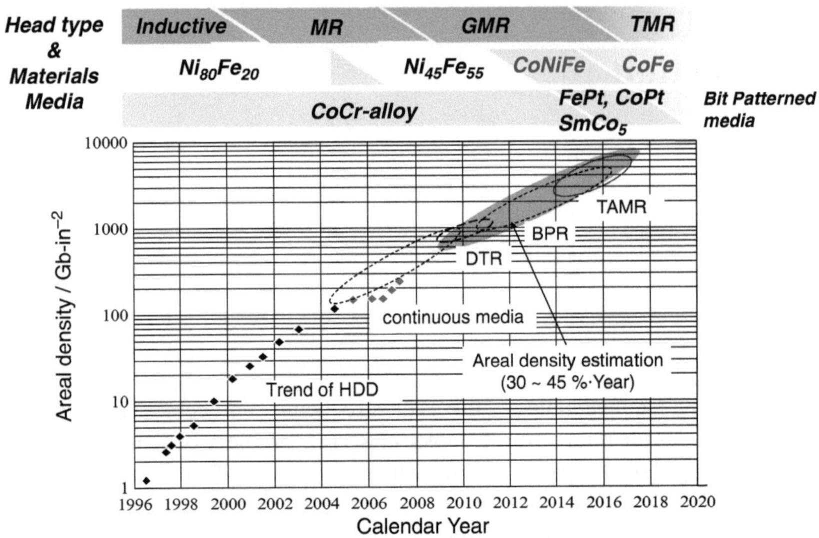

Fig. 1.1 Trend of increasing areal density of hard-disk drives (HDD) with time

One approach to meet these requirements was to prepare the film with the structure composed of nanocrystal domains dispersed in an amorphous matrix. This structure is called "nanocrystalline soft/hard magnetic material". Although the preparation method using a heating-quenching technique was reported, it was not suitable as a magnetic head manufacturing process because of the damage caused in other head components. We proposed an electrochemical method consisting of a new concept that is different from the thermal method. The electrodeposited CoNiFe film led to the development of an ultra small GMR head. In this section, we first introduce the new concept based on electrochemical nanotechnology, which was developed through research on electro- and electroless-deposited CoNiFe alloy films. Subsequently we introduce its application to the fabrication of cusp-field single-pole type (CF-SPT) head, which is considered to be promising as a next generation magnetic head, and recent research on the enhanced concept for the magnetic domain control of films.

The energy density of portable batteries will be considered next. The changing trend of energy devices is illustrated in Fig. 1.2 with portable battery energy devices in mind. The idea of electrochemical nanotechnology has been adopted for developing energy systems for portable electric devices to enhance their energy density. Currently, Li ion battery is used for high-end electric devices such as cellular phones, portable computers, and PDAs. The demand for higher capacity and higher power for the energy systems comes from the requirement for future enhanced IT technologies and for the welfare of our society. While the energy density of a battery pack is limited primarily by the materials used in the battery, the nanometer-scale design of the electrode leads to an improved capacity and power output of the battery. The conventional electrode materials are prepared at a high temperature.

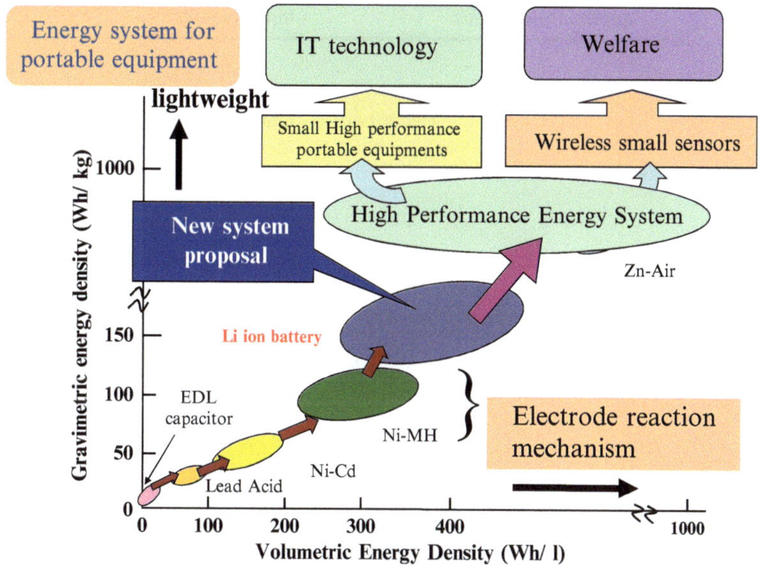

Fig. 1.2 Trend of energy devices

The homogeneity in structure and composition of those materials are advantageous for mass production of the batteries. On the other hand, electrochemical methods for the preparation of new electrode materials permit the control of the phase, the amount of impurities/additives, and the nanometer-scale structure of the product. Controlling those aspects of materials led to the development of Sn–Sx, Ni–Sn, and mesoporous Sn as new candidates for the electrode material of future Li batteries. With these technologies, it is anticipated that further progress will be made in the performance of batteries in the future. Electrochemical nanotechnology is also powerful for developing fuel cells. New materials and new processes for synthesizing catalysts by wet procedures based on electrochemical nanotechnology have been proposed. Furthermore, with MEMS technologies, the construction and operation of miniaturized fuel cells have been demonstrated.

Finally, we describe results of studies of bioanalysis and bioelectrochemistry based on nanotechnology. For example, magnetic nanoparticles offer some attractive possibilities for applications in the fields of bioanalysis and biomedicine as shown in Fig. 1.3 [7]. Magnetic nanoparticles can be used as a magnetic marker for the detection of biorelevant materials such as a cell (10–100 μm), a virus (20–450 nm), a protein (5–50 nm), or a gene (2 nm wide and 10–100 nm long). These materials are comparable in size to magnetic nanoparticles. Also, iron-oxide nanoparticles of magnetite (Fe_3O_4) and maghemite (γ-Fe_2O_3), which are reported to be nontoxic to the human body, are applied as a magnetic carrier for drug delivery systems (DDS), as a heating element in hypothermia, and as a contrasting agent for magnetic resonance imaging (MRI). In these cases, magnetic properties of iron-oxide nanoparticles, which are very different from those of the bulk material, are utilized for

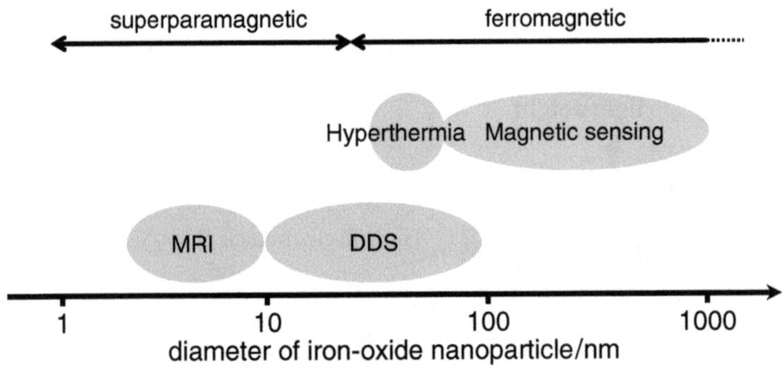

Fig. 1.3 Bioapplications of iron-oxide nanoparticles

biomedical treatment. Thus, special magnetic properties in addition to the biologically advantageous size of magnetic nanoparticles are utilized in their bioanalytical and medical applications such as bioassays, biomedicine, ultrasensitive biodetection, and bioimaging. This book introduces the present status of nanotechnologies for bio-chip applications, electrochemical field effect transistor (FET) micro pH and biosensors, electrochemical and magnetic technologies for bioapplications, and nano bioelectrochemical interfacing.

References

1. Masuko N, Osaka T, Fukunaka Y (eds) (1993) New trends and approaches in electrochemical technology. Kodansha and VCH, Tokyo and Weinheim
2. Masuko N, Osaka T, Ito Y (1996) Electrochemical technology: innovation and new development. Kodansha and Gordon & Breach, Tokyo and Amsterdam
3. Osaka T, Datta M (eds) (2000), New trends in electrochemical technology, Energy storage systems for electronics, Vol. 1. Gordon & Breach, Amsterdam
4. Schultze JW, Osaka T, Datta M (eds) (2002) New trends in electrochemical technology, Electrochemical microsystem technologies, vol 2. Taylor & Francis, London and New York
5. Datta M, Osaka T, Schultze JW (eds) (2005) New trends in electrochemical technology, Microelectronic packaging, vol 3. CRC Press, Boca Raton, London, New York and Washington, D. C.
6. Shacham-Diamand Y, Osaka T, Datta M, Ohba T (eds) (2009), Advanced nanoscale ULSI interconnects: fundamentals and applications. Springer, New York
7. Osaka T, Matsunaga T, Nakanishi T, Arakaki A, Niwa D, Iida H (2006) Syntheis of magnetic nanoparticles and their application to bioassays. Anal Bioanal Chem 384:593–600

Part I
Nanotechnology for Energy Devices

Chapter 2
Nanotechnologies for Li Batteries

Hitomi Mukaibo and Toshiyuki Momma

2.1 Introduction

Batteries store chemical energy, which is converted into electric energy by electrochemical reactions. Those that cannot be used once the electric energy is totally discharged are called primary cell. Batteries that can induce chemical reaction by charging electric energy to reactivate the electrode material and be used many times are called secondary batteries.

Due to the rapid growth of electrical and mechanical integration technologies such as wireless telecommunications, emerging integrated optoelectronic circuits, and rapidly growing microelectromechanical systems (MEMS), secondary batteries that are mounted on semiconductor tips are attracting attention throughout the world. In Table 2.1, conventional batteries have been classified according to their application. The expected power of the microbatteries is between 10^2 and $10^3\,\mu W\,cm^{-2}$ in the temperature range −20 to 80°C, and a capacity of up to $10^3\,\mu Ah\,cm^{-2}$ with a required operating voltage range of 2–3 V . The number of cycles depends on applications: primary cell may be sufficient for smart cards whereas applications in aerospace require more than 10^4 cycles. The battery proximity to microelectronic components is required to prevent any liquid leakage. The thickness of the battery should not exceed 0.3–3 mm including packaging. The battery surface area depends on the power requirement and may vary from 10^{-2} to 20 cm^2 [1].

Within the different types of secondary batteries such as the lead–acid battery, cadmium–nickel battery, and nickel–hydroxide battery, the Li battery has outstanding features of high capacity (realizes smaller size and lighter weight of the battery itself), high operating voltage (allows large power with few batteries; realizing smaller size and lighter weight of the appliances), no memory effect (allows intermittent charging), higher safety (stable against overcharging and high temperature),

H. Mukaibo and T. Momma (✉)
Faculty of Science and Engineering, Waseda University, 3-4-1 Okubo, Shinjuku-ku, Tokyo, 169-8555, Japan
e-mail: momma@waseda.jp

T. Osaka et al. (eds.), *Electrochemical Nanotechnologies*,
Nanostructure Science and Technology, DOI 10.1007/978-1-4419-1424-8_2,

Table 2.1 Classification of batteries according to application. [2]

Applications	Capacity/mAh	Type
Sensors, CMOS memories, credit cards, implantable medical devices, MEMS	200	Microbatteries
Electric watches, calculators, medical devices	200	Miniature batteries
Power tools, toys, radio, portable TVs, flash lights	2,000	Portable batteries
Cars, tractors, trucks, electrical vehicles	50,000	SLI batteries

and long cycle life. The application of this battery to the microbattery system is one of the key developments for today's market requirements.

Solid-state batteries which have been studied for about forty years are attractive electrochemical systems because these batteries avoid the problems involved in classical batteries with liquid electrolyte: i.e., ease of utilization, resistance to shocks and vibrations, absence of possible pollution due to liquid electrolyte, thermal stability, absence of self discharge, and possible miniaturization. The use of conventional thin-film technology for the formation of solid-state microbatteries offers various advantages [3]: (1) thin-film technology is widely used in advanced microelectronics, (2) lower electrical resistance in the transverse direction of the layers can be achieved by the thinning of its thickness, (3) the interface resistance could be reduced by improved electrode–electrolyte interface contact achieved by the thin-film technology, which provides clean surface of the compound and very good adhesion between layers, (4) possible moisture problems are avoided by the deposition of fast ionic conductor in vacuum chambers, (5) convenient substrate materials such as silicon wafers can be used and (6) encapsulation of the battery can be achieved by the deposition of an insulating layer on the top of the device.

Naturally, such conditions require different criteria for the thin-film microbatteries cell design compared to the larger batteries systems. This section will briefly review the achievements reported on the design and fabrication of Li microbatteries.

2.2 Structure of Microbatteries

A chemical battery consists of a cathode and an anode electrode, separated by an electrolyte. The schematic figure of the construction of a solid-state micro battery is represented in Fig. 2.1. Once these electrodes are connected externally, the discharge of the battery occurs and the chemical reactions proceed simultaneously at both electrodes. Electrons flow through the external circuit, and ions transfer through the electrolyte solution counterbalancing the charge deviation. The charge transfer is schematically represented in Fig. 2.2.

The amount of electrical energy produced from this reaction is expressed either by per unit of weight (Wh kg^{-1}) or by per unit of volume (Wh l^{-1}). This is a function

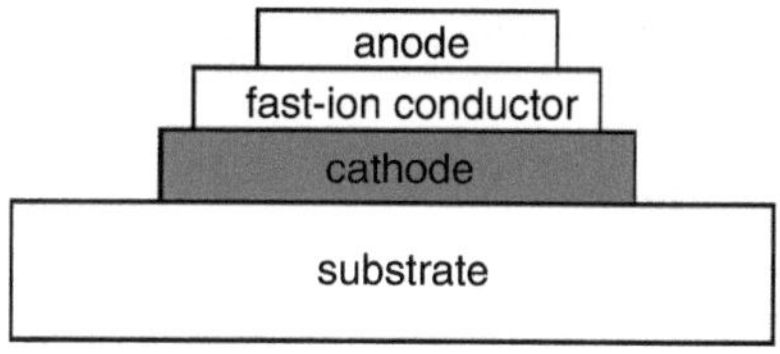

Fig. 2.1 Schematic representation of the construction of a solid-state microbattery. Reprinted from [4], with permission from Elsevier

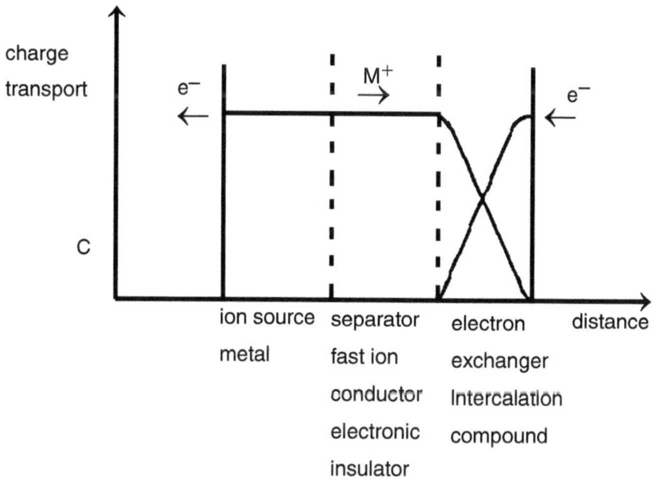

Fig. 2.2 Charge transfer in a solid-state battery. Reprinted from [4], with permission from Elsevier

of the cell voltage (V) and capacity (Ah kg^{-1} or Ah l^{-1}), both of which are linked directly to the chemistry of the system.

$$[\text{Wh kg}^{-1}] = [\text{V}] \times [\text{Ah kg}^{-1}] \tag{2.1}$$

The voltage of the battery cell is defined as the difference between the reaction potential of cathode and anode electrode. The capacity is the amount of electrons stored in or released from the electrodes.

2.3 Fabrication Technologies for Microbatteries

Thin film deposition technologies are widely used for the fabrication of microbatteries. They can be divided into two main categories: Physical Vapor Deposition (PVD) and Chemical Vapor Deposition (CVD).

2.3.1 Physical Formation Processes of Thin Films [2, 5]

2.3.1.1 Evaporative Methods

To control the properties of the fabricated thin films, it is necessary to operate under vacuum to minimize the interaction between residual gases and the surface of growing films. The thin films can be obtained in the crystalline or amorphous state by controlling the concentration of the vapor. The process of the film formation by this method involves the following three stages; (1) evaporation or sublimation of the charge to form vapor, (2) transfer of atoms or molecules from evaporation source to the substrate, and (3) condensation of vapor on the substrate.

For evaporative or sublimation process, correct selection of evaporation method, the evaporation source, and the evaporation temperature is required to surmount the attractive intermolecular forces existing within the starting material. The parameters depend primarily on the materials used and the film purity required. Indirect resistance heating, flash evaporation , and electron beam heating techniques are used for this purpose.

Flash-Evaporation Method

Flash evaporation is another technique for the deposition of films whose constituents have different vapor pressures. Small quantities of the constituents in the desired ratio are continuously dropped at a predetermined rate from a vibration feeder into a sufficiently heated crucible or boat so that they are evaporated instantly. The temperature must be sufficiently high to evaporate the less volatile material. Figure 2.3 shows the flash-evaporation apparatus used to grow the different layers that compose a microbattery: metallic contacts, cathode, electrolyte, lithium anode deposited under vacuum of 10–100 mPa pressure. This system has two vacuum chambers. Vacuum chamber A is devoted to evaporation of In-Se films while lithium and glass films are formed in chamber B. Two interlock mechanisms are used: the first one transfers the grown In-Se film to the chamber B and the second interlock system is utilized to carry lithium pieces from an inert-gas glove box to the evaporation boat.

Reactive Evaporation Method

This process involves a chemical reaction between the evaporated constituent and the residual gas atmosphere. The technology of reactive evaporation is applied in all cases where direct evaporation of a chemical compound is not possible because of thermal dissociation or very low vapor pressure. Stoichiometric oxide films can be obtained with a relatively high controlled oxygen partial pressure and a slow metal atom condensation rate.

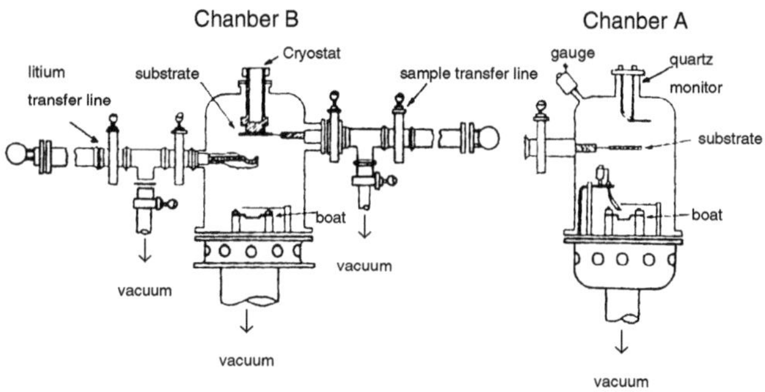

Fig. 2.3 Schematic representation of a flash evaporation system used to fabricate microbatteries. Reprinted from [3], Copyright Elsevier

Condensation and Film Formation

In condensation there exists a relation between a critical deposition rate and the substrate temperature.

Even at room temperature, most deposited films are in a nonequilibrium state and contain vacancies, dislocations, stacking faults, and grain boundaries. The movement of atoms in and on the surface layers is the key factor to approach equilibrium state of the deposited film. The most important parameter controlling the mobility of atoms in a solid film is diffusion. Therefore a better-ordered solid film is formed when the condensation process occurs closer to the melting point of the film material. This can be achieved, for example, by increasing the substrate temperature. In addition to influencing surface mobility and ordering processes, the substrate temperature will also affect the grain size. In contrast, films deposited on cooled substrate favor the formation of amorphous structures.

2.3.1.2 Sputtering Methods

An important advantage of the sputtering technique over evaporative methods is that high quality films with good adhesion can be obtained. However, it has a drawback of low (10–100 nm min^{-1}) deposition rate.

This process takes place in a vacuum chamber where the starting pressure is 10^{-4} Pa or lower to prevent contamination of the deposited films. The working pressure is then achieved with the working gas (generally argon). The sputtering process is the bombardment of the negative target by the accelerated positive ions produced upon gas discharge. They cause ejection mainly of neutral particles by pulse transfer phenomena, and the ejected particles fly through the working gas and condense on the substrate. Figure 2.4 shows a schematic drawing of a general set-up for the sputtering method. Two electrodes are installed in the chamber. One of them, the

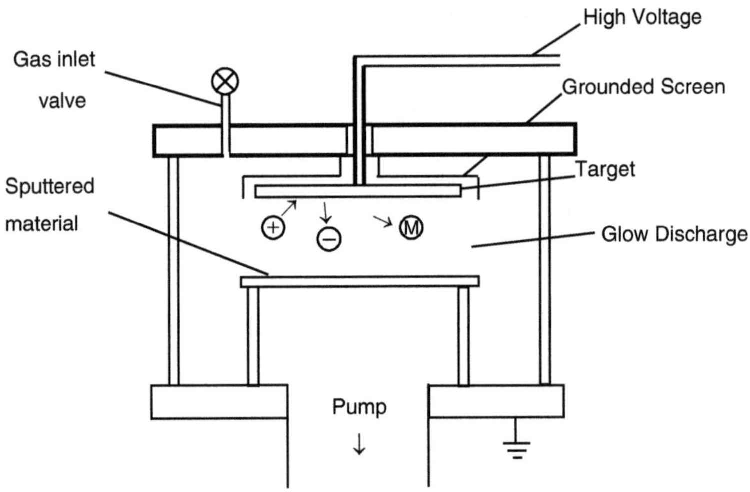

Fig. 2.4 Simple diode-type sputtering system. Reprinted from [5], with permission from Elsevier

target, serves as the source of material for the films to be produced and is at a negative potential. A substrate holder is situated opposite the target which can be earthed or applied to a floating potential. This holder can be heated or cooled, according to the desired property of the deposited thin film.

It is possible to use a high frequency electric field instead of a direct current (d.c.) field explained above. This process is called the radio-frequency (r.f.) sputtering. The electrons oscillate in the plasma and gain sufficient energy for the higher ionizing collisions. This process leads to the formation of a pulsating negative charge on the target surface, allowing an efficient ion bombardment only on the target electrode. This method is mainly used for depositing thin films from insulating targets such as glass.

Another way to increase ionization efficiency of the electrons during gas discharge is to apply a magnetic field perpendicular to the target surface, where the electrons will be forced in spiral paths parallel to the target surface. This magnetron set-up can be used for both d.c. and r.f. discharges and gives very high sputtering yield.

2.3.1.3 Ion Plating Methods

Ion plating method is a combination of the evaporation process and the sputtering technique. Although the apparatus is similar to that used in evaporation, the substrate holder is electrically insulated and the substrates are biased negatively so that an electric field exists between the target source and the substrate. Under adequate gas pressure and voltage gradient, a glow discharge is generated, an argon or a reactive gas (O_2, N_2, etc.) atmosphere. Evaporation is performed in the presence of the argon gas discharge and material ions are formed and accelerated in the electric field so that condensation and film formation take place under the influence of ion bombardment. This technique gives highly dense films (often the same density as the bulk material), good adhesion of the film to the substrate, and a high evaporation rate.

2.3.2 Chemical Formation Processes of Thin Films [2, 5]

2.3.2.1 Chemical Vapor Deposition Methods

In the Chemical Vapor Deposition (CVD) methods, the starting material undergoes specific chemical reactions at the hot surface of the substrate to form thin layers of the desired material. The reaction can be stimulated by various energy sources, e.g. plasma, giving plasma enhanced CVD (PECVD), or a laser, giving laser CVD.

Several conditions are necessary to achieve this process. Firstly, the reactants must be volatile and stable so that they can be transported to and from the deposition zone. Secondly, the solid product should have low vapor pressure under the deposition condition. The film properties (such as the film thickness and purity) depend strongly on the process parameters such as the reactor geometry, gas flow rate, gas composition, substrate temperature, and pressure. The CVD performed under low pressure, is especially referred to as the low pressure CVD (LP-CVD).

2.4 System and Cell Designing for Li Microbattery Fabrication

Conventional batteries are 2-D cells with a parallel (or pseudo-parallel) arrangement of planar cathode and anode separated by an electrolyte (Fig. 2.1). On the other hand, three-dimensional (3D) configuration (such as those discussed in section 2.6) offers decreased diffusion lengths and increased cell capacity due to the higher surface area under the same areal footprint (i.e., square footage) on the surface of the device. Such structure offers the merit of obtaining both high energy density and high power density. In order to maximize energy and power density, 3-D microbatteries will comprise a large number of closely packed cathodes and anodes (as in Fig. 2.9). However, in contrast to a 2-D battery, in which uniform current density is naturally obtained over the surface of the electrodes, the current density in 3-D microbattery is inherently non-uniform. In general, non-uniform currents result in poor utilization of the electrode materials, and are thus associated with lower cell efficiencies, reduced electrode stability due to non-uniform stresses, and non-uniform heat dissipation [6–8]. The simulations of the current and potential distributions demonstrate the tremendous effect of the electrode geometry on the current distribution (Figs. 2.5, 2.6, and 2.7) [9].

Fabrication of an array of such Li microbatteries improves the limited capacity and current rating of the microbattery and enables their use in miniaturized systems which require higher capacities and voltages than what a single microbattery can

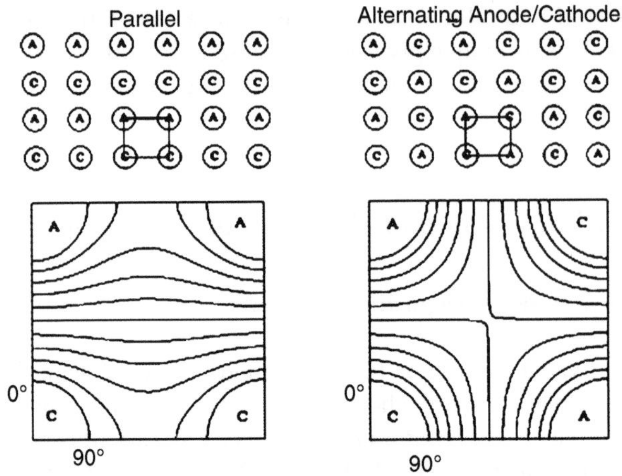

Fig. 2.5 Schematic diagram of 3-D cylindrical battery arrays in parallel row (*left*) and alternating anode/cathode (*right*) configurations. (*Bottom panels*) Isopotential lines between cathode (C) and anode (A) for unit battery cells. Reprinted from [9], with permission from Elsevier

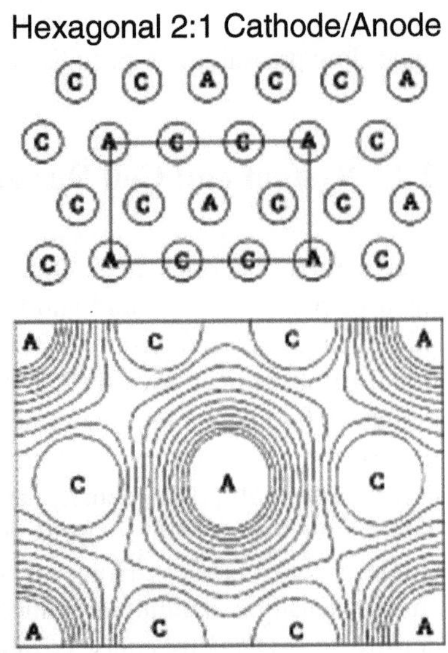

Fig. 2.6 (*Top panel*) Schematic diagram of hexagonal 2:1 cathode/anode battery array. (*Bottom panel*) Isopotential lines between cathode (C) and anode (A) for unit battery cell. Reprinted from [9], with permission from Elsevier

provide. In such cases, the development of the design criteria of the management systems such as the switch matrices that enable the control of multiple microbatteries to charge and discharge as designated [10] is also a critical issue.

For power sources with reduced size, increased flexibility, longer lifetime, and increased reliability, there are also some approaches concerning the hybridization of the Li microbatteries with energy conversion devices (i.e. solar cells and

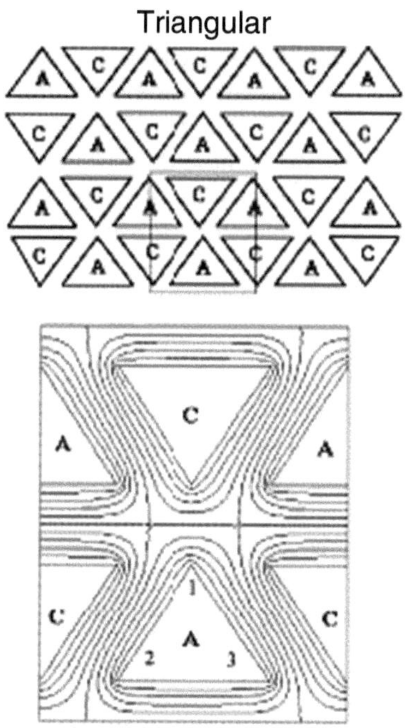

Fig. 2.7 (*Top*) Schematic diagram of triangular battery array. (*Bottom*) Isopotential lines between cathode (C) and anode (A) for unit battery cell. Reprinted from [9], with permission from Elsevier

piezo-electric generator) [11, 12]. This is claimed to be particularly feasible for autonomous sensors that do not have a direct connection to a host for either power or communication.

2.5 Li Metal Microbatteries

Li metal is attractive as an anode material for batteries as it is the most electronegative metal (–3.04 V vs. SHE) that benefits the high cell voltage, as well as the lightest (equivalent weight $M=46.94$ g mol^{-1}, and density $\rho=40.53$ gcm^{-3}) which facilitates the design of lighter and smaller batteries.

The rechargeable thin film battery based on a lithium anode was first designed in the 1980s [13–16]. In 1983 Levasseur et al. proposed a lithium battery where 1 μm thick lithium borosilicate electrolyte film was deposited onto a positive electrode made from a thin pressed pellet of TiS_2. This cell could deliver 10 μA for 30 h with an average voltage of 2 V, but the system had poor rechargeability [16]. An evaporated thin film of lithium served as the negative electrode. During the same period, Kanehori et al. reported the fabrication of a battery having phosphosilicate electrolyte and a TiS_2 cathode [14, 17]. They fabricated 1–3 μm thick film of TiS_2 film by CVD at a substrate temperature of 450°C using a gas composed of $TiC1_4$

and H_2S diluted with helium. An amorphous lithium phosphosilicate electrolyte and a lithium negative electrode were evaporated on this layer. The CVD process allows the TiS_2 crystals to orient with the c axis parallel to the substrate plane, resulting in a good reversible intercalation of the lithium into the TiS_2 electrode. This cell gives a capacity of more than 100 μA h cm^{-2}, and by controlling the current density and the depth of discharge, it retained its capacity for 100–1,500 cycles [17].

The project developed at the Eveready Battery Company (EBC) reported thin film solid state Li/oxysulfide glass/TiS_2 microbatteries fabricated by sputtered depositions [18, 19]. It has been found that the metallic lithium anode reacts with this electrolyte to form a high resistive layer, which is assumed to be Li_2S. To avoid direct contact between lithium and the electrolyte, an intermediate LiI layer is deposited by vacuum evaporation (Fig. 2.8). Although the LiI layer limits the cathodes which can be used due to its restricted stability window of about 2.8 V and the ionic conductivity falls from 2×10^{-5} S cm^{-1} to 2×10^{-6} S cm^{-1}, excellent performance over thousands of charge-discharge cycles was obtained. Several of the Eveready cells have undergone more than 10,000 cycles at current densities of up to 100 μA cm^{-2} with little change in performance, while others stored at room temperature for nearly two years have retained 98% of their initial voltage.

2.6 Li Ion Microbatteries

2.6.1 Fundamental Principles of Li ion Battery System

When Li^+ is reduced to metallic Li on the Li metal anode surface during the recharge of the cell, it forms a deposit more porous than the original metal (*i.e.* the dendritic growth). Hence the area of contact between the Li metal and the electrolyte increases with cycling. Since this interface is not thermodynamically stable, the cells become increasingly sensitive to thermal, mechanical, and electrical abuse.

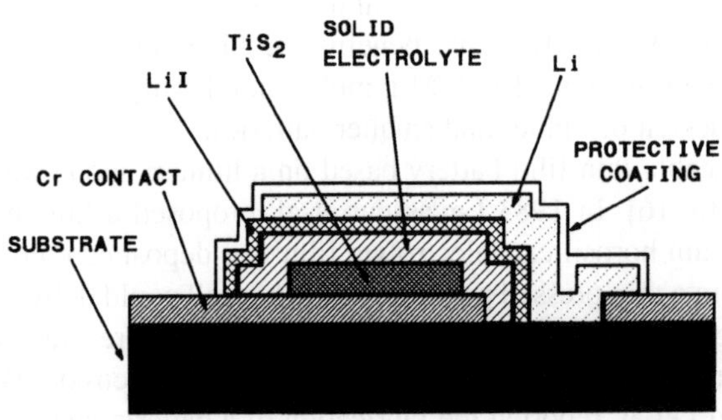

Fig. 2.8 Cross section of the thin film EBC microbattery configuration. Total thickness about 10 μm. Reprinted from [18], with permission from Elsevier

One of the approaches to circumvent these safety issues concerning the use of Li metal was to substitute the metallic Li for layer-structured Li ion intercalating materials. Because of the presence of Li in its ionic rather than metallic state, Li ion cells solve the dendrite problem and are, in principle, inherently safer than Li-metal cells.

The discovery and research of the highly reversible, low-potential Li^+ intercalation–deintercalation process in carbonaceous material attracted strong interest as the anode material for Li ion batteries [20–24], finally leading to the creation of C anode / $LiCoO_2$ cathode rocking-chair cell commercialized by Sony Corporation in 1991 [25]. This type of Li ion cell, having a voltage exceeding 3.6 V (three times that of conventional alkaline systems) and gravimetric energy densities as high as 120–150 Wh kg^{-1} (two to three times those of usual Ni-Cd batteries), is found in most of current high-performance portable electronic devices.

2.6.2 Strategy for 3-D Integrated Microbattery

Innovative advancements in microbatteries are intimately linked to the availability of new materials and the development of novel battery designs. In recent years there has been the realization that improved battery performance can be achieved by reconfiguring the electrode materials currently implied in 2-dimentional (2-D) batteries into 3-D architectures [26]. The general strategy of this approach is to design cell structures that maximize power and energy density per unit area and realize high-rate discharge capabilities. This goal can be achieved from a 3-D matrix of electrodes (in a periodic array or an aperiodic ensemble, see Fig. 2.9) with short transport lengths and large energy capacity within the footprint area. For example, according to White et al., in 3-D microbattery with electrode arrays of 50:1 aspect ratio (height/width ratio), the expected capacity may become larger by 3.5 times than the conventional 2-D battery design with the same areal footprint [9].

2.6.3 Micromachining Process for 3-D Li Ion Microbattery

Micromachining methods such as photolithographic patterning and selective etching that evolved from the realm of integrated circuits have enabled the fabrication of 3-D Microsystems.

Two very different micromachining methods are applied for the fabrication of 3-D electrode arrays. One approach named carbon MEMS or C-MEMS is based on the pyrolysis of photoresists. The use of photoresist as the precursor material is a key consideration, since photolithiography can be used to pattern these materials into appropriate structures. The second approach involves the micromachining of silicon molds that are then filled with electrode material. Successful construction of both anode and cathode electrode arrays has been demonstrated using these microfabrication methods.

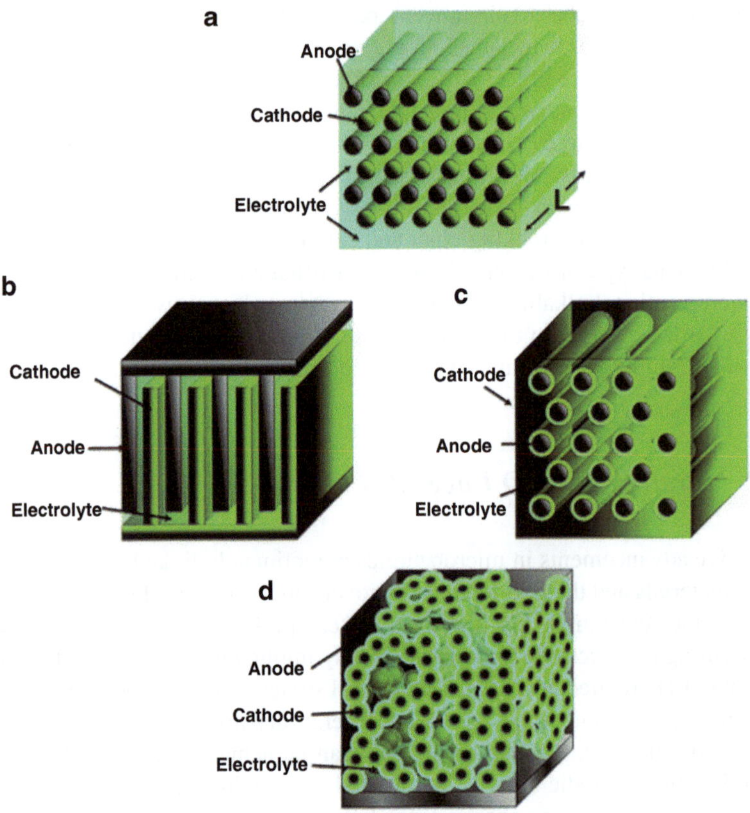

Fig. 2.9 Examples of prospective 3-D architectures for charge-insertion batteries: (**a**) array of interdigitated cylindrical cathodes and anodes; (**b**) interdigitated plate array of cathodes and anodes; (**c**) rod array of cylindrical anodes coated with a thin layer of ion-conducting dielectric (electrolyte) with the remaining free volume filled with the cathode material; (**d**) aperiodic "sponge" architectures in which the solid network of the sponge serves as the charge insertion cathode, which is coated with an ultra thin layer of ion-conducting dielectric (electrolyte), and the remaining free volume is filled with an interpenetrating, continuous anode. Reprinted from [26] with permission. Copyright 2004, American Chemical Society

2.6.3.1 Carbon MEMS (C-MEMS)

In Carbon-Microelectromechanical system (C-MEMS), carbon interdigitated electrodes are fabricated from positive photoresist spin coated on a silicon substrate. This photoresist is patterned by photolithography and pyrolyzed in an inert environment at high temperature to form the carbon electrode [27–30]. In a more recent work, laser excitation has been used to both pyrolyze the film and to write the electrode pattern. [31]. Figure 2.9 shows an example of the 3-D microbattery that has been actually fabricated using the C-MEMS process [32].

Galvanostatic measurements on the C-MEMS array in a half-cell with lithium as both the counter and reference electrode [32] show a large irreversible capacity loss on first discharge followed by good cycling properties consistent with the behavior of conventional bulk coke electrodes. The lithium capacity normalized to the foot print area of the electrode array is 0.125 mAh cm^{-2}. This value is nearly twice that of an unpatterned pyrolyzed film of SU-8 photoresist [32]. The higher capacity is due to the greater active volume, contributed by the carbon posts over the footprint area (Fig. 2.10).

2.6.3.2 Micromachining of Silicon Molds

A second approach for fabricating electrode arrays has involved micromachining of silicon molds, which, for example, could be filled with electrode material by colloidal processing methods. In contrast to C-MEMS, this fabrication approach is suitable for both anodes and cathodes. The process flow for such electrode array fabrication is depicted in Fig. 2.11. Recent works by D. Golodnitsky et al., reports on the manufacturing of 3-D rechargeable Li ion microbattery cell applying this silicon mold method [33, 34]. The cell has a sandwich-like structure of conformal thin-film electrodes, electrolyte, and current collectors. One cell with a roughly 1 μm thick cathode ran at C/10–2 C charge/discharge rates at room temperature for

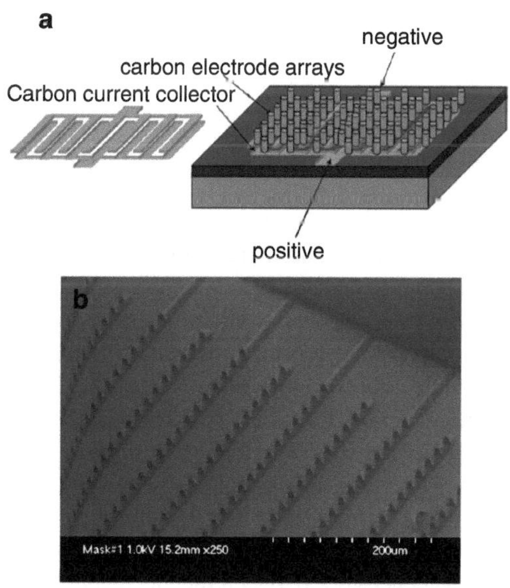

Fig. 2.10 Design of C-MEMS 3-D microbattery and a typical SEM of low aspect ratio C-MEMS battery arrays. Both the electrodes and contact fingers are made of carbon. Reproduced from [32] with permission from ECS – The Electrochemical Society

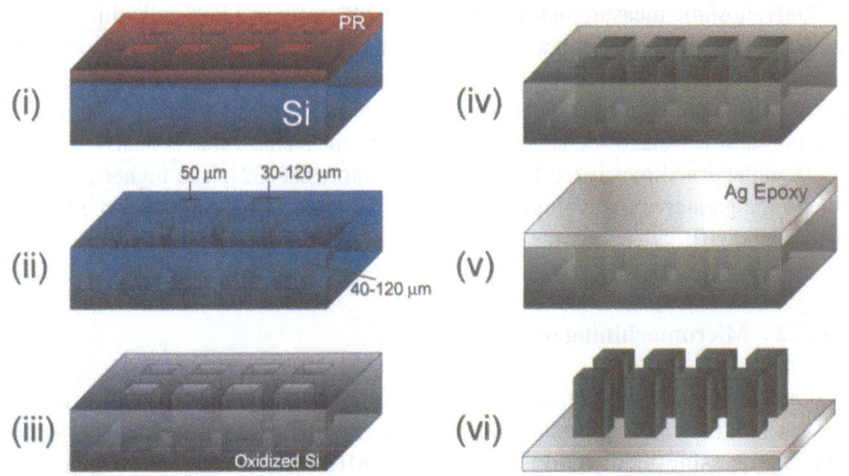

Fig. 2.11 Processing flow for 3-D electrode array fabrication using silicon micromachining with colloidal filling of the electrode material. The six steps are identified as the following: (i) patterned photoresist (PR) on silicon substrate, (ii) PR removal after DIRE micromachining, (iii) insulate silicon mold by oxidation, (iv) colloidal electrode filling material centrifuged into the mold, (v) silver epoxy added to provide mechanical stability and electrical contact, (vi) the electrode flipped over and released from the mold by immersion in a TEAOH solution. Reprinted with permission from [26]. Copyright 2004, American Chemical Society

200 cycles with 0.2 % per cycle capacity loss and about 100 % Faradaic efficiency. The cell exhibited a capacity of 2 mAh cm^{-2}, which is about 30 times higher than the capacity of a similarly built planar (2-D) cell with the same foot print and same cathode thickness [33].

The micromachining of silicon molds is one key factor that determines the dimensions of the electrode array and therefore the energy of the 3-D battery. The use of electrochemical etching in HF to fabricate high-aspect ratio trenches in silicon was introduced in 1990 [35]. Recently, Chamran et al. reported the use of this technique for 3-D microbattery fabrication [36]. Silicon etched by the photo assisted process can yield an array of holes of 5 µm in diameter with depths of ~100 µm and the center to center distance between holes being 10 mm. Calculations indicate that 3-D batteries constructed from such dimensions will be capable of providing over 2 mWh in a 5 mm^3 package, an appropriate size and energy for powering MEMS devices [26].

As a future development, realization of monolithic rocking chair cells may be one option [1]. A common network between the electrolyte and electrode materials would provide a continuous matrix for the lithium cation transfer from the electrode to the electrolyte or vice versa, and would consequently reduce the interfacial charge transfer resistance (Fig. 2.12).

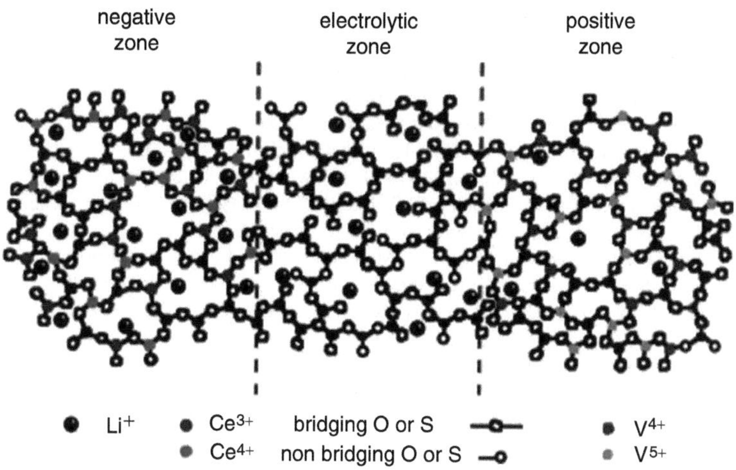

Fig. 2.12 Towards a glassy monolithic solid state cell: a common network is used for the electrolyte and electrode materials. The electrodes contain two different redox couples using two different transition metals, and are separated by an electrolytic zone obtained by the addition of a network modifier. Reprinted from [1], with permission from Elsevier

References

1. Souquet JL, Duclot M (2002) Solid State Ionics 148:375
2. Julien C, Nazri G-A (1994) Solid state batteries: materials design and operation. Kluwer Academic Publishers, Massachusetts, USA
3. Julien C (1994) In: Pistoia G (ed) Lithium batteries new materials, developments and perspectives. In Industrial Chemistry Library. Volume 5. Elsevier Science, Amsterdam, Netherlands
4. Balkanski M (2000) Solar energy mater. Solar Cells 62:21
5. Levasseur A, Menetrier M, Dormoy R, Meunier G (1989) Mat Sci Eng B3:5
6. West AC, Matlosz M, Landolt D (1991) J Electrochem Soc 138:728
7. Orazem M, Newman J (1984) J Electrochem Soc 131:2857
8. Mao Z, White RE, Jay B (1991) J Electrochem Soc 138:1615
9. Hart RW, White HS, Dunn R, Rolison DR (2003) Electrochem Comm 5:120
10. Sukumar V, Alahmad M, Buck K, Hess H, Li H, Cox D, Zghoul FN, Jackson J, Terry S, Blalock B, Mojarradi MM, West WC, Whitacre JF (2004) J Power Sourc 136:401
11. Koeneman PB, Busch-Vishniac IJ, Wood KL (1997) J Microelectromech Syst 6:355
12. Harb JN, LaFollette RM, Selfridge RH, Howell LL (2002) J Power Sourc 104:46
13. Meunier G, Dormoy R, Levasseur A (1989) Mater Sci Eng B3:19
14. Kanehori K, Matsumoto K, Miyauchi K, Kudo J (1983) Solid State Ionics 9–10:1445
15. Creus R, Sarradin J, Astier R, Pradel A, Ribes M (1989) Mater Sci Eng B3:109
16. Levasseur A, Kbala M, Hagenmuller P, Couturier G, Danto Y (1983) Solid State Ionics 9–10:1439
17. Miyauchi K, Matsumoto K, Kanehori K, Kudo T (1983) Solid State Ionics 9–10:1469
18. Jones SD, Akridge JR (1995) J Power Sourc 54:63
19. Jones SD, Akridge JR (1993) J Power Sourc 43–44:505
20. Hèrold A (1955) Bull Soc Chim 187:999
21. Guèrard D, Hèrold A (1975) Carbon 13:337

22. Besenhard JO, Fritz HP (1974) J Electroanal Chem 53:329
23. Besenhard JO (1976) Carbon 14:111
24. Mohri M, Yanagisawa N, Tajima Y, Tanaka H, Mitate T, Nakajima S, Yoshida M, Yoshimoto Y, Suzuki T, Wada H (1989) J Power Sourc 26:545
25. Nagaura T, Tozawa K (1990) Prog Batt Solar Cells 9:209
26. Long JW, Dunn B, Rolison DR, White HS (2004) Chem Rev 104:4463
27. Kinoshita K, Song X, Kim J, Inaba M (1999) J Power Sourc 81–82:170
28. Kim J, Song X, Kinoshita K, Madou M, White B (1998) J Electrochem Soc 145:2314
29. Kostecki R, Song XY, Kinoshita K (2000) J Electrochem Soc 147:1878
30. Ranganathan S, McCreery R, Majji SM, Madou M (2000) J Electrochem Soc 147:277
31. Kostecki R, Song XY, Kinoshita K (2002) Electrochem Solid-State Lett 5:E29
32. Wang C, Taherabadi L, Jia G, Madou M, Yeh Y, Dunn B (2004) Electrochem Solid-State Lett 7:A435
33. Nathan M, Golodnitsky D, Yufit V, Strauss E, Ripenbein T, Shechtman I, Menkin S, Peled E (2005) J Microelectromech Syst 14:879
34. Golodnitsky D, Yufit V, Nathan M, Shechtman I, Ripenbein T, Strauss E, Menkin S, Peled E (2006) J Power Sourc 153:281
35. Lehmann V, Foll H (1990) J Electrochem Soc 137:653
36. Chamran F, Christophersen M, Kim C-J, In Abstracts of the 204th Meeting of the Electrochemical Society, Electrochemical Society: Pennington, NJ, 2003; Abst. #1292

Chapter 3
Nanotechnologies for Fuel Cells

Jong-Eun Park, Takahiro Shimizu, and Tetsuya Osaka

3.1 Introduction

Fuel cells, which convert chemical energy directly into electrical energy with high efficiency and low emission, are currently attracting interest because of their huge potential for power generation in stationary and portable devices, transport applications, and sustainable energy production. Direct methanol fuel cells (DMFCs) are especially attractive as portable power sources because of characteristics such as simple construction, easy operation, liquid fuel, and high efficiency [1, 2].

DMFCs use methanol as fuel and generate electricity through the electrochemical reaction of methanol in the presence of catalyst. They are similar to polymer electrolyte membrane fuel cells (PEMFCs) as both types of cells use a polymer membrane as the electrolyte. The structure of DMFCs is simple because the system does not require fuel reforming steps to take protons out of alcohol or fossil fuel, which are required for hydrogen–oxygen fuel cells [3]. In addition, methanol, commonly used alcohol, has the advantage of high energy density, offering the consumer the potential for longer operating time and system with increased functionality.

Table 3.1 shows the thermodynamic energy densities and practical energy densities for several battery systems and fuel cells. Fuel cells potentially offer 5–10 times greater energy densities than rechargeable batteries [4].

Recently, there has been a growing interest in miniaturizing DMFCs and exploring their potential as a long-life and charge-free power source for portable and mobile

J.-E. Park
Faculty of Science and Engineering, Waseda University, 3-4-1 Okubo, Shinjuku-ku, Tokyo, 169-8555, Japan; Samsung Electro-Mechanics Co., Ltd.

T. Shimizu
Faculty of Science and Engineering, Waseda University, 3-4-1 Okubo, Shinjuku-ku, Tokyo, 169-8555, Japan; Fuel Cell Nanomaterials Center, University of Yamanashi

T. Osaka (✉)
Faculty of Science and Engineering, Waseda University, 3-4-1 Okubo, Shinjuku-ku, Tokyo, 169-8555, Japan
e-mail: osakatets@waseda.jp

T. Osaka et al. (eds.), *Electrochemical Nanotechnologies*,
Nanostructure Science and Technology, DOI 10.1007/978-1-4419-1424-8_3,

Table 3.1 Comparison of thermodynamic energy densities and practical energy densities for rechargeable batteries and fuel cells

Rechargeable batteries	Electrochemical reaction	Energy density (W h/L)	Practical energy density (W h/L)
Li-ion battery	$C_6 + 2LiCoO_2 \rightarrow LiC_6 + 2Li_{0.5}CoO_2$	1,350	360
Ni-MH	$NiOOH + MH \rightarrow Ni(OH)_2 + M$	988	205
Ni-Cd battery	$2NiOOH + Cd + 2H_2O \rightarrow 2Ni(OH)_2 + Cd(OH)_2$	439	120
Fuel Cells			
PEMFC	$H_2 + 1/2O_2 \rightarrow H_2O$ (gaseous H_2 and air)	2.8	1.4
PEMFC	Liguid H_2 and air	4,632	2,316
DMFC	$2CH_3OH + 3O_2 \rightarrow 2CO_2 + 4\ H_2O$ (100% MeOH)	4,767	1,906
DMFC	2 M MeOH (7%) in water	334	134

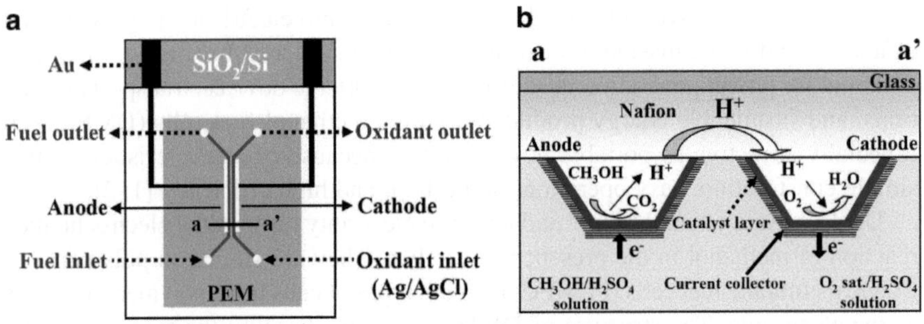

Fig. 3.1 Illustration of a micro-direct methanol fuel cell: (**a**) top view of a planer cell and (**b**) cross-sectional view with working principle

electronic devices [5, 6]. For this purpose, micro electro mechanical system (MEMS) technology is a promising approach for fabricating micro-devices or systems integrated on a silicon substrate such as miniaturized fuel cells, the so-called micro-fuel cells, i.e., micro-direct methanol fuel cells (μ-DMFCs). Micro power sources are important for future integrated MEMS devices, such as micro-sensors, micro-actuators, and single-chip communication devices. Medical application is also a fertile field for micro power source: if implantable, they are critical for medical devices such as cerebrospinal fluid shunt pumps and micro-insulin pumps. Thus, μ-DMFCs with various degrees of micro fabrication have been reported in the literature [7–15].

This chapter reviews selected topics related to μ-DMFCs and MEMS technologies for portable power sources; emphasis is on the results of recent studies carried out to clarify the factors affecting the performance of μ-DMFC as a new power source.

3.2 Concept and Design of Micro-Fuel Cell

An example of the micro-direct methanol fuel cell (μ-DMFC) is illustrated in Fig. 3.1. The cell is equipped with two parallel micro channels to supply fuel ($CH_3OH/H_2SO_4/H_2O$) and oxidant (saturated $O_2/H_2SO_4/H_2O$) [16–18]. At the anode, where Pt–Ru is employed as catalyst, the methanol reacts with water to produce electrons, carbon dioxide, and protons. The protons transfer from the anode to the cathode through an electrolyte membrane that covers over the micro channels. At the cathode, where Pt is employed as catalyst, oxygen is reduced and then it recombines with the protons to form water. In order to enhance the proton conductivity between the electrode and the membrane, sulfuric acid aqueous solution is employed as a supporting electrolyte that does not only carry the fuel and oxidant but also reject the carbon dioxide produced during the electro-oxidation of methanol. In short, the overall process in the methanol/air fuel cell is the direct conversion of methanol to electric power, only with carbon dioxide and water as byproducts.

The concept of this novel structure is that the anodic and cathodic micro-channels are arranged in parallel on a planer single silicon substrate, unlike the arrangement in the conventional bipolar design. The parallel-planar electrodes along the channels allow for increased power generation as a result of the large interface between solid electrode and liquid fuel/oxidant. The catalyst layers directly supported on the micro channels also contribute to highly efficient current collection. In addition, the planer design does not require a conventional hot-pressing process to form a membrane electrode assembly (MEA) because the electrolyte membrane is placed on the patterned electrode. Therefore the design realizes a future possibility that this micro-patterned cell can be easily piled up to form a tiny energy device.

3.3 Fabrication of the Micro-DMFC

The micro DMFC as mentioned above is fabricated onto a silicon (100) substrate through a series of steps tailored from MEMS techniques [16–18]. This procedure is shown in Fig. 3.2. A 500 nm layer of silicon dioxide, formed on the polished (110) oriented silicon (p-type, 1–10 Ωcm, 200 μm thick) substrate is used and the surface is cleaned with acidic solution (H_2SO_4:H_2O_2). Then a pattern was made on the surface by photolithography (Fig. 3.2a), followed by buffered hydrogen fluoride (BHF) etching (Fig. 3.2b). A slope type channel with an angle of 54.7 degrees to the surface was formed by the anisotropic chemical etching process in a 30 wt% aqueous KOH solution at 80°C. The Si (111) plane was etched selectively, while the (100) plane remained stable (Fig. 3.2c). [19, 20] To complete the fabrication pattern, the feed-holes were etched by deep reactive ion etching (D-RIE) process (Fig. 3.2d). The holes have diameter of 1 mm diameter with a hole spacing of 8 mm and 10 mm horizontally and vertically, respectively. The width and the depth of the microchannel are 100 μm and 50 μm, respectively, while the clearance between channels is 50 μm. The silicon dioxide layer was removed by BHF etching and the silicon layer was cleaned by the SPM method (Fig 3.2e). Subsequently, a 100 nm layer of silicon oxide was grown thermally (dry oxidation at 1,100°C for 12 min) on the entire surface of the silicon wafer (Fig. 3.2f). Finally, a photosensitive resistor for printed wiring boards was applied to prepare for selective deposition of metal film using the spray coating method and lithography (Fig. 3.2g, h). Titanium/gold (Ti/Au) layer for current collectors was formed by electron-beam deposition and the lift-off method (Fig. 3.2i, j). A 100 nm gold layer was deposited in an electron beam evaporation chamber, preceded by a 20 nm Ti layer just beneath the gold to promote adhesion.

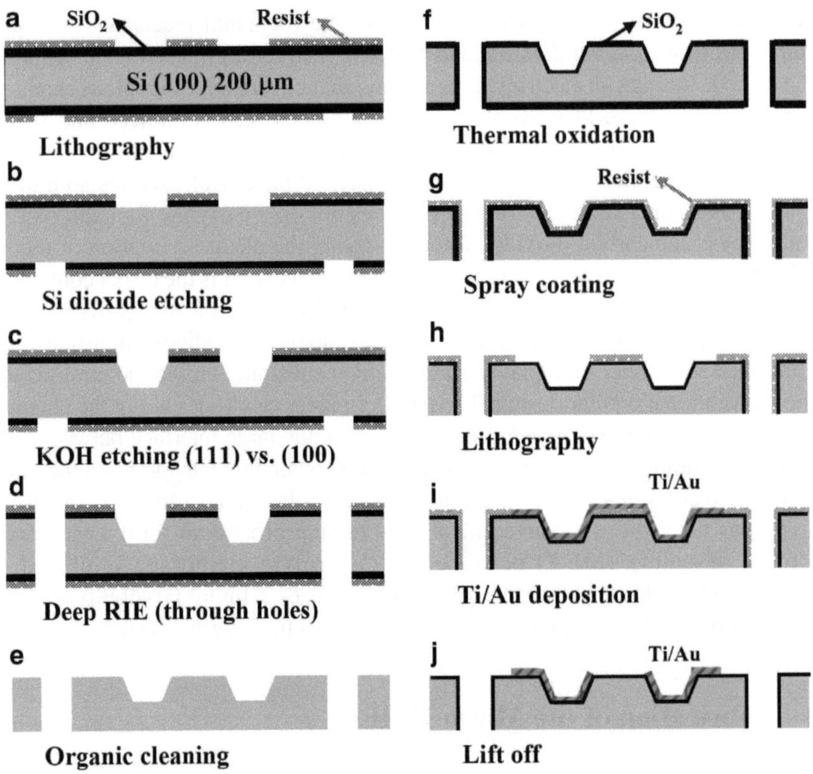

Fig. 3.2 Process steps for the fabrication of the micro-DMFC chip on a silicon wafer

3.4 Formation of Catalysts and Cell Assembly

The catalyst layers of the anode and the cathode were prepared by electroplating either PtRu or Pt, respectively onto the Ti/Au electrodes. Table 3.2 shows the electroplating process [18]. The electroplating process produced a uniform coating of the surface exposed sidewalls, and bottom of the microchannel electrodes (Fig. 3.3a). The structure and surface morphology of PtRu and Pt showed spherical shape with a uniform size distribution (Fig. 3.3b and c).

The channels, which had catalyzed electrodes on the surfaces, were covered with Nafion 112 (thickness: 50 μm, equivalent weight: 1,100 $g\,mol^{-1}$, ionic conductivity: 0.083 $S\,cm^{-1}$) to provide ionic conductivity between the anode and the cathode. The Nafion membrane was pressed with a glass plate to avoid solution leakage (Fig. 3.4a). Voltage-current measurements were performed at room temperature with a mass flow control system of fuel and oxidant as shown in Fig. 3.4b. The fuel and oxidant solutions were supplied to the electrodes with the micro-syringe pumps from the outlet of each channel. The flow rate of both the fuel and oxidant solutions was 80 $\mu L\ min^{-1}$. Composition of the fuel solution was 2M methanol solution

Table 3.2 Bath compositions and electroplating conditions for the electroplating of Pt and PtRu catalysts

	$H_2PtCl_6\cdot$ $6H_2O$/mM	$RuCl_3\cdot$ $3H_2O$/mM	$(CH_3COO)_2$ Pb/mM	Mass loading/ mg cm^{-2}	Pulse deposition[a] (i_p=50 mA/cm^2)
Pt	20	–	0.5	2.4	t_{on}=100 ms, $_{off}$=500 ms
PtRu	20	20	–	2.85	T_{charge}: 2 C/cm^2

[a]i_p Peak current density, t_{on} On-time, t_{off} Off-time, T_{charge} Applied charge

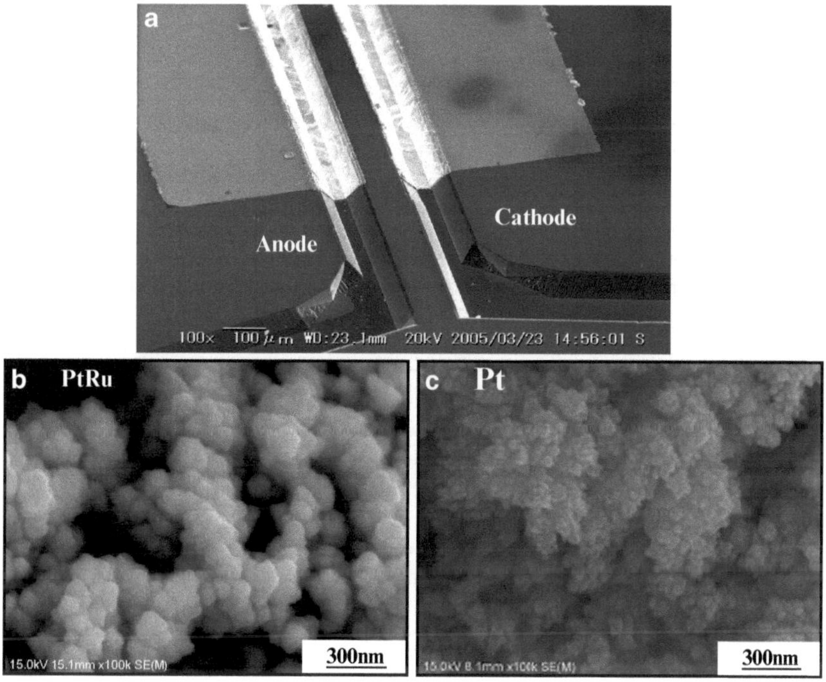

Fig. 3.3 SEM images of (**a**) catalysts layer in micro-DMFCs channels, (**b**) PtRu at the anode, and (**c**) Pt at the cathode

containing 0.5 M H_2SO_4, while that of the oxidant solutions was 0.5 M H_2SO_4 solutions saturated with oxygen.

3.5 Evaluation of the Performance of a μ-DMFC with Oxygen Saturated Solution as oxidant

Figure 3.5 shows a plot of cell voltage and power density for the micro-DMFC at the flow rate of 80 μL min^{-1}. A maximum power density of 1.4 mW cm^{-2} at 0.25 V was obtained for the 2 M methanol-saturated oxygen. In the voltage-current curve, an abrupt drop of the cell voltage at approximately 8 mA cm^{-2} was confirmed, suggesting

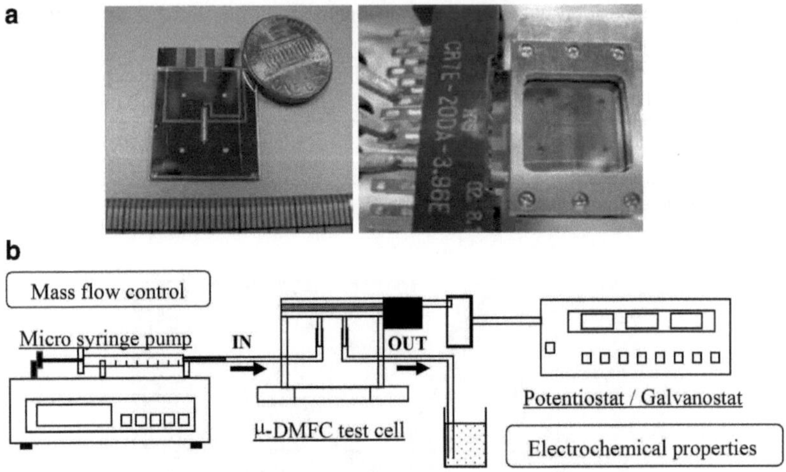

Fig. 3.4 A prototype of micro-DMFC cell: (**a**) pictures of a planer cell (left) and its holder with terminal pins (right), and (**b**) a diagram of experimental apparatus for electrochemical measurements [16, 17]. Reproduced with permission of the Electrochemical Society of Japan and Elsevier

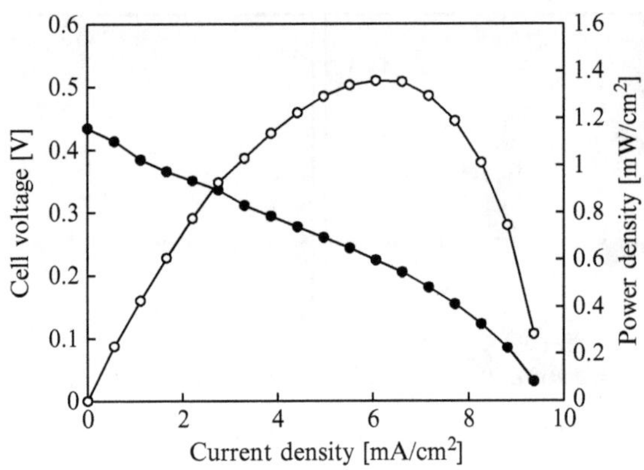

Fig. 3.5 Voltage-current curve (•) and power-current curve of the whole cell. (○) Oxidant solution was O_2 saturated aqueous solution containing 0.5 M H_2SO_4

an existence of concentration polarization at the anode and/or the cathode. In order to clarify which side of the electrode mainly causes the significant loss of cell voltage, individual potential profiles of the anode and cathode during cell polarization were examined as shown in Fig. 3.6. The results indicated that the gradient of the curves significantly changed only at the cathode in the high current density region. The drop of the cathode potential above 8 mA cm^{-2} was similar to the voltage drop of the whole cell shown in Fig. 3.5. Therefore, the main cause of the concentration polarization at high current region was attributed to the cathode. It was partly because the concentration of oxygen in water (approximately 1.5 mM) was remarkably

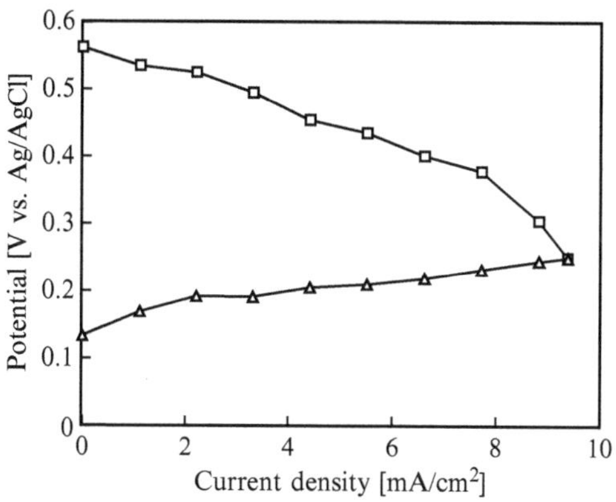

Fig. 3.6 Potential-current curves of each electrode (anode: Δ, cathode: □) in the cell. Oxidant solution was O_2 saturated aqueous solution containing 0.5 M H_2SO_4

small compared with that of methanol (2 M), which also indicated the larger concentration polarization at the cathode. Taking these into consideration, Fig. 3.5 and 3.6 demonstrate the oxygen transport limit at the cathode, for the μ-DMFC was operated with the oxygen dissolved into the fuel solution [21]. It can be concluded that the amount of oxidant supplied to the electrode was a critical issue for μ-DMFC. This conclusion is also supported by the literature describing that the maximum of power density could be limited by the concentration polarization [22, 23]. If the concentration of oxygen and/or oxidant can be maintained enough for the reaction, then a significant improvement will be expected in the μ-DMFC performance.

3.6 Thermodynamic Aspects of a Fuel Cell

The performance of a DMFC is directly related to oxygen concentration when the cell performance is cathode-limited as described previously: the higher the concentration, the better the performance of the cell. Therefore, an alternative to use oxygen is considered to get higher performance; one solution is to use hydrogen peroxide or other peroxides as the oxidant. Figure 3.7 shows the thermodynamic potentials expected for acidic single electrolyte methanol/O_2 and methanol/H_2O_2 systems at room temperature [24]. The acidic single electrolyte is the most extensively researched methanol/O_2 system because it is employed in polymer electrolyte membrane (PEM) fuel cells [25–27]. The maximum thermodynamically attainable voltage from these systems is 1.20 V; this voltage was calculated using 0.5 M H_2SO_4 (pH 0) for the acid and assuming a concentration of 1.0 M for methanol and oxygen. On the other hand, the maximum thermodynamically attainable voltage from these systems is 1.74 V when the voltage is calculated using 0.5 M H_2SO_4 (pH 0) for the acid and assuming a concentration of 1.0 M for methanol and hydrogen peroxide. Experimentally, the open circuit voltage (OCV) for a methanol/

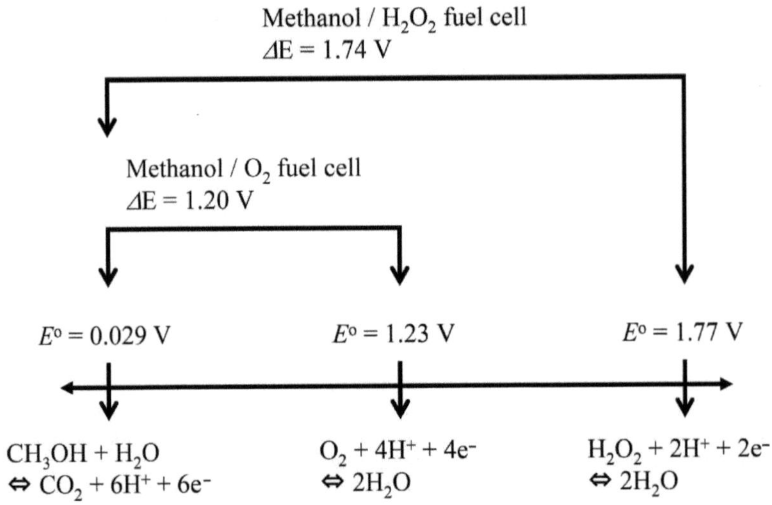

Fig. 3.7 Thermodynamic potentials of methanol/O_2 and methanol/hydrogen peroxide in acid systems [24]. Potentials vs. are referred against standard hydrogen electrode (SHE) at 25°C

O_2 (Air) fuel cell is ca. 0.5–0.6 V [28, 29]. This deviation is primarily caused by the large overpotential associated with the reduction of O_2, but the decrease also can be attributed to other resistances in the fuel cell itself, including solution resistance and electrical contact resistance.

3.7 Evaluation of the Performance of a µ-DMFC with Hydrogen Peroxide

To increase the cathode potential, hydrogen peroxide was examined as an oxidant according to the concentration of oxygen and the electron number of reactions [30–32]. The concentration of oxygen saturated aqueous solution is roughly 1.5 mM at room temperature. Thus, the same volume of the 3 mM hydrogen peroxide solution possibly generates almost the same electron quantity as the quantity generated by the oxygen dissolved solution, and the 30 mM hydrogen peroxide solution possibly generates ten times as much electron quantity as that generated by the oxygen dissolved solution [21]. Figure 3.8 shows the micro-fuel cell polarization curves taken of two different oxidants (dissolved O_2 and various H_2O_2 in 0.5 M H_2SO_4 solution) with 2 M methanol. A maximum power density of 8.5 mW cm^{-2} was obtained for the 2 M methanol-30 mM hydrogen peroxide solution. This result was much better than that of oxygen, 1.4 mW cm^{-2}. The increase in power density with hydrogen peroxide as the oxidant can be attributed to a significant increase of current density within the potential range of fuel cells. In the current region higher than 80 mA cm^{-2}, the voltage drop gradient of hydrogen peroxide increased, while the

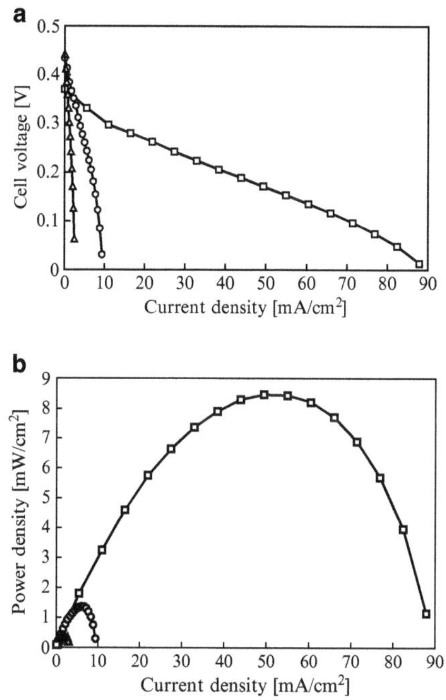

Fig. 3.8 Performance of the μ-DMFC with various oxidant solutions (**a**) voltage-current curves and (**b**) power-current curves Oxidant solution was O_2 saturated (○), 3 mM H_2O_2 (Δ), or 30 mM H_2O_2 (□) aqueous solution containing 0.5 M H_2SO_4

extent was modest compared with the case of oxygen saturated solution at the same cell voltage. The better performance of the μ-DMFC operated with hydrogen peroxide could be attributed to a higher amount of oxidant, indicating that the performances of electrode and membrane were sufficient for high power operation.

As for the effect of concentration of the oxidant, we speculate that the difference of concentration polarization between oxygen and 3 mM hydrogen peroxide can be attributed to the difference between their diffusion coefficients or the concentration decrease of hydrogen peroxide caused by non-electrochemical decomposition. This is because the total electron quantities of the oxygen saturated solution and the 3 mM hydrogen peroxide solution are almost the same. [21, 33]

In order to confirm from another aspect that the resulting cell was cathode-limited and to clarify the mechanism of the increase of cell performance with hydrogen peroxide as an oxidant, individual potential profiles of the anode and the cathode during cell polarization were examined. Figure 3.9 shows the cathode current density increased in the following order: 3 mM hydrogen peroxide <oxygen <30 mM hydrogen peroxide. Also, the application of 30 mM hydrogen peroxide reduced the concentration polarization on the cathode. However, the anode currents did not change with the variation of oxidant solutions, which means the cell performance is cathode-limited. In addition to the changes of current, the open circuit voltages

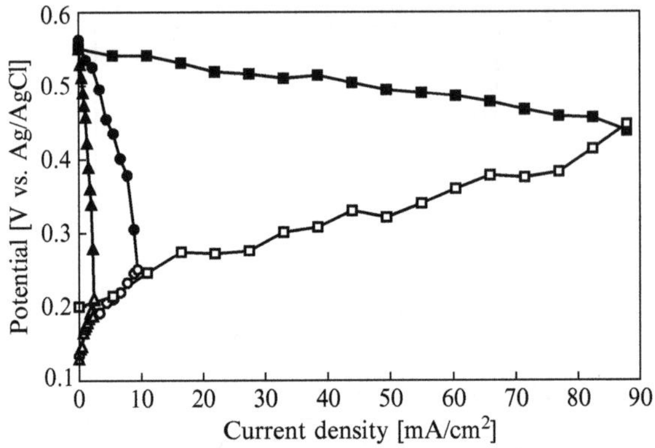

Fig. 3.9 Potentials against current density for the anode and the cathode corresponding to Fig. 3.8a. Oxidant solution was O_2 saturated (anode: ○, cathode: ●), 3 mM H_2O_2 (anode: Δ, cathode: ▲), or 30 mM H_2O_2 (anode: □, cathode: ■) aqueous solution containing 0.5 M H_2SO_4

(OCVs) were also different as shown in Fig. 3.8. In the case of oxygen and 3 mM hydrogen peroxide, the OCV was approximately 0.44 V, while it was approximately 0.37 V in the case of 30 mM hydrogen peroxide. We hypothesized that this decrease of OCV was the result of oxidant crossover from the cathode to the anode. Based on the fact that the open circuit potentials of the cathode were almost the same for all the solutions, the OCVs should be determined by those of the anode. The open circuit potentials of the anode were 0.13 V for oxygen and 3 mM hydrogen peroxide, and 0.2 V for 30 mM hydrogen peroxide as shown in Fig. 3.9). The concentration of the fuel was 2 M and the same for all the experiment, consequently the increase of the anode potential could be explained by the mixed potential due to the crossed-over oxidant to the anode in addition to the overpotential stemmed from polarization resistance. From all the above considerations, we can conclude that given present catalyst technology, methanol/hydrogen peroxide cells are applicable primarily in low-power, long-duration systems.

3.8 Conclusion

Microfabrication processes have been used successfully to form micro-fuel cells on silicon wafers. Aspects of the design, materials, and forming of a micro-fabricated methanol fuel cell have been presented. The processes yielded reproducible, controlled structures that performed well for liquid feed, direct methanol/O_2 saturated solution (1.4 mW cm^{-2}) and direct methanol/H_2O_2 systems (8 mA cm^{-2}). In addition to optimizing micro-fuel cell operating performance, there are many system-level issues to be considered when developing a complete micro power system. These issues include: electro-deposition procedure, catalyst loading, channel depth, oxidants supply, and system integration. The micro-fabrication processes that have been developed with the progress of MEMS technology provide new prospects for the design of micro-fuel cells. This MEMS technology may improve the performance of the fuel cells of various sizes. It is also expected that the improvement of MEMS-based small reformers will make possible high power density and multi-fuel micro-fuel cells.

References

1. Ren X, Wilson MS, Gottesfeld S (1996) J Electrochem Soc 143:L12
2. Kim D, Cho EA, Hong SA, Oh IH, Ha HY (2004) J Power Sourc 130:172
3. Lamy C, Lima A, LeRhun V, Delime F, Coutanceau C, Leger JM (2002) J Power Sourc 105:283
4. Maynard HL, Meyers JP (2002) J Vac Sci Technol B 20:1287
5. Kelley CS, Deluga GA, Smyrl WH (2000) Electrochem Solid-State Lett 3:407
6. Kelley CS, Deluga GA, Smyrl WH (2002) AIChE J 48:1071
7. Yamazaki Y (2004) Electrochimica Acta 50:663
8. Wainright JS, Savinell RF, Liu CC, Litt M (2003) Electrochim Acta 48:2869
9. Lu GQ, Wang CY, Yen TJ, Zhang X (2004) Electrochimica Acta 49:821
10. Yu J, Cheng P, Ma Z, Yi B (2003) Electrochimica Acta 48:1537
11. Min K, Tanaka S, Esashi M (2002) Electrochemistry 70:924
12. Hayase M, Kawase T, Hatsuzawa T (2004) Electrochem Solid-State Lett 7:A231
13. Yu J, Cheng P, Ma Z, Yi B (2003) J Power Sourc 124:40
14. Yen TJ, Fang N, Zhang X, Lu GQ, Wang CY (2003) Appl Phys Lett 83:4056
15. Lu GQ, Wang CY, Yen TJ, Zhang X (2004) Electrochim Acta 49:821
16. Motokawa S, Mohamedi M, Momma T, Shoji S, Osaka T (2004) Electrochem Commun 6:562
17. Motokawa S, Mohamedi M, Momma T, Shoji S, Osaka T (2005) Electrochemistry 73:346
18. Motokawa S, Obata H, Mohamedi M, Momma T, Shoji S, Osaka T (2005) Electrochemistry 73:352
19. Philipsen HGG, Kelly JJ (2005) J Phys Chem B 109:17245
20. Kovacs GTA, Maluf NI, Petersen KE (1998) Proceedings of the IEEE 86:1536
21. Narita E, Lawson F, Han KN (1983) Hydrometallury 10:21
22. Choban ER, Spendelow JS, Gancs L, Wickowski A, Kenis PJA (2005) Electrochimica Acta 50:5390
23. Choban ER, Waszczuk P, Kenis PJ (2005) Electrochem and Solid-State Lett 8:A348
24. Bard AJ, Faulkner LR (2001) Electrochemical methods; fundamentals and applications, 2nd edn. John Wiley, New York
25. Savadogo O (2004) J Power Sourc 127:135
26. Larminie J, Dicks A (2000) Fuel cells explained. John Wiley, West Sussex, England
27. Haile SM (2003) Acta Mater 51:5981
28. Shimizu T, Momma T, Mohamedi M, Osaka T, Sarangapani S (2004) J Power Sourc 137:277
29. Shimizu T, Mohamedi M, Momma T, Osaka T (2006) Electrochemistry 74:326
30. Prater DN, Rusek JJ (2003) Appl Energy 74:135
31. Bewer T, Beckmann T, Dohle H, Stolten JM (2004) J Power Sourc 125:1
32. Hasegawa S, Shimotani K, Kishi K, Watanabe H (2005) Electrochem Solid-State Lett 8:A119
33. Narayanan SR, Valder TI, Chun W (2002) United States Patent, 6,485,851

Chapter 4
Nanotechnology for Material Development on Future Energy Storage

Kiyoshi Kanamura, Hirokazu Munakata, and Kaoru Dokko

4.1 Introduction

Various kinds of energy devices have been developed as power sources for portable electronic devices and electric vehicles. Fuel cell, rechargeable lithium ion battery, and super capacitor are the most interesting devices, and they have been extensively studied to improve their electrochemical performance around the world [1, 2]. In these electrochemical devices, chemical energy is directly converted to electric energy through charge transfer process occurring at an interface between electrode and electrolyte. The electrochemical reactions take place at the interface and their reaction rates strongly depend on the nature of interface consisting of electrode and electrolyte materials. In some case, the electrode reaction is so slow that the electrode reaction kinetics should be carefully investigated in order to improve charge transfer reaction rate. On the other hand, the slow electrode reaction can be technically overcome by a large interface area for the electrode reaction, leading to an improvement of apparent reaction rate. For example, the true surface area of the porous electrode used in practical battery and fuel cell is much larger than that of flat electrode. When the surface area is 100 times larger than that of flat electrode, the apparent electrode reaction rate is also 100 times. However, this is too simple to estimate the advantage of the porous electrode. The porous electrode has so many problems that the reaction rate may not become 100 times [3]. Figure 4.1 shows the electrode reaction occurring on flat electrode and porous electrode. In the case of the flat electrode, the electrode reaction takes place uniformly on an entire electrode surface. On the other hand, the electrode reaction taking place on the porous electrode surface has a distribution of electrode reaction rate depending on its porous nature and a kind of electrode material. For example, both electronic and ionic conductivities of porous electrode are very important properties to establish an electrochemical interface and to realize apparently high charge transfer rate. One of the key technologies for porous

K. Kanamura (✉), H. Munakata, and K. Dokko
Graduate School of Urban Environmental Sciences, Tokyo Metropolitan University, 1-1 Minami-Ohsawa, Hachioji, Tokyo, 192-0397, Japan

T. Osaka et al. (eds.), *Electrochemical Nanotechnologies*,
Nanostructure Science and Technology, DOI 10.1007/978-1-4419-1424-8_4,

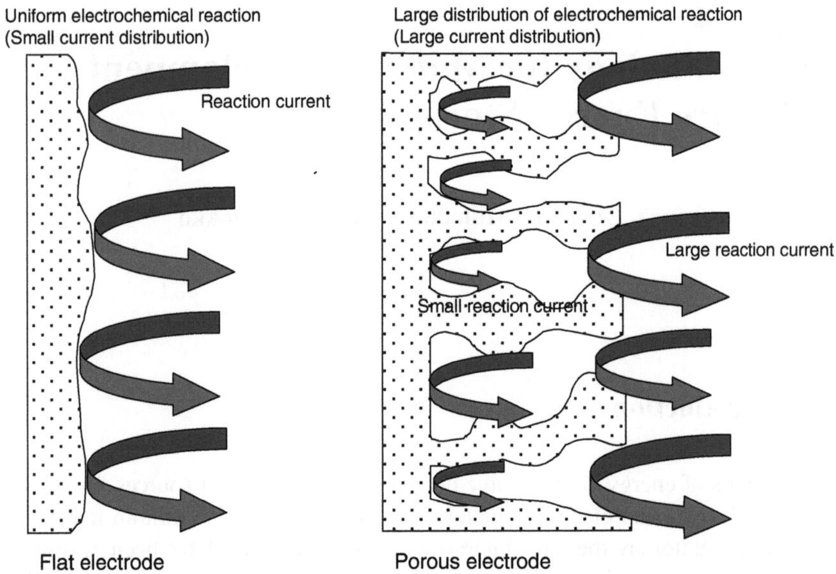

Fig. 4.1 Schematic illustration of an electrochemical reaction taking place on flat and porous electrodes

electrodes used in electrochemical energy conversion system is a fabrication process of porous electrode with three-dimensionally ordered porous structures. Recently, three-dimensionally ordered macroporous materials have been extensively studied on various application fields, such as catalyst, photonic material, sensor, and so on [4–11]. At first silica porous materials have been prepared by using colloidal crystal templating method. This study has inspired a lot of scientists working in the field of material science. So far, many kinds of macroporous materials, such as zirconia, titania, carbon, and so on, have been successfully prepared and applied to various applications. In this section, three applications of three-dimensionally ordered materials to electrochemical energy conversion systems are introduced.

4.2 Rechargeable Lithium Ion Battery

Three-dimensionally ordered macroporous ceramic with high Li^+ ion conductivity was prepared by colloidal crystal templating method using monodispersed polystyrene beads [12]. Monodispersed polystyrene beads with 3 μm diameter were dispersed in water and then filtrated by using a membrane filter under a small pressure difference. After this treatment, polystyrene beads were accumulated on the membrane filter with closed pack structure, as shown in Fig. 4.2. Then, the membrane consisting of accumulated polystyrene beads was removed from the membrane filter and put on a glass substrate. After drying at room temperature, the

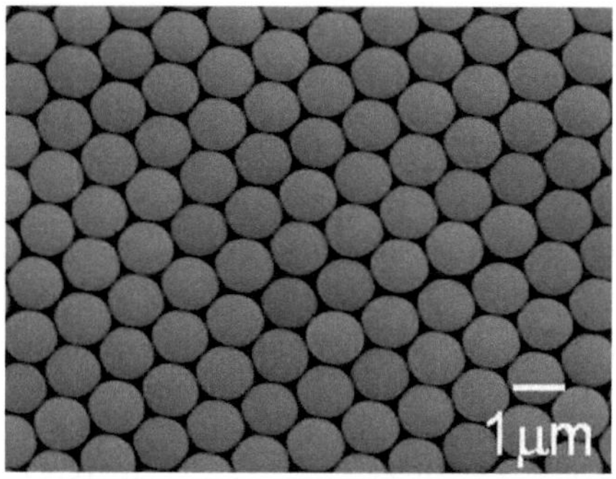

Fig. 4.2 Scanning electron micrograph of three-dimensionally ordered monodispersed polystyrene beads

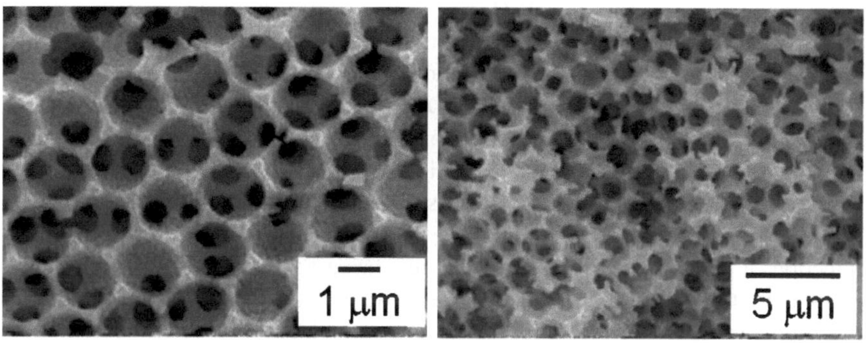

Fig. 4.3 Scanning electron micrographs of surface and cross section for prepared three-dimensionally ordered macroporous $Li_{0.35}La_{0.55}TiO_3$

polystyrene beads were heated at 110°C for 1 h to connect each polystyrene bead to improve the mechanical strength of the membrane consisting of polystyrene beads. Using this membrane as template, three-dimensionally ordered macroporous ceramic membrane was prepared.

A sol or nanoparticle for Li^+ ion conductive ceramic is necessary for preparing the three-dimensionally ordered macroporous structure of Li^+ ion conductive ceramic. For example, a sol for $Li_{0.35}La_{0.55}TiO_3$ has been prepared by using LiOH, $La(CH_3COO)_3$, and $Ti(C_3H_7O)$ as starting materials [13]. The membrane consisting of polystyrene beads was immersed in this sol under reduced pressure. Through this vacuum impregnation process, the sol was put into free volume between polystyrene beads and was converted to gel. The composite membrane consisting of polystyrene and gel was heated at 450°C to remove polystyrene and convert the gel to amorphous ceramic. Finally, the porous amorphous ceramic was crystallized by a heat treatment at 1,000°C. Figure 4.3 shows the scanning electron micrographs of surface and cross section of the prepared, three-dimensionally ordered, macroporous $Li_{0.35}La_{0.55}TiO_3$. This ceramic has a highly uniform porous structure and a high ionic conductivity of 4×10^{-4} S cm^{-1}. Each macropore is connected by small connecting through pores. This membrane is an electrolyte for all solid state electrode system. When this porous material is filled with active materials for rechargeable lithium ion battery, all ceramic electrode system is constructed. In order to fill active material, nanoparticle or sol for active material has to be prepared. For example, a sol for $LiCoO_2$ active material was obtained from $Li(C_3H_7O)$ and $Co(CH_3COO)_2$ as starting materials [14]. This sol was injected into three-dimensionally ordered macroporous $Li_{0.35}La_{0.55}TiO_3$ ceramic membrane by vacuum impregnation method. The injected sol was converted to the Li-Co-O gel during the following drying process. The composite membrane consisting of Li-Co-O gel and three-dimensionally ordered macroporous $Li_{0.35}La_{0.55}TiO_3$ was heated at 450°C to form amorphous $LiCoO_2$ inside of pores. This process should be repeated to increase amount of $LiCoO_2$ injected into three-dimensionally ordered macropores.

Finally, the composite membrane consisting of Li-Co-O gel and three-dimensionally ordered macroporous $Li_{0.35}La_{0.55}TiO_3$ was heated at 700–800°C to crystallize amorphous $LiCoO_2$. In the course of this heat treatment, electrolyte material might react with $LiCoO_2$. However, X-ray diffraction pattern of this composite membrane shows no reaction between $Li_{0.35}La_{0.55}TiO_3$ electrolyte and $LiCoO_2$ active material. Figure 4.4 shows the scanning electron micrograph of the prepared composite membrane consisting of $LiCoO_2$ and $Li_{0.35}La_{0.55}TiO_3$. Some of pores were fully occupied by $LiCoO_2$, but other pores were still open. This means that the impregnation of $LiCoO_2$ is not perfect. Probably, nanoparticle of $LiCoO_2$ is more suitable for injection of active material into macropores. From a density measurement of this membrane, the occupancy of $LiCoO_2$ was estimated to be 60%. Figure 4.5 shows the discharge and charge curves of all ceramic electrode system consisting of $LiCoO_2$ and $Li_{0.35}La_{0.55}TiO_3$.

The obtained curves were similar to those reported in literature [15]. This indicates that all solid state $LiCoO_2$ electrode functions as cathode of rechargeable lithium ion battery at relatively high discharge and charge rates. The thickness of

Fig. 4.4 Scanning electron micrograph of three-dimensionally ordered composite membrane consisting of $Li_{0.35}La_{0.55}TiO_3$ and $LiCoO_2$

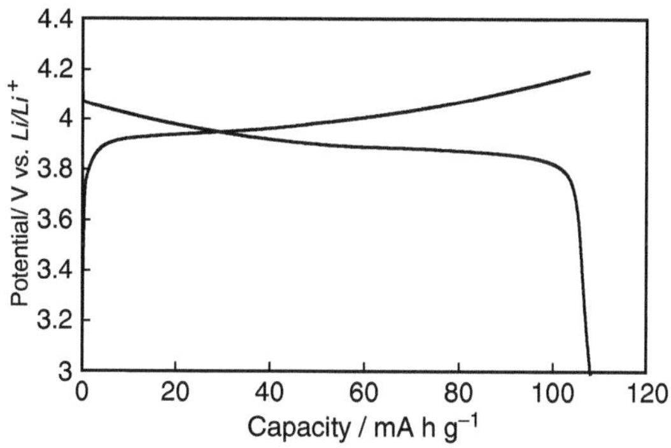

Fig. 4.5 Discharge and charge curves of three-dimensionally ordered composite electrode system consisting of $Li_{0.35}La_{0.55}TiO_3$ and $LiCoO_2$

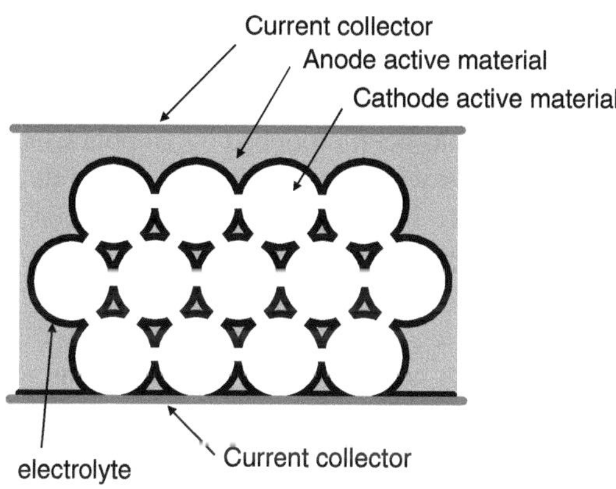

Fig. 4.6 Schematic illustration of 3D battery

this electrode was 100 μm which was thicker than that of standard electrode. Nevertheless, the discharge capacity of 108 mAh g^{-1} was obtained in this test. Thus, the three-dimensionally ordered macroporous solid electrolyte matrix is very useful in constructing all solid state rechargeable lithium ion battery. This is due to highly uniform structure which provides excellent ionic and electronic conductivities of all solid state electrode system.

In the above paragraphs, three-dimensionally ordered solid electrolyte was prepared for a construction of all solid state rechargeable lithium ion battery. On the other hand, various kinds of three-dimensionally ordered structures consisting of electrochemically active material have been prepared in order to realize extremely high rate capability of rechargeable lithium ion batteries with liquid electrolyte. In addition, such types of active materials have been prepared to establish three-dimensional batteries which is a new category of future lithium ion battery.

Figure 4.6 shows the schematic illustration of Three-Dimensional Battery (3D battery) consisting of three-dimensionally ordered macroporous materials. This is a very interesting concept for battery technology, and some groups are researching on this subject [16]. When 3D battery is constructed, the energy density and power density of rechargeable lithium ion batteries will be surprisingly improved.

4.3 Polymer Electrolyte Membrane Fuel Cell

As you know, the polymer electrolyte membrane fuel cell (PEMFC) has gained much attention around the world, due to its excellent characteristics, such as clean energy conversion system, high power density, and high energy density. In fuel cells, chemical energy is directly converted to electric energy with high efficiency in theory. This high efficiency is very attractive for electric vehicle and also for electronic devices, such as portable computer, cellular phone, and other small devices [2]. In fact, research and development of PEMFC are now very active. One of the key materials in PEMFC is a proton conductive membrane. The most famous membrane is Nafion® which has been produced by Dupont Company. Nafion® is a kind of perfluorinated polymer with ion exchange groups in its side chains, so that it is chemically and thermally more stable than hydrocarbon polymer [17, 18]. Nafion® easily swells by water or methanol absorption, leading to low stability of PEMFC operation and cross over flow of fuel [19]. Both problems result in lowering of energy conversion efficiency and power of fuel cell. Another important point is the cost of a membrane. Of course, perfluorinated polymer is more expensive than hydrocarbon polymer. In order to overcome these disadvantages, some new polymer electrolyte membranes have been prepared by using nanoparticles and nano-structured materials. Silica nanoparticles have been used as fillers for polymer membrane to realize high mechanical strength and thermal stability [20, 21]. In the case of polymer electrolyte membrane, the same method has been applied in order to improve mechanical strength and suppress the expansion of polymer by absorption of methanol or water. Another interesting method of suppressing the expansion of proton conductive polymer is the pore filling membrane technology [22, 23]. In this case, a mechanically strong polymer membrane is used as porous matrix and highly proton conductive polymer is utilized as proton conductive phase filled in pores. Many kinds of polymers have been used in various practical fields. Some of the polymers have high mechanical strength and no absorption of methanol and water. Polyethylene or polypropylene is classified as this type of polymer. Using this kind of polymer, nano-porous material has been prepared and applied to pore filling membrane. Figure 4.7 shows the schematic illustration of pore filling membrane for proton conductive polymer applied to PEMFC. The polymer for matrix should have a mechanical strength to prevent an expansion of polymer. The polymers filled in this matrix only have high proton conductivities, so that their high mechanical strength is not necessary due to mechanical hardness of matrix polymer. The same concept has been applied to composite membrane consisting of

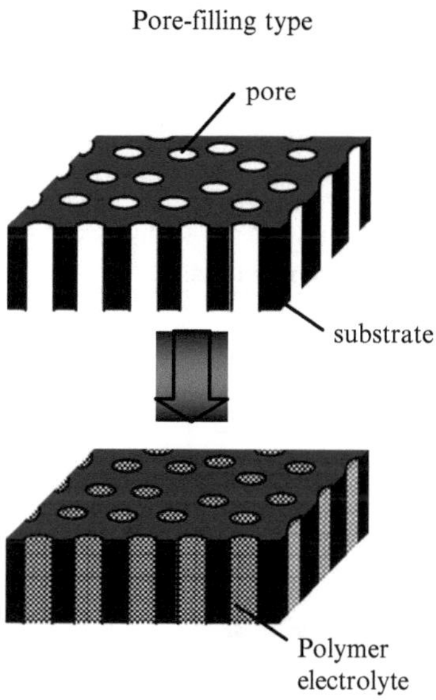

Fig. 4.7 Schematic illustration of pore filling polymer membrane

porous silica and proton conductive polymer [24]. There are many kinds of porous silica. Some silica matrix has been prepared and applied to proton conductive material. One such interesting material is sulfonated nano-porous silica.

Silica has a lot of OH groups on its surface which can react with silane coupling agents. Using this reaction scheme, the surface of silica has been modified for various applications. The nano porous silica with surface modification has been prepared by sulfonation of surface silanol groups, which exhibits relatively high proton conductivity at high temperatures without moisture. However, its proton conductivity is still lower than that of Nafion®. Another interesting material is macroporous silica with proton conductive polymer. Some research groups have reported a preparation of three-dimensionally ordered macroporous silica using a colloidal crystal templating method [25, 26]. Most of macroporous silica materials are too small to be applied to real materials, such as electrolyte membranes for fuel cells. In order to prepare a large size porous silica matrix with reasonable uniformity of macropores, the preparation method has been modified as shown in Fig. 4.8. In this preparation process, polystyrene beads (1,000 nm–100 nm in diameter) were mixed with silica beads (100 nm–10 nm in diameter) to prepare a stable suspension containing composite particles between polystyrene beads and silica beads. This suspension was filtrated with membrane filter equipment and both beads were accumulated on the membrane filter with ordered structure. After the filtration, the filter with thin membrane consisting of both beads was carefully removed and dried. After the drying process, the filter was removed to separate the membrane consisting of both beads. This membrane was heated at 450°C and then 800–1000°C in order to remove polystyrene beads and sinter colloidal silica particles. During the

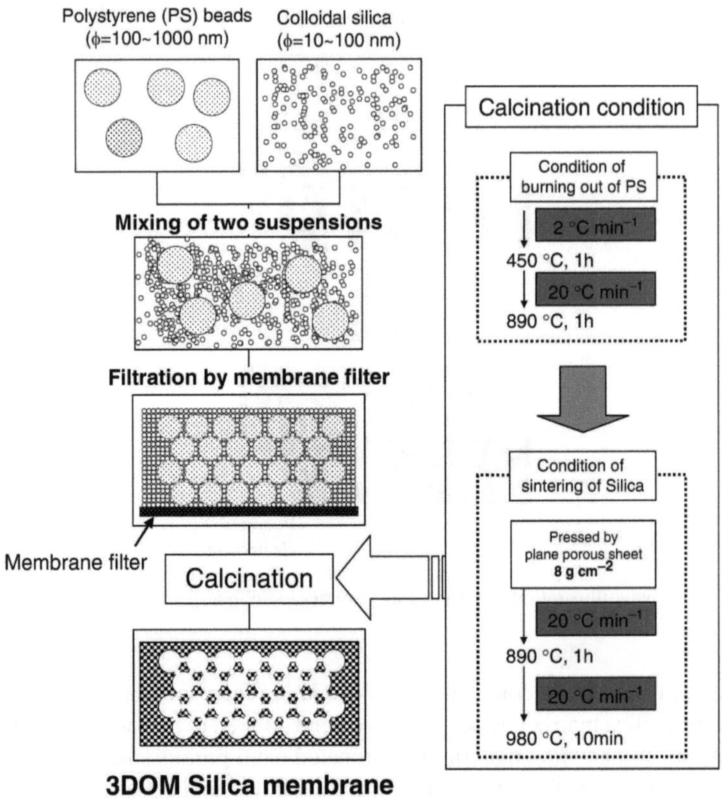

Fig. 4.8 Preparation process of three-dimensionally ordered macroporous silica membrane

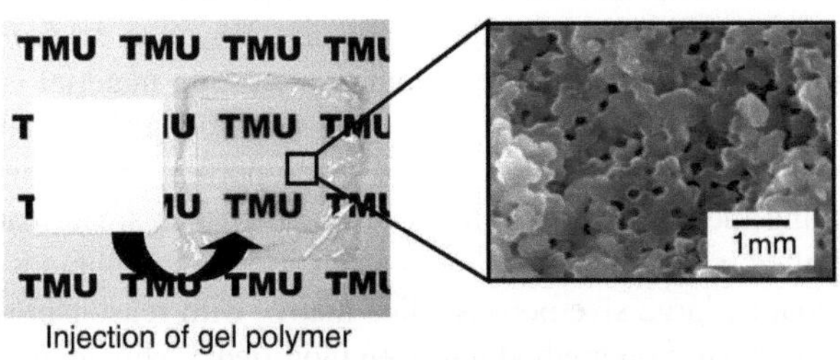

Fig. 4.9 Photograph of composite membrane between 2-acrylamido-2-methyl propane sulfonic acid and three-dimensionally ordered macroporous silica

sintering process, a proper weight was added to the porous silica membrane to obtain a flat silica membrane. After this heat treatment process, a proton conductive polymer was injected into all pores in the porous silica membrane. In Fig. 4.9, a photograph of the obtained composite membrane and a scanning electron micrograph of the composite membrane are shown.

Table 4.1 Proton conductivity and methanol permeability of the composite membrane with three-dimensionally ordered macroporous silica matrix. Measurement temperature: 30°C, Methanol concentration: 10 mol dm^{-3}

Membranes	Conductivity (S cm^{-1})	Permeability (10^{-6} cm^{2}s^{-1})	φ=conductivity/ permeability (10^4 S cm^{-3}s)
AMPS	0.24	5.4	4.4
Nafion®	0.062	3.3	1.9
Silica composite	0.081	0.82	9.9

Here, 2-acrylamido-2-methyl propane sulfonic acid (AMPS) and water were used to form proton conductive gel polymer in pores of macroporous silica membrane. A polymerization of monomer took place inside pores of porous silica matrix. After the polymerization, all pores were completely occupied by AMPS polymer. Both proton conductivity and methanol permeability of this composite membrane are summarized in Table 4.1 with data for Nafion® membrane. The proton conductivity of the composite membrane was slightly higher than that of Nafion, even though the nonconductive silica matrix occupied 30% of the composite membrane. This is due to a higher proton conductivity of AMPS polymer. On the other hand, methanol permeability of the composite membrane was strongly suppressed compared with original AMPS polymer and also smaller than that for Nafion® membrane. From these data, it can be said that the composite membrane is useful to improve the performance of proton conductive polymer. This technology can be applied to various kinds of composite membrane. Figure 4.10 shows the photograph and scanning electron micrograph of the composite membrane consisting of three-dimensionally ordered macroporous polyimide and proton conductive polymer and macroporous polyimide membrane. Three-dimensionally ordered macroporous polyimide was prepared by using the silica template with opal structure. In this case, mono-dispersed silica spherical beads can be used as template material. A suspension of mono dispersed silica beads was ordered by using filtration process or solvent evaporation process. A polyamic acid or other kinds of precursor for polyimide was injected into free volume between silica particles and then polymerized to prepare the composite membrane between silica and polyimide. After that, the silica particles were removed by an etching process with HF solution to obtain three-dimensionally ordered macroporous polyimide.

The membrane shown in Fig. 4.10 was prepared using this three-dimensionally ordered macroporous polyimide obtained according to the above process with AMPS polymer. The proton conductivity and methanol permeability of the composite membrane are summarized in Table 4.2. The proton conductivity of the composite membrane was higher than that of Nafion® and the methanol permeability of the composite membrane was slightly lower than that of Nafion®. Both tendencies are good for membrane for direct methanol fuel cell. In this way, three-dimensionally ordered macroporous materials are suitable for matrix of soft proton conductive polymer with higher proton conductivity.

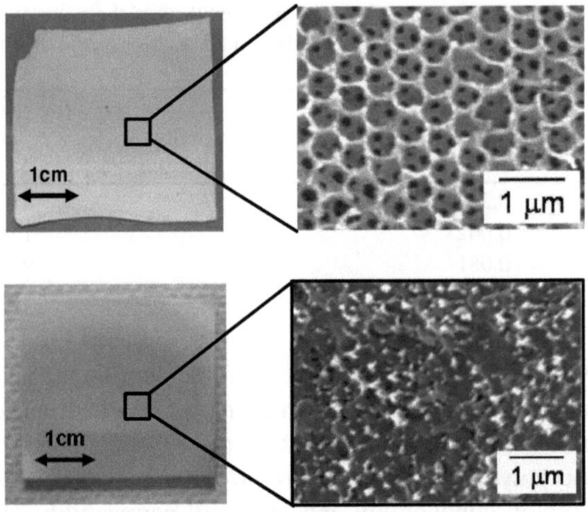

Fig. 4.10 Photograph and scanning electron micrograph of three-dimensionally ordered macroporous polyimide and composite membrane consisting of macroporous polyimide and proton conductive polymer

Table 4.2 Proton conductivity and methanol permeability of the composite membrane with three-dimensionally ordered macroporous polyimide matrix. Measurement temperature: 30°C, Methanol concentration: 10 mol dm^{-3}

Membranes	Conductivity (S cm^{-1})	Permeability ($10^{-6} cm^2 s^{-1}$)	φ = conductivity/ permeability (10^4 S cm^{-3} s)
AMPS	0.24	5.4	4.4
Nafion	0.062	3.3	1.9
Polyimide composite	0.12	1.3	9.0

4.4 Capacitor Material

Capacitor has been extensively investigated due to its high rate discharge and charge performance. Electrochemical double layer capacitor (EDLC) is now more interesting especially due to its relatively high capacity and high rate capability [27]. Usually, materials with very high surface area have been utilized to EDLC and their porous nature optimized to obtain high electric capacity. Most of the standard materials are carbon with macro, meso, and micro pores. The surface area of such kind of carbons ranges from 1,000 $m^2 g^{-1}$ to 2,000 $m^2 g^{-1}$. This is more or less very large surface area. However, electrochemical active pore is limited by the size of the pore. In general, micropores in carbon material cannot be utilized in EDLC devices. Mesopores are the most important in carbon materials to realize high electric capacity of EDLC [28]. So far, various kinds of methods and activation processes of carbon materials have been proposed to optimize pore size distribution which is a key issue for high electric capacity and high rate capability [29, 30]. For controlling pore size distribution in

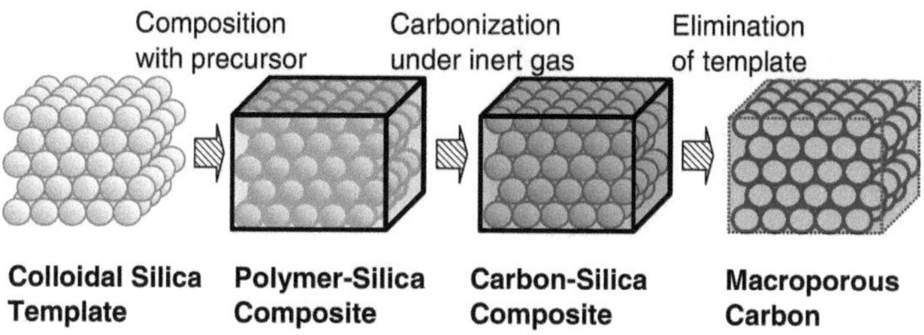

Fig. 4.11 Three-dimensionally ordered macroporous carbon by using monodispersed silica template and polyimide

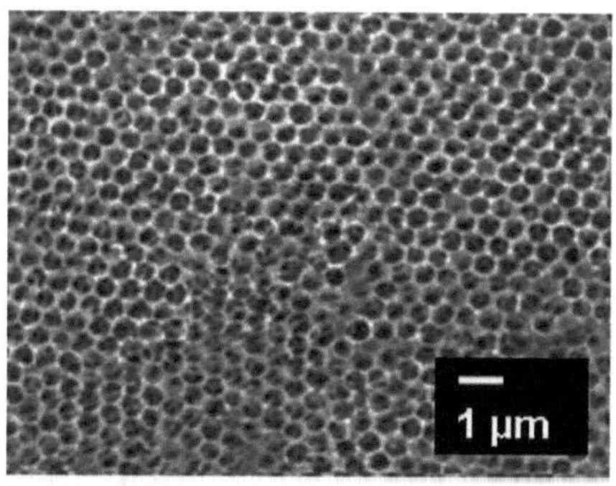

Fig. 4.12 Scanning electron micrograph of three-dimensionally macroporous carbon prepared by three-dimensionally macroporous polyimide

carbon materials, three-dimensionally ordered macroporous structure has been applied to EDLC material. Three dimensionally ordered macroporous carbon can be prepared by using colloidal crystal template consisting of silica mono-dispersed particles [31]. Figure 4.11 shows the schematic illustration of the preparation procedure for three-dimensionally ordered macroporous carbon. Firstly, mono-dispersed silica particles were assembled in ordered structure, which was an opal one, using membrane filter equipment. After that, the obtained silica membrane was slightly sintered to improve mechanical strength of the assembled silica particles. Then, organic compound, such as furfural alcohol, sugar, and precursor of polymers, was injected into the free space of silica matrix to prepare composite membrane consisting of organic compound and silica. This composite membrane was heated at a proper temperature of 800–1,000°C, under inert atmosphere to convert organic compound to carbon. After this heat treatment, the membrane consisting of carbon and silica was obtained. Finally, silica matrix was removed by using etching process with HF aqueous solution. Figure 4.12 shows the scanning electron micrograph of the prepared three-dimensionally ordered macroporous carbon which is synthesized from polyimide polymer.

The size of membrane was larger than 1 cm^2 and through-hole between macropores was clearly observed. This material has been applied to materials for EDLC and exhibits an interesting electrochemical property. The surface area of this carbon was much higher than that calculated from the size of the macropore, indicating

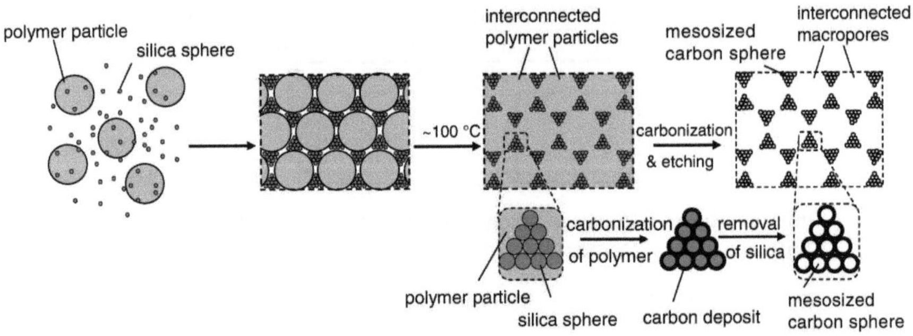

Fig. 4.13 Preparation process of three-dimensionally ordered macroporous carbon with controlled pore size distribution by using monodispersed polystyrene and silica beads

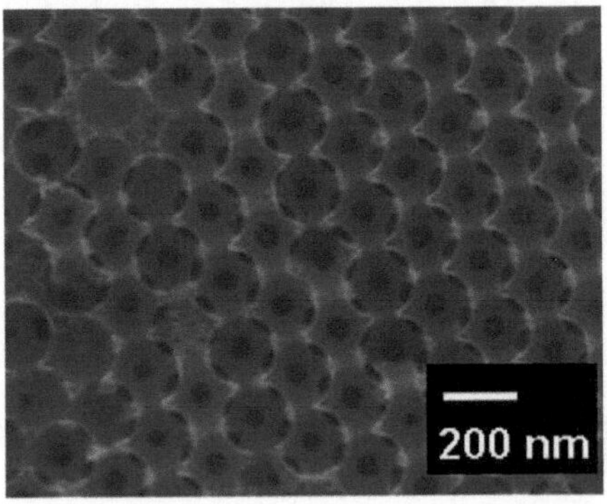

Fig. 4.14 Scanning electron micrograph of three-dimensionally ordered macroporous carbon with controlled pore size distribution

that both meso and micro pores are present in this carbon. Probably, such small pores exist in the walls of macropores. This is one of the interesting materials. Another interesting process for preparation of three-dimensionally ordered porous carbon is shown in Fig. 4.13. Polystyrene beads of relatively large size (for example 200 nm) and silica beads of small size (10–20 nm) are mixed and filtrated on membrane filter to prepare a composite between polystyrene beads and colloidal silica. In this process, the size of polystyrene beads and colloidal silica is vice versa. The composite membrane was heated at 450°C and then heated at 1,000°C under inert atmosphere. During the heat treatment, polystyrene beads were melted and penetrated into free space between colloidal silica particles. Probably, the colloidal silica particles are covered with melted polymer. After the heat treatment, three-dimensionally ordered macroporous material was obtained, which consisted of colloidal silica and carbon. This composite macroporous membrane is treated with HF aqueous solution to remove the silica part. Finally, the three-dimensionally ordered macroporous carbon was obtained. Figure 4.14 shows the scanning electron micrograph of the prepared carbon. A porous nature of this carbon is measured by B.E.T. method.

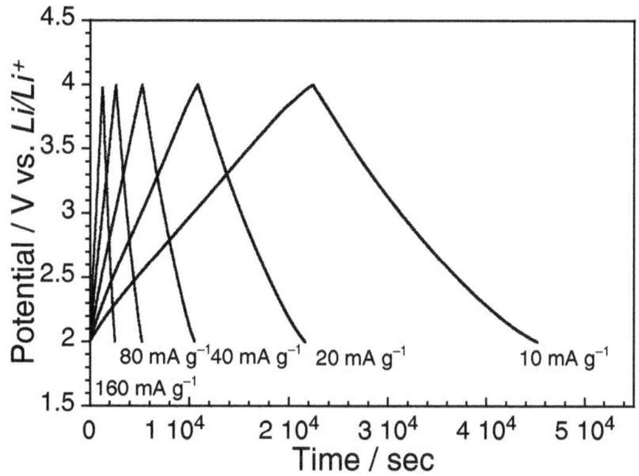

Fig. 4.15 Discharge and charge curves of three-dimensionally ordered macroporous carbon with controlled pore size distribution in Et_4NBF_4/PC at various current densities

This carbon membrane consisted of 10–20 nm pores and 200 nm pores. The pore size of 200 nm corresponds to the size of polystyrene used in this preparation process and the pore size of 10–20 nm reflects the size of colloidal silica. This carbon had high surface area, which was 1,004 $m^2 g^{-1}$. Such high surface area is suitable for capacitor application.

Figure 4.15 shows the discharge and charge curves of the prepared three-dimensionally ordered macroporous carbon with controlled meso pores in Et_4NBF_4/PC at various currents. From the results, the electric capacity of this carbon was estimated to be 90 $F g^{-1}$.

4.5 Summary

Three-dimensionally ordered porous materials have been applied to electrochemical energy conversion systems, such as lithium battery, fuel cell, and electrochemical double layer capacitor. Based on this technique, functional materials for other applications can be produced. The advantages of three-dimensionally ordered materials are based on micro or nano size ordered pores.

Such a structure provides highly uniform porous nature. This is the most important point. For example, the uniform reaction may occur in all the pores. A higher mechanical strength may be realized from ordered structure. Especially, two-dimensional electrochemical reactions are strongly enhanced by using three-dimensionally ordered porous materials, as mentioned above. In other words, two-dimensional electrochemical reactions are converted to pseudo three-dimensional ones. This procedure is useful for practical applications. Presently, preparation of three-dimensionally ordered macroporous materials is not so easy due to low mechanical strength and the presence of minor defects. More extensive research will be carried out in the near future.

References

1. Tarascon JM, Armand M (2001) Nature 414:359
2. Savadogo O (1998) J New Mater Electrochem Syst 90:47
3. Newman J, Electrochemical Systems, Third Edition, John Wiley and Sons, New York
4. Stein A, Schroden RC (2001) Curr Opin Solid State Mater Sci 5:553
5. Holland BT, Blanford CF, Stein A (1998) Science 281:538
6. Subramanian G, Manoharan VN, Thorne JD, Pine DJ (1999) Adv Mater 11:1261
7. Velev OD, Kaler EW (2000) Adv Mater 12:531
8. Bartlett PN, Baumberg JJ, Birkin PR, Ghanem MA, Netti MC (2002) Chem Mater 14:2199
9. Jiang P, Cizeron J, Bertone JF, Colvin VL (1999) J Am Chem Soc 121:7957
10. Jiang P, Hwang KS, Mittleman DM, Bertone JF, Colvin VL (1999) J Am Chem Soc 121:11630
11. Zakhidov AA, Baughman RH, Iqbal Z, Cui C, Khayrullin I, Dantas SO, Marti J, Ralchenko VG (1998) Science 282:897
12. Dokko K, Akutagawa N, Isshiki Y, Hoshina K, Kanamura K (2005) Solid State Ionics 176:2345
13. Kanamura K, Mitsui T, Rho YH, Umegaki T (2002) Key Engineering Materials 228–229:285
14. Rho Young Ho, Kanamura K, Umegaki T (2003) J Electrochem Soc 150(1):A107
15. Ohzuku T, Ueda A (1994) J Electrochem Soc 141:2972
16. Long JW, Dunn B, Rolison DR, White HS (2004) Chem Rev 104:4463
17. Gierke TD, Munn GE, Wilson FC (1981) J Polym Sci Polym Phys Ed 19:1687
18. Heitner-Wirguin C (1996) J Membr Sci 120:1
19. Ren X, Springer TE, Zawodzinski TA, Gottesfeld S (2000) J Electrochem Soc 147:466
20. Jung DH, Cho SY, Peck DH, Shin DR, Kim JS (2002) J Power Sourc 106:173
21. Bauer F, Willert-Porada M (2004) J Membr Sci 233:141
22. Yamaguchi T, Miyata F, Nakao S (2003) J Membr Sci 214:283
23. Zhou J, Childs RF, Mika AM (2005) J Membr Sci 254:89
24. Kikukawa T, Kuraoka K, Kawabe K, Yamashita M, Fukumi K, Hirao K, Yazawa T (2005) J Membr Sci 259:161
25. Velev OD, Jede TA, Lobo RF, Lenhoff AM (1998) Chem Mater 10:3597
26. Cassagneau T, Caruso F (2002) Adv Mater 14:34
27. Conway BE (1999) Electrochemical Super Capacitors. Kluwer Academic/Plenum Publisers, New York
28. Salitra G, Soffer A, Eliad L, Cohen Y, Aurbach D (2000) J Electrochem Soc 147:2486
29. Endo M, Maeda T, Takeda T, Kim YJ, Koshiba K, Hara H, Dresselhaus MS (2001) J Electrochem Soc 148:A910
30. Shiraishi S, Kurihara H, Shi L, Nakayama T, Oya A (2002) J Electrochem Soc 149:A855
31. Frackowiak E, Delpeux S, Jurewicz K, Szostak K, Cazorla-Amoros D, Beguin F (2002) Chem Phys Lett 361:35

Chapter 5
Micro/Nano Fabrication Technologies and Micro Flow Devices for Future Energy Devices

Shuichi Shoji, Masanori Ishizuka, Hironobu Sato, Takahiro Arakawa, and Jun Mizuno

Keywords μDMFC • 3-D microstructures • Microelectrodes • Micro fluidic devices • Microvalve

5.1 Introduction

In recent years, 3-D microstructures have been utilized to realize high performance micro devices and systems in Micro Electro Mechanical Systems (MEMS) [1]. Silicon has been widely used in MEMS field and micro/nano fabrication methods based on photolithography have been developed. In the case of 3-D micro structures fabricated using bulk Si, the Deep Reactive Ion Etching (DeepRIE) as well as the conventional anisotropic wet etching using alkaline solutions are now commonly used. For high performance energy devices, downsizing and 3-D structuring are indispensable. Since fine metallization methods on 3-D microstructures are requested, the uniform photoresist coating over highly topographic surfaces is one of the key technologies. Electrodepositon of photoresist is also excelled in the formation of uniform photoresist coating on 3-D structures, but it can be applied only on a conductive layer [2]. For commonly used processes, spray coating method is advantageous in many cases.

In order to realize low cost fabrication of micro energy devices, plastic micro/nano fabrications based on molding are attractive methods. Hot embossing, injection molding, and casting have been commonly used for this purpose [1]. However, conventional methods have to be refined to fabricate micro scale plastic structure. Plastic micro/nano fabrications called polymer MEMS technologies have been developed in recent years. For structuring the plastic devices and systems, bonding of plastic parts becomes very important. In order to realize high performance

S. Shoji (✉), M. Ishizuka, H. Sato, T. Arakawa, and J. Mizuno
School of Science and Engineering, Waseda University, 3-4-1 Okubo, Shinjuku-ku, Tokyo, 169-8555, Japan
e-mail: shojis@waseda.jp

T. Osaka et al. (eds.), *Electrochemical Nanotechnologies*,
Nanostructure Science and Technology, DOI 10.1007/978-1-4419-1424-8_5,

energy devices, increment of surface area of the anode and cathode per unit area is effective. 3-D metal microelectrodes have many advantages in this case. On the other hand, miniaturization of the direct methanol fuel cell (DMFC) is one of the trends in energy devices. Since precise flow control of the fuel solution and oxidant solution is required in microfabricated device, high performance microvalves that can control the flow of the order of μL or nL are also indispensable.

In this session, micro direct methanol fuel cells (DMFC) fabricated by Si MEMS and plastic MEMS are described. For examples of micro/nano fabrication technologies, fine micro plastic fluidic chips fabricated by molding and low temperature bonding are reported. The 3-D metal micro mesh electrode fabricated by simple UV photolithography is described next. For an example of micro flow devices, the pneumatically actuated microvalve and the microreactor having multireagent inlets are introduced.

5.2 DMFC Fabricated by Si MEMS

A new concept DMFC based on well considered electrochemical background and MEMS process was designed and fabricated [3]. The structure is shown in Fig. 5.1. The main feature of this device is the planer structure and the integration of anodes and cathodes side by side. This device has two micro channels for introducing new fuel solution and oxidant solution to the reaction area continuously to achieve high power generation. Another feature is that the inclement of the electrode surface area compared to the planer electrode used in the conventional bipolar type μDMFC.

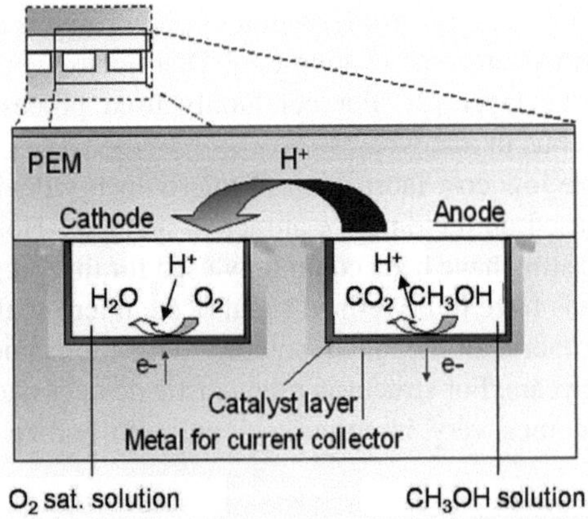

Fig. 5.1 Concept and principle of the DMFC fabricated by Si MEMS. [4] Reprinted from Proc. of Transducers '05, Houjou H, Motokawa S, Ishizuka M, Mizuno J, Momma T, Osaka T, Shoji S. Metallization on 3-D microstructures using spray coating for high performance micro direct methanol fuel cell (μDMFC), 1437, 2005, with permission from IEEE

The fuel solution is fed into the anode side channel. Methanol reacts at the anode and releases electrons, protons, and carbon dioxide. At the cathode, molecular oxygen reacts with proton being transported through the PEM (Polymer Electrolyte Membrane) from the anode and produces water. The electrons travel through the external circuit to the cathode. Power generation is performed by the above oxidation–reduction reaction principles. In order to realize micro-mini power sources, large numbers of DMFC should be integrated serially on a substrate. The structure of the proposed DMFC is suitable for this application.

Fabrication process of the DMFC is shown in Fig. 5.2. The simple structure enables easy fabrication. Chip size of the fabricated DMFC is 20 mm × 25 mm. A (100)-silicon wafer of 200 μm thickness is used as the substrate. 0.5 m silicon dioxide is formed thermally and patterned to form through-holes and channels. (a)–(b) The channels of the front side are etched by KOH from top side (c) and through-holes are etched by deep-RIE from the backside (d). The width of the channel and the distance between two channels are 100 μm. A silicon dioxide layer is grown thermally on the entire exposed surface of the silicon wafer for the isolation layer (e).

Figure 5.2f and g shows the metallization process on the 3-D microstructures. By spray coating, the photoresist film is formed on the 3-D topographic structures and patterned by photolithography. (f) Ti/Au for current collectors is formed by electron-beam deposition and subsequent lift-off process. (g) Catalyst layers of the anode and cathode are formed by electroplating Pt-Ru for the anode and Pt for the

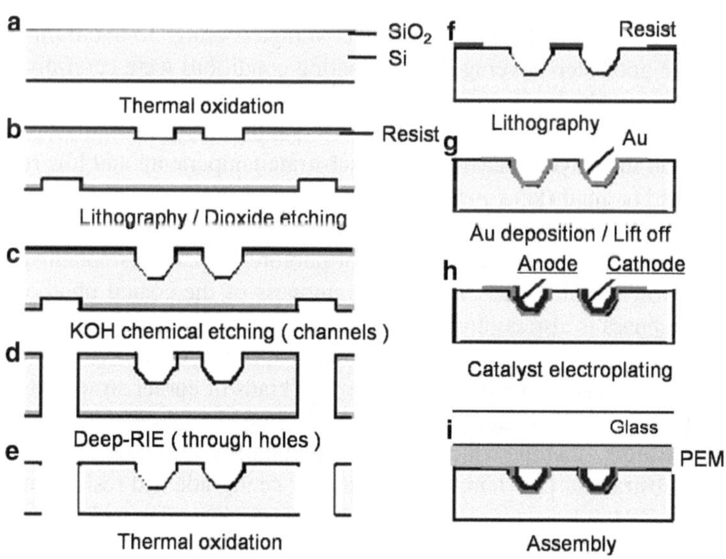

Fig. 5.2 Fabrication process flow of the DMFC chip. [4] Reprinted from Proc. of Transducers'05, Houjou H, Motokawa S, Ishizuka M, Mizuno J, Momma T, Osaka T, Shoji S. Metallization on 3-D microstructures using spray coating for high performance micro direct methanol fuel cell (μDMFC), 1437, 2005, with permission from IEEE

cathode, onto the Ti/Au 3-D electrodes. (h) Dupont™ Nafion112 is used as PEM. The thickness of Nafion112 membrane is 50 μm. Finally the DMFC is assembled by placing the Nafion112 membrane between the silicon substrate and the glass cover. These three layers are clamped mechanically (i).

In the above process, the metallization on 3-D microstructures is the key technology. Poor step coverage by spin coating for this 3-D structure is a big problem. Spray coating is indispensable to achieve good step coverage. Some papers on spray coating for 3-D structures were published [5, 6]. However, further optimization of the coating conditions is required to apply this method in actual use [4].

As the spray coating system, we utilized an EVG 101 spray coater, which was developed by the Electronic Vision Group [7]. In this system, the photoresist particles are generated by the ultrasonic spray nozzle. The distribution of particles is micrometer in size and the mean diameter of the particles is around 20 μm. AZP4620 positive photoresist (Clariant Corp.) of 40.9 % solid content and a viscosity of 442.4 cSt is used for the spray coating experience. To obtain the proper particle size distribution, it is necessary to dilute photoresist so that the viscosity becomes smaller than 20 cSt. During the spray coating process, the sample substrate is rotating with very low spin speed (60 rpm) in order to deposit photoresist to the whole substrate and minimize the centrifugal force.

The key point to performing good step coverage on 3-D microstructure is to dry quickly after the photoresist particles are sprayed to the surface. If drying speed is slow, the sprayed particles link together and large photoresist droplets are formed. The droplets flow down the sloped microstructure and get stacked at the cavity. These phenomena cause poor step coverage even in the spray coating method. As a result, these problems lead to poor step coverage coating. To avoid this problem and improve good step coverage, spray coating conditions were carefully studied.

Three factors which influence the step coverage of photoresist are the substrate surface temperature, the resist dilution ratio (solid content of diluted photoresist solution), and the solvent material. High substrate temperature and low resist dilution ratio lead to good step coverage. On the other hand, selection of solvent is also important. A solvent with low boiling point and high vapor pressure is suitable for good step coverage. Quick drying speed of photoresist particles obtains good step coverage. However it increases surface roughness of the coated photoresist. The surface roughness is also controlled by optimization of spray conditions. Trade off exists between the drying speed and the linking speed of photoresist particles.

The step coverage of the photoresist for two kinds of corner structures is shown in Figs. 5.3 and 5.4. These are SEM photomicrographs of cross-sectional views of the silicon wafers. Acetone which shows high volatility was chosen for the solvent material. Substrate surface temperature was 65° centigrade and resist dilution ratio was 1:10 (AZP4620: acetone). Good step coverage was obtained on the vertical (Fig. 5.4) as well as on the obtuse-angled corner (Fig. 5.3).

3-D lithography was performed by using the contact mask aligner (Karl Suss, MA6/BA6). The channel formed by anisotropic wet etching was chosen for the channel fabrication of the μDMFC because UV light can be irradiated to the sidewall. The photograph of the fabricated μDMFC is shown in Fig. 5.5. The performance of

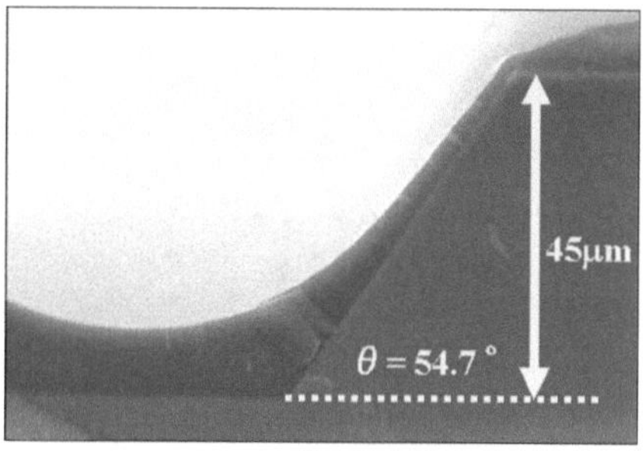

Fig. 5.3 Cross-sectional view (SEM) of photoresist coating on the channel formed by KOH anisotropic etching. [4] Reprinted from Proc. of Transducers'05, Houjou H, Motokawa S, Ishizuka M, Mizuno J, Momma T, Osaka T, Shoji S. Metallization on 3-D microstructures using spray coating for high performance micro direct methanol fuel cell (μDMFC), 1437, 2005, with permission from IEEE

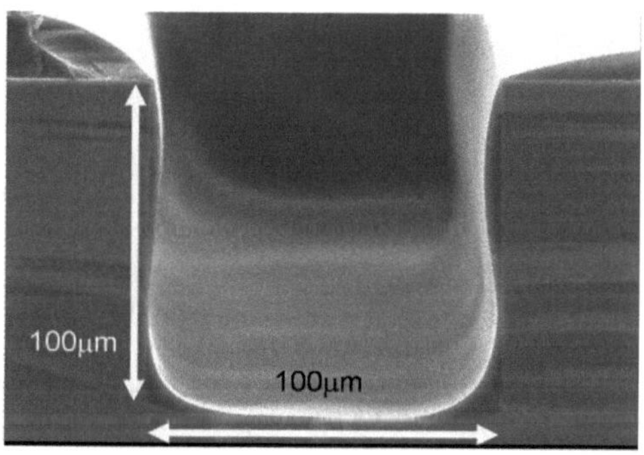

Fig. 5.4 Cross-sectional view (SEM) of photoresist coating on the channel formed by deep-RIE. [4] Reprinted from Proc. of Transducers'05, Houjou H, Motokawa S, Ishizuka M, Mizuno J, Momma T, Osaka T, Shoji S. Metallization on 3-D microstructures using spray coating for high performance micro direct methanol fuel cell (μDMFC), 1437, 2005, with permission from IEEE

the DMFC was tested at room temperature using 2 M methanol/0.5 M sulfuric acid/H_2O as the fuel and O_2-sat/0.5 M sulfuric acid/H_2O as the oxidant. The O_2 saturated solution was prepared by using oxygen bubbling into 0.5 M sulfuric acid/H_2O solution. The supply of fuel and oxidant was made by a micro syringe pump connected to the fabricated DMFC unit.

The power generation characteristics are shown in Fig. 5.6. Fuel flow rate is 5 μL min^{-1} and oxidant flow rate is 30, 40, and 50 Lmin^{-1}. There is a tendency for the performance to improve, as the oxidant flow velocity is increased. This is because the amount of supply of O_2 increases. The open circuit voltage of 548 mV and the maximum power density of 0.72 mW cm^{-2} have been obtained at room temperature with 2 M methanol. This power generation experiment shows that the DMFC with 3-D electrodes succeeds in actual electrical power generation.

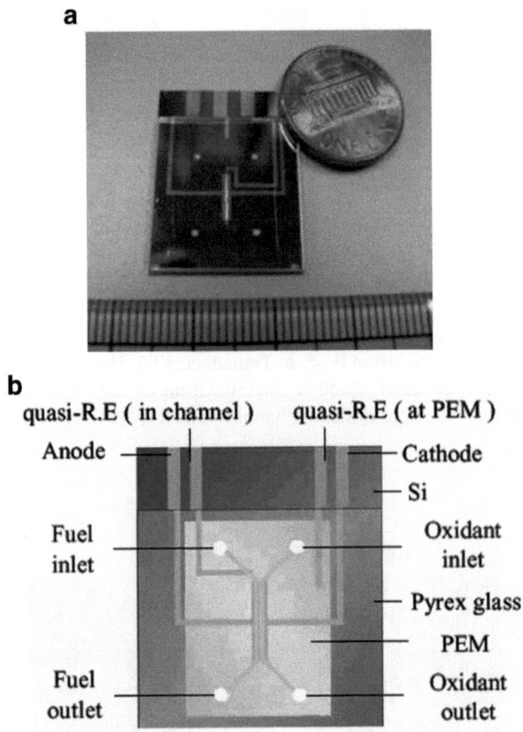

Fig. 5.5 Photograph of the fabricated DMFC (**a**) Over view (**b**) Top view. [4] Reprinted from Proc. of Transducers'05, Houjou H, Motokawa S, Ishizuka M, Mizuno J, Momma T, Osaka T, Shoji S. Metallization on 3-D microstructures using spray coating for high performance micro direct methanol fuel cell (μDMFC), 1437, 2005, with permission from IEEE

The open circuit voltage (548 mV) of the DMFC is almost same or a little higher than that of the other DMFC.

5.3 DMFC Fabricated by Plastic MEMS

μDMFCs reported so far were fabricated on the silicon or glass substrates [3, 6]. Considering the cost reduction, polymer micro devices were requested in actual use [8, 9]. A plastic DMFC, which has similar structure (Fig. 5.1), was then fabricated. Cyclo olefin polymer (COP) was chosen as the polymer material because of its high chemical durability and easy metallization of the electrodes. In order to achieve high power density, 3-D electrodes which increase the catalyst surface area keeping the total size were formed on the plastic 3-D structure.

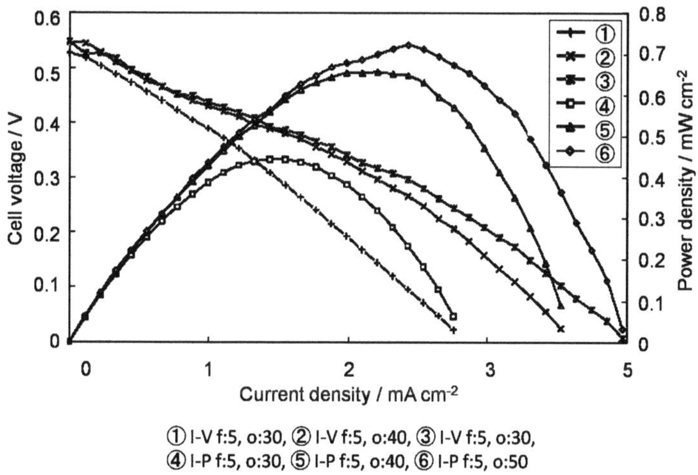

Fig. 5.6 Power generation characteristic of the fabricated DMFC

A Ni mold master was fabricated by silicon anisotropic etching and successive Ni electroplating. Figure 5.7 is the photomicrograph of the Ni mold. Hot embossing of the COP substrate was performed using hot embossing equipment (EVG, EVG520). The optimal embossing condition is 140°C and 2000 N press strength. To realize metal electrodes on a 3-D substrate, spray coating method as shown above was utilized for coating photoresist on the obtuse corner of the plastic micro channels [4]. The 3-D Au electrodes were formed by EB evaporation and lift-off process. Pt and Pt–Ru were deposited as the catalyst layers by electroplating.

The photograph of the plastic DMFC is shown in Fig. 5.8. The open circuit voltage of 237 mV and the maximum power density of 0.04 mW cm^{-2} have been obtained at room temperature with 2 M methanol.

5.4 Micro/Nano Fabrication Technologies

5.4.1 *Fine Plastic Fluidic Microchip*

A fabrication method for plastic microchip which has fine micro channel structures was developed by combining hot embossing and direct bonding techniques using PMMA [8]. For the fabrication of plastic microchips, thermal direct bonding method has been used. However, the deformation of the channel occurred during bonding and shallow micro channels of less than 10 μm were not fabricated. This was caused by the applied pressure under the heat around the glass transition temperature (T_g) of PMMA. Dimension control of about a few μm for the micro-channels was

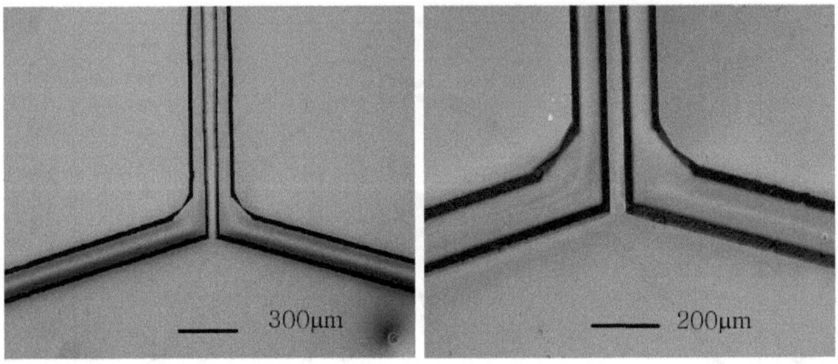

Fig. 5.7 Photomicrographs of the Ni mold master (*left*) and hot embossed plastic substrate (*right*)

Fig. 5.8 Photograph of the plastic DMFC

realized by the low temperature plasma activated bonding as well as the optimization of hot embossing conditions. In order to bond at lower than T_g, pretreatment was applied by oxygen plasma before bonding [10]. A PMMA microchip, which has shallow dams of about 5 μ gaps for efficient capillary electrophoresis, was fabricated.

Figure 5.9 shows the design of the microchip. Dimensions of the micro-channel are 50 μm wide and 19 μm deep. It has six ports (diameter: 1 mm) and two dams

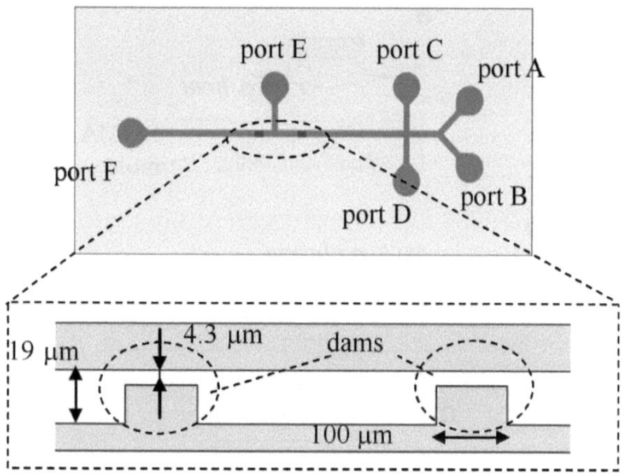

Fig. 5.9 The design of the plastic microchip; (**a**) whole view and (**b**) cross section of two dams. [10] Reprinted from Int. Conf. on Miniaturized Systems for Chemistry and Life Sciences: μTAS 2006, Shinohara H, Mizuno J, Kitagawa F, Otsuka K, Shoji S, Fabrication of Highly Dimension Controlled PMMA Microchip by Hot Embossing and Low Temperature Direct Bonding, 158, 2006, with permission from the Society for Chemistry and Micro-Nano Systems

(50 μm wide and 4.3 μm gap). The fabrication process of the microchip is shown in Fig. 5.10. The Si mold is fabricated by conventional photolithography and Deep-RIE methods. Mold lubricant (EGC-1720 from Sumitomo 3 M, Ltd.) is coated on the surface of the mold for easy demolding. The micro-channel region is fabricated by PMMA (COMOGLASS from Kuraray Co., Ltd.) hot embossing using EVG520HE of EV Group Co. (Fig. 5.10a).

Next, six ports are drilled in appropriate positions on the micro-channel plate. The surface of the micro-channel plate and the PMMA lid are then activated by an oxygen plasma activation bonding system (EVG810LT from EV Group Co.) (Fig. 5.10b). Finally, the two PMMA structures are directly bonded. The typical bonding conditions are -temperature of 85°C, pressure of 3 MPa, and annealing time of 3 min (Fig. 5.10c).

SEM micrographs of the micro-channel regions are shown in Fig. 5.11. Relative dimension differences between the Si mold and the embossed PMMA are smaller than 1%. This result indicates successful transfer from the mold to the PMMA plate.

5.4.2 Comb-Shaped 3-D Micromesh Metal Electrodes

The fabrication methods of the 3-D micromesh Ni structures were developed by modifying the multi-angle exposure of UV photpresist [11–13]. An inverse-micromesh frame was fabricated by multiple inclined backside exposures, and Ni was electroplated using the inverse-micromesh frame as a mold (Fig. 5.12). In this process, a

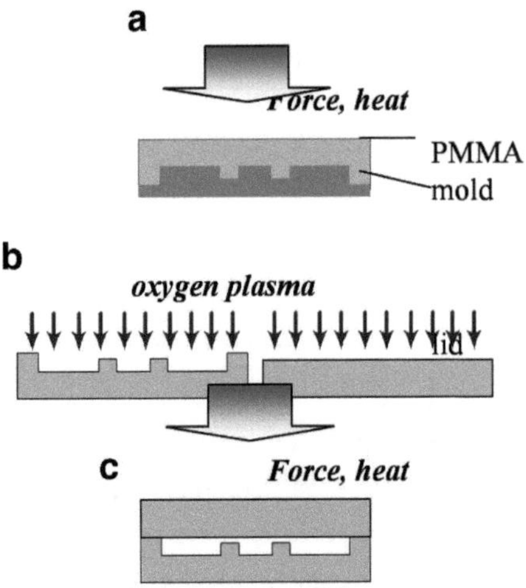

Fig. 5.10 Fabrication process of the plastic micro fluidic chip. [10] Reprinted from Int. Conf. on Miniaturized Systems for Chemistry and Life Sciences: μTAS 2006, Shinohara H, Mizuno J, Kitagawa F, Otsuka K, Shoji S, Fabrication of Highly Dimension Controlled PMMA Microchip by Hot Embossing and Low Temperature Direct Bonding, 158, 2006, with permission from the Society for Chemistry and Micro-Nano Systems

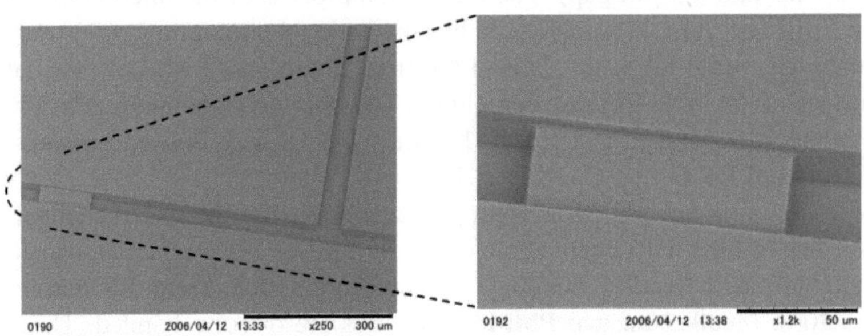

Fig. 5.11 SEM photomicrographes of the micro-channel with dam structure. [10] Reprinted from Int. Conf. on Miniaturized Systems for Chemistry and Life Sciences: μTAS 2006, Shinohara H, Mizuno J, Kitagawa F, Otsuka K, Shoji S, Fabrication of Highly Dimension Controlled PMMA Microchip by Hot Embossing and Low Temperature Direct Bonding, 158, 2006, with permission from Society for Chemistry and Micro-Nano Systems

ring-shaped Au appeared by over exposure of LA900 around the Cr/Au openings and was used as a seed layer for electroplating. In the next stage, micromesh structures on the planar comb-shaped electrodes were fabricated. We refined the above-mentioned fabrication method (Fig. 5.12) used for insulating between counter electrodes. To insulate electrodes, two individual Cr/Au patterns were formed on the topside of the glass substrate. However, if the topside patterned Cr/Au layer was used as a mask

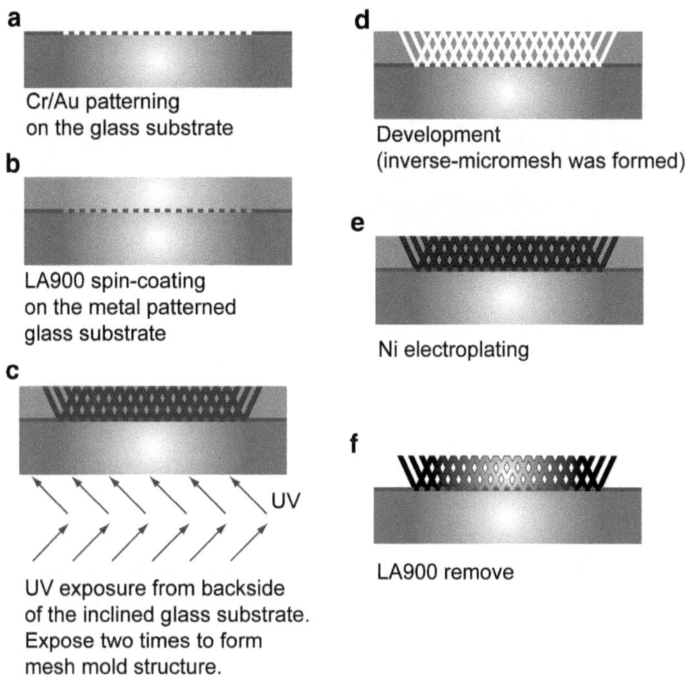

Fig. 5.12 Fabrication process of the 3-D metal micro meshes. [12]

during inclined backside exposure, unexpected Au as well as ring-shaped Au appeared by over exposure of LA900 at the edge of the planar electrodes and Ni was electroplated. To solve this problem, we combine the conventional techniques and two novel techniques of the backside Cr patterning and protecting resist.

Cr patterns are formed on the backside of the glass substrate in addition to surface Cr/Au patterns. Cr patterning on both sides enables to define the exposure areas. For finer patterns, exposure areas cannot be defined perfectly due to the influence of UV diffraction at the edge of backside Cr patterns even if Cr shadow layers are formed on both sides of a glass substrate. For this reason the negative resist (CA3000) was formed to avoid formation of Ni walls between metal electrodes. Two additional techniques enabled the realization of the 3-D micromesh metal structures formed on the planar electrodes on the glass surface.

By adding the backside Cr patterning and the protecting resist to the conventional techniques, comb-shaped 3-D micromesh Ni structures were fabricated as shown in Fig. 5.13. The gap between the micromeshes is 30 μm, the height of the mesh structure is 20 μm, and the diameter of the inclined micropillar is about 3 μm. The inverse-micromesh mold was filled with Ni using a ring-shaped Au as a seed layer for electroplating. Adequate metal is required on the Ni surface for actual use.

Fig. 5.13 Photograph of the electrode chip and magnified SEM micrograph of the metal micro meshes [12]

5.5 Microvalve

Microvalve has been gaining attention in the MicroTAS dealing with nanolitter or picolitter volume biochemicals due to low cost, straightforward fabrication process, and transparency of optical measurement. Pneumatically actuated microvalve made of Polydimethylsiloxane (PDMS) is one of the realistic devices for this purpose. [14, 15] However, conventional fabrication method of the pneumatically-actuated PDMS valves using a molding method provides sharp corners and weak PDMS-to-PDMS bonding strength. The sharp corner causes crucial leakage and diffusion pathway of chemicals. When the membrane valve is installed into a microchannel for miniaturization, sufficient adhesion force is required, in particular, between layers to resist an applied air pressure.

A microchannel valve having smooth surface and the stacked PDMS structures were fabricated (Fig. 5.14). Underneath the microchannel, a membrane and a pressure applying chamber are formed. The depth of microchannel of 50 μm and the membrane thickness of 60 μm were considered. For a 300 μm × 500 μm square membrane, it was numerically identified that more than a 150 kPa pressure was needed to obtain the deflection of 50 μm, which can close the microchannel perfectly.

In microfabrication, two critical points mentioned above were focused. Firstly, the smooth surface channel was obtained from PDMS replication of the convex rounded PMMA channel master, which was transferred from the concave round etched channels on a glass substrate. The molded PDMS layer was attached to the soft-cured PDMS film spin-coated on a silicon substrate. After hard cure, two PDMS layers were firmly bonded, implying the monolithic bonding process [16]. Based on the interfacial fracture toughness, the maximum bonding strength was evaluated by measuring the threshold air pressure of the adhesion test specimen fabricated with various bonding conditions. The PDMS microvalve with the smooth surface was fabricated by the optimized bonding condition. Figure 5.15 shows the cross-sectional view of the fabricated microvalve.

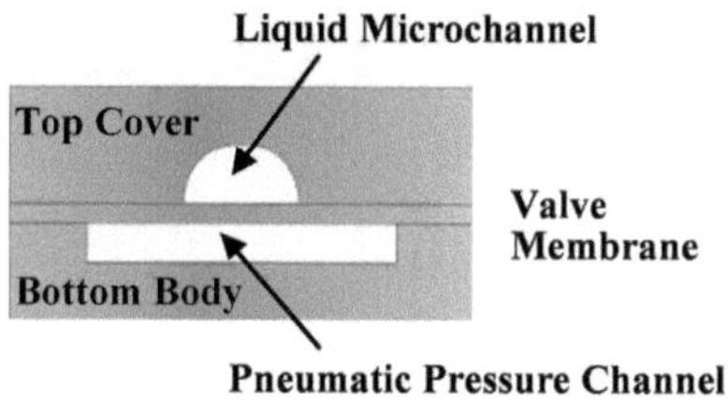

Fig. 5.14 Schematic of the pneumatic microvalve

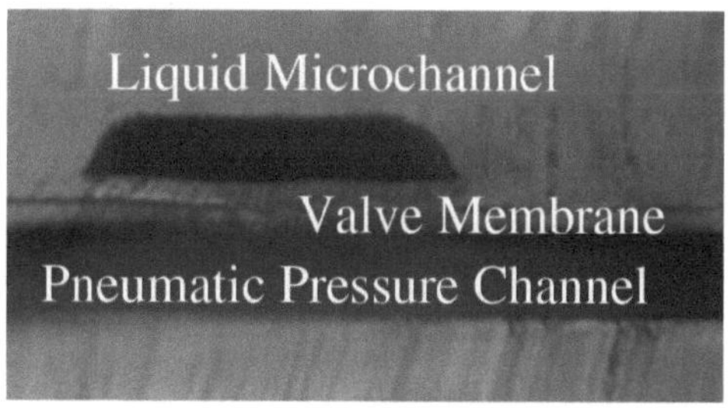

Fig. 5.15 Cross-sectional view of the fabricated PDMS microvalve

A microreactor having 4 reagent inlets was fabricated as the top view and is shown in Fig. 5.16. The PC controlled applied pressure system for driving the multi-inlets micro reactor was used as illustrated in Fig. 5.17. We characterized response time of the microchannel valve by applying stepped air pressure. A high speed CCD camera analyzed the valve Open–Close modes. It was determined that 190 ms were taken to bulge membrane perfectly, indicated to valve Open mode. The volume of the variable valve part was estimated to be about 7.5 nL. The smooth surface channel valve showed perfect sealing during the Close mode. Quick response within 500 ms was expected for sequential chemical reactions with this system. The PTFE coated PDMS microvalves were also fabricated considering the chemical durability. [17]

5.6 Conclusions

High efficient micro energy devices are realized by the increment of the electrode surface areas keeping the chip size. Use of the 3-D electrodes instead of the planer electrodes is quite effective for this purpose. To realize a metal pattern on 3-D structures, spray coating method was utilized and the optimum condition to achieve good step coverage was observed. Good step coverage of photoresist was obtained on the obtuse corner formed by anisotropic wet etching of (100) silicon and even

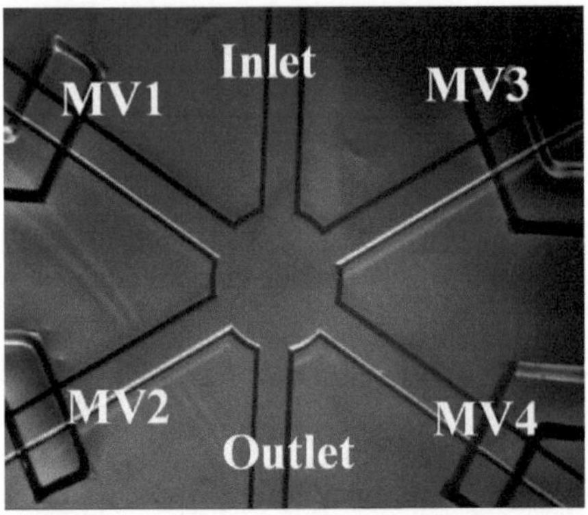

Fig. 5.16 Top view of microreactor having 4 reagent inlets

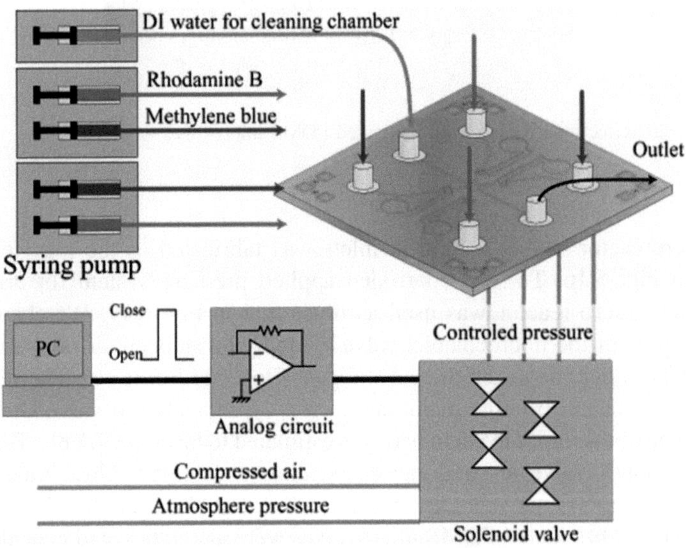

Fig. 5.17 PC control system for driving the multi-inlets micro reactor

vertical corner formed by deep RIE. We applied this method to fabricate a high performance μDMFC. The open circuit voltage of 548 mV and the maximum power density of 0.72 mW cm^{-2} were obtained at room temperature with 2 M methanol. A MEMS based μDMFC was also realized on the COP substrate. Hot embossing is utilized to fabricate the polymer micro channels, and spray coating method is applied to fabricate 3-D micro electrodes structure on the polymer substrate. This structure can apply to a μDMFC array chip which has numbers of cells in series.

A PMMA microchip, which has shallow dams of about 5 μm gaps, was fabricated and evaluated. Accurate dimension control of the micro-channels was realized by not only optimizing hot embossing conditions but also employing low temperature plasma activated bonding. The plastic MEMS fabrication technologies including bonding are indispensable for low cost energy devices. The 3-D metal micromesh structures fabricated by multi-angle UV exposure and electroplating increase the surface area of the metal electrodes. This kind of 3-D structure is useful in achieving high energy density of future μDMFC.

In order to improve the performance of μDMFC, precise flow control of the fuel solution and oxidant solution is quite important. The pneumatic microvalves are one choice for this application. However, we have to choose the materials of the microvalve structures considering the chemicals. In actual design of the high performance μDMFC, fluidic simulation tools in micro meter scale channels for micro flow control play important roles.

References

1. Shoji S (2000) Micromachining for biosensors and biosensing systems. In: Yong VC, Ngo TT (eds) Biosensors and their applications. Kluwer Academic/Plenum Pub, New York, pp 225–241
2. Heschel M, Boustra S (1997) Conformal coating by photoresist of sharp corners of anisotropically etched through-holes in silicon. Tech Dig Transducers'97, vol 1, pp 209–212
3. Motokawa S, Momamedi M, Momma T, Shoji S, Osaka T (2004) MEMS-based design and fabrication of a new concept micro direct methanol fuel cell. Electrochem Commun 6:562–565
4. Houjou H, Motokawa S, Ishizuka M, Mizuno J, Momma T, Osaka T, Shoji S (2005) Metallization on 3-D microstructures using spray coating for high performance micro direct methanol fuel cell (μDMFC). In: Proceeding of Transducers'05, Seoul, pp 1437–1440
5. Singh VK, Sasaki M, Song JH, Hane K (2003) Heating effect on photoresist in spray coating technique for three-dimensional lithography. Jpn J Appl Phys 42:4027–4030
6. Pham NP, Boellaard E, Burghartz JN (2004) Photoresist Coating Methods for the Integration of Novel 3-D RF Microstructures. J Microelectrochem Syst 13(3):491–499
7. Wieder B, Brubaker C, Glinsner T, Kettner P, Nodes N (2002) Spray coating for MEMS, NEMS and Micro systems, Pacific Rim Workshop on transducers an micro/nano technologies
8. Mizuno J, Honda T, Glinsner T, Ishizuka M, Edura T, Tsutsui K, Ishida H, Shoji S, Wada Y (2004) Fabrications of micro-channel devices by hot emboss and direct bonding of PMMA. In: International conference on MEMS, Nano, and Smart systems, Banff, pp 26–29
9. Mizuno J, Ishida H, Farrens S, Dragoi V, Shinohara H, Suzuki T, Ishizuka M, Glinsner T, Lindner F P, Shoji S (2005) Cyclo-olefin polymer direct bonding using low temperature plasma activation bonding. In: Proceeding of Transducers'05, Seoul, pp 1346–1349
10. Shinohara H, Mizuno J, Kitagawa F, Otsuka K, Shoji S (2006) Fabrication of highly dimension controlled PMMA microchip by hot embossing and low temperature direct bonding. In: International conference on miniaturized systems for chemistry and life sciences: μTAS 2006, Tokyo, pp 158–160.
11. Sato H, Kakinuma T, Go JS, Shoji S (2004) In-channel 3-D micromesh structures using maskless multi-angle exposure and their microfilter application. Sens Actuators A 111:87–92
12. Sato H, Yoshimine K, Otsuka T, Shoji S (2007) Interdigitated array 3D micromesh electrodes for electrochemical sensors. J Micromech Microeng 17:909–914

13. Sato H, Houshi Y, Otsuka T, Shoji S (2004) Fabrication of polymer and metal three-dimensional micromesh structures. Jpn J Appl Phys 43:8341–8344
14. Thorsen T, Maerkl SJ, Quake R (2002) Microfluidic large-scale integration. Science 298:580–584
15. Jeon NL et al (2002) Design and fabrication of integrated passive valves and pumps for flexible polymer 3-dimensional microfluidic systems. Biomed Microdevices 4(2):117–121
16. Arakawa T, Go JS, Jeong EH, Kawakami S, Takenaka K, Mori M, Shoji S (2004) 3-dimensional nano volume PDMS microreactor equipped with pneumatically-actuated in-channel membrane valves. In: International conference on miniaturized systems for chemistry and life sciences: μTAS2004, Malmo, pp 381–383
17. Kanai M, Uchida D, Sugiura S, Shirasaki Y, Go JS, Nakanishi H, Funatsu T, Shoji S (2003) PDMS microfluidic devices with PTFE passivated channels. In: 7th International conference on micro total analysis systems: μTAS2003, vol 1, Squaw Valley, pp 429–432

Part II
Nanotechnologies for Magnetic Storage Devices

Chapter 6
Magnetic Heads

Tokihiko Yokoshima

6.1 Introduction

Figure 6.1 shows how rapidly the areal density of hard disk drives (HDD) has been increasing over the past 20 years [1]. Several critical innovations were necessary to bring about such rapid progress in the field of magnetic recording [2]. One of the most significant innovations from the viewpoint of material improvement was the electrodeposition of permalloy ($Ni_{80}Fe_{20}$), which was introduced by IBM in 1979 as the core material of a thin-film inductive head to increase the magnetic recording density [3]. After the introduction of the magneto-resistive (MR) element as the read head and the electrodeposited permalloy as the write head by IBM in 1991 [4], the rate of increase in the recording density of HDDs jumped from 30% per year to 60% per year. Recently, a giant magneto-resistive (GMR) element has been used for the read element instead of the MR element. The rate of increase in the recording density jumped to over 100% per year in 1999, which is an *incredible* rate of increase. Since 2002, however, the rate of increase has decreased to 30%; thus, new innovations are required to maintain the rate of increase. In 2004, the practical use of perpendicular magnetic recording instead of longitudinal magnetic recording was announced [5]. This system is a critical innovation for developing high-performance HDD systems with high-recording density. The design of the magnetic recording head was changed because of the change of the recording system.

The development of a new magnetic recording head with higher performance and smaller dimensions is a key requirement for realizing high-density magnetic recording. Furthermore, to meet the demand for the rapid increase in magnetic recording density, the soft magnetic film in the core must have low magnetostriction, λ_s, high electrical resistivity, ρ, high thermal stability, low film stress, and high corrosion resistance, as well as high saturation magnetic flux density, B_s. The recording heads, using higher magnetic materials with high value of B_s can write to a high-H_c

T. Yokoshima (✉)
High Density Interconnection Group, Nanoelectronics Research Institute (NeRI), National Institute of Advanced Industrial Science and Technology (AIST), Tsukuba Central 2, 1-1-1 Umezono, Tsukuba, Ibaraki, 305-8568, Japan
e-mail: t.yokoshima@aoni.waseda.jp

T. Osaka et al. (eds.), *Electrochemical Nanotechnologies*,
Nanostructure Science and Technology, DOI 10.1007/978-1-4419-1424-8_6,

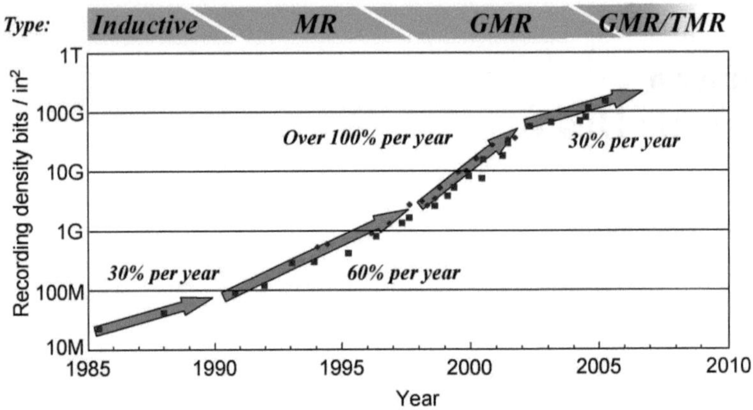

Fig. 6.1 Areal density of hard disk drives has been increasing over the past 20 years [1]

Fig. 6.2 SEM image and a schematic of a GMR head [1]

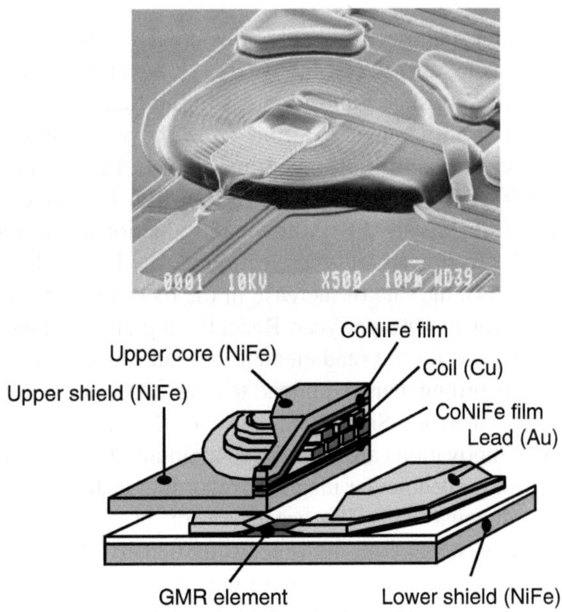

medium because of the high magnetic field generated. On the other hand, for higher-frequency recording, the soft magnetic film should have higher resistivity while maintaining a high value of B_s to suppress the eddy-current loss. Therefore, B_s and ρ of the soft magnetic thin film used for the recording head should be much higher than those of the conventional $Ni_{80}Fe_{20}$ permalloy, for which B_s = 9–10 kG and ρ = 15–20 μΩcm. So, permalloy with the composition of $Ni_{45}Fe_{55}$ was used instead of $Ni_{80}Fe_{20}$. Recently, CoNiFe has been used instead of permalloys. However, the development of a new magnetic recording head with smaller dimensions is strongly required for realizing a higher-density magnetic recording device. An SEM image and a schematic of a GMR head are shown in Fig. 6.2 [1].

6.2 Thin-Film Head

6.2.1 *Mechanism of Magnetic Recording*

Figure 6.3 shows system of magnetic recording of HDD. A recording bit is a small area of the recording layer, and the bit is magnetized. The magnetized bit pattern is made by magnetic flux, which is formed by an exciting current in the coil of the recording head. The recording bit is read by the read-head element by measuring the magnetic flux from the recording layer on the medium, which is rotated. Ferromagnetic material is used as the core of the magnetic-electric converter. Generally, soft magnetic materials that have high permeability (μ) and low coercivity (H_c) are used. Contact heads, which are in contact with the medium, and flying heads, which are not in contact with the medium, can be used for the magnetic recording. Magnetic tapes and floppy disks use contact heads, and HDDs use flying heads. The distance between the surface of the flying head and the medium is less than 10 nm, and the distance is maintained by an air flow between the medium and the slider on which the head is installed.

6.2.2 *Thin-Film Inductive Head*

A magnetic recording head that uses magnetic flux flowing from and into the coil for read–write operations, is called an "inductive head". Figure 6.4 shows a schematic illustration of an inductive head.

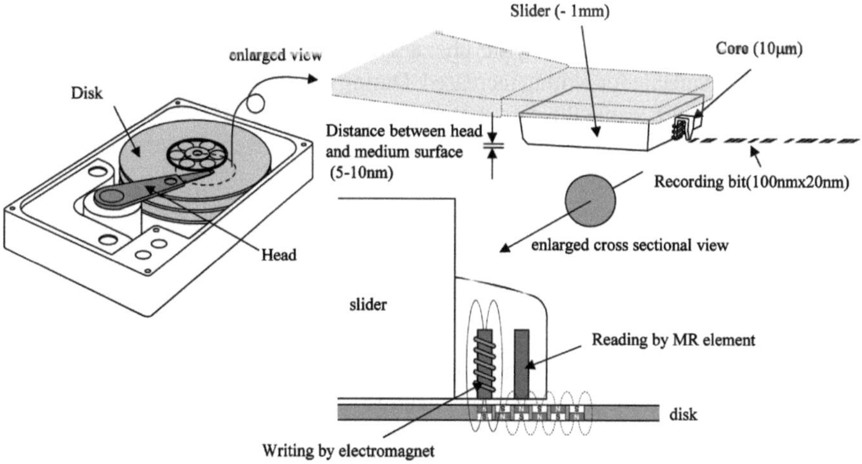

Fig. 6.3 System of magnetic recording of HDD

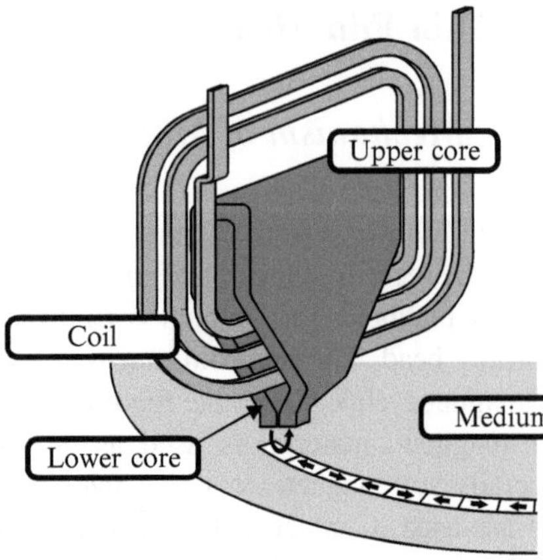

Fig. 6.4 Schematic illustration of an inductive head

In the case of bit writing, the magnetization of a small area of the magnetic layer is changed to produce a recording bit by the leakage of magnetic flux through a gap of the core, resulting from magnetic flux produced from coils by an exciting current. In the case of bit reading, an electric signal generated by electromagnetic induction in the coil is read by measuring the leakage of magnetic flux through the coil from the gap to the core. In 1960, inductive heads with a ferrite core and Cu wire coils were used. In 1979, a thin-film inductive head was developed by IBM Corporation [3], and the recording density increased rapidly as a result of this innovation. The development of the earlier inductive heads had involved the machine-tooling of the ferrite and the manual twisting of Cu-wire. However, the thin-film inductive head was not developed mechanically but by sputtering, evaporating, and plating. Since the core and the coil were produced by a thin-film process involving photolithography without using machine tools, a thin and fine core and coil could be realized. Thus, the read–write characteristics improved rapidly, and the rate of production increased dramatically. Since the core material was changed from ferrite to a metal-thin film, a high B_s was achieved. As a result, the read–write characteristics improved dramatically and the magnetic flux available for writing increased. During that time, soft magnetic permalloy ($Ni_{80}Fe_{20}$) thin film became the core material in practical use. The B_s values of ferrite and $Ni_{80}Fe_{20}$ permalloy are 3–5 kG and B_s=9–10 kG, respectively. A variety of methods have been used for realizing the head core.

6.2.3 *Magnetoresistive (MR) Head*

In 1991, IBM Corporation developed an MR head for practical use: a thin-film head with magnetoresistive element. The magnetoresistance effect is the phenomenon in which the resistance of a magnetic material changes with the magnetic field. The effect is also called an anisotropic magnetoresistive effect (AMR) to distinguish it from the giant magnetoresistive effect (GMR). The MR device in the head is used for reading, and a common inductive head is used for writing. A head in which a soft magnetic thin

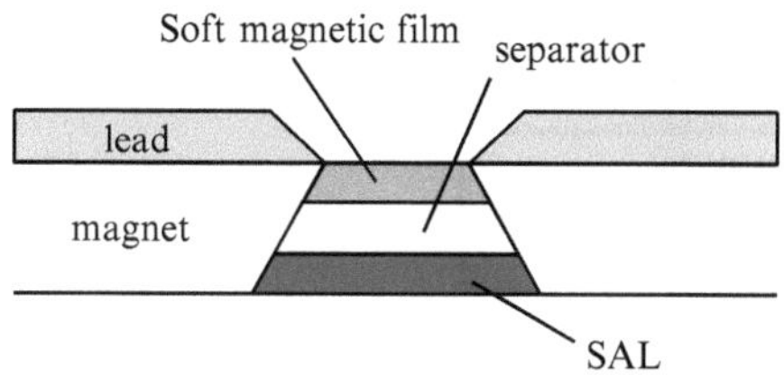

Fig. 6.5 Structure of an MR element

film is used, both as a lower core and an upper shield of the MR element, is called a merged-type MR head. Figure 6.5 shows the structure of an MR element.

When the direction of the current that flows in a thin film is the same as the direction of magnetization of the thin film, the resistance of the film increases. On the other hand, when the direction of magnetization of a thin film and the direction of the current that flows in the film are perpendicular, the resistance of the film decreases. Since the voltage in an MR element changes upon changing the soft magnetic thin films in the MR element due to the leakage of flux from the recording bit on the recording layer, the recording bit can be detected. The output signal is proportional to the change in magnetic flux density with respect to time because the electromagnetic induction of the inductive head is used for writing. The relative speed of the head relative to the medium must be high. On the other hand, the output signal depends on magnetic flux density in the MR element. Thus, the relative speed of the head to the medium, i.e., the rotation speed, does not have to be high.

A request of properties of the inductive head could be altered because the read and write elements were divided. Since the intuitive element is not used as the read element, the high sensitivity of the inductive element is unnecessary. Therefore, the number of turns of coil could be decreased, and furthermore, the recording head could become smaller. The soft magnetic thin films with very low hysteresis loss, that is the films with very low coercivity, is not required. So, some soft magnetic films, except for commonly used permalloy, could be used for core materials. Now, high B_s soft magnetic thin films are in practical use for realizing high-recording density.

In 1999, a GMR head, applying a giant magnetoresistive effect instead of an MR element came into practical use. The GMR effect is of very high sensitivity compared with the MR effect, and this resulted in the rate of increase in recording density jumping to over 100% per year. In 2005, a TMR head applying the Tunneling Magnetoresistance(TMR) effect came into practical use. TMR and GMR heads have been in practical use ever since.

6.2.4 Development of Thin-Film Head

The process flow for the thin-film head is shown in Fig. 6.6 [6]. The Cu coils, Cu lead, Au pad, and the core of the soft magnetic thin films in the head are prepared by electrodeposition. These metal parts require a high-aspect ratio. Also, the method

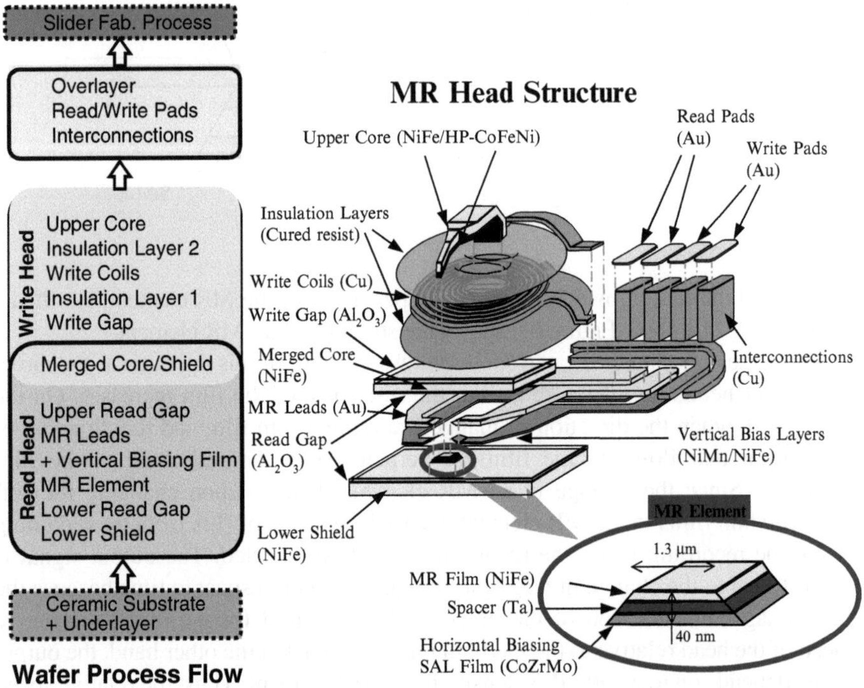

Fig. 6.6 Process flow for the thin-film head

of electrodeposition with photolithography is advantageous for selective deposition and the easy formation of thick films. In particular, the head core of the magnetic thin films is very fine with a width of less than 1 μm. Electrodeposition has a serious problem with the distribution of current density because the distributions of film thickness and metal composition occur often. pH modification near the deposition surface occurred by a side reaction that generated H_2. The alloy composition was changed markedly by this slight modification of pH during electrodeposition. In particular, the electrodeposition of NiFe alloy is very easily changed by pH modification during electrodeposition. Furthermore, obtaining a constant film composition is very difficult in Fe-based plating because of "anomalous codeposition". It is necessary to maintain a deviation of film composition of less than 0.1% because the magnetic properties of permalloy films depend on the film composition.

This serious problem was solved by two new plating developments: the paddle plating cell system and the frame plating method. Figure 6.7 shows a schematic illustration of the paddle plating cell system [7]. This method of plating surfaces allows the formation of films with constant composition because the plating solution near the deposition surface is agitated by a "paddle". This agitation also removes bubbles on the deposition surface and unwanted products formed during electrodeposition. Effective agitation suppresses pH modification during electrodeposition, thus ensuring a constant film composition. Figure 6.8 shows the process

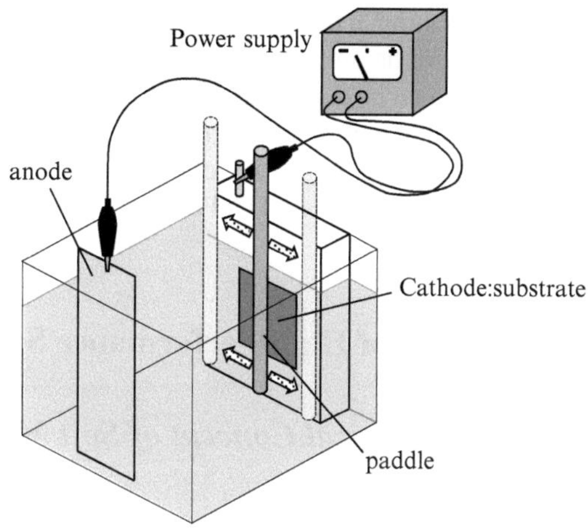

Fig. 6.7 Schematic illustration of the paddle plating cell system

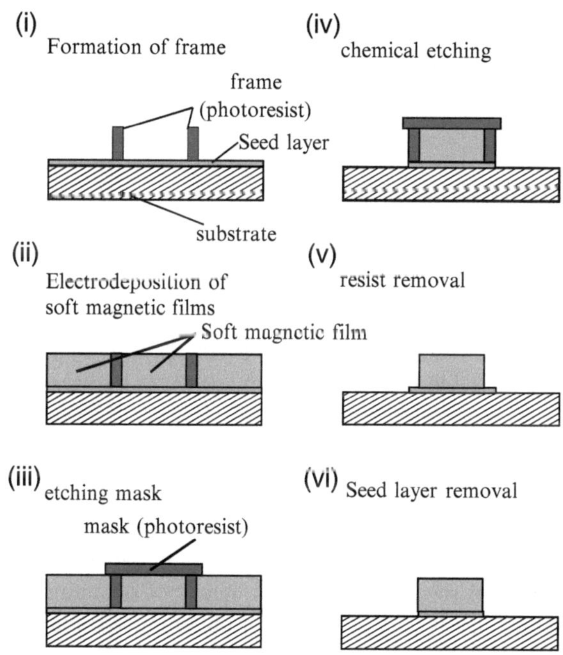

Fig. 6.8 Process flow of the frame plating method

flow of the frame-plating method [8]. This method suppresses the distribution of the current density on a fine pattern. In this method, a line resist pattern called a "frame" is used to suppress the distribution of current density at the edge of the pattern. After electrodeposition, the metal pattern in the frame pattern is covered with a resist pattern. The film deposited outside the frame pattern is removed by chemical etching. Thus, a metal pattern is obtained inside the frame pattern. Since a resist pattern having a narrow width and a small area is used, distribution of current density can be suppressed. The composition of the film pattern is consistently uniform, so that permalloy with homogeneous magnetic properties can be obtained. These methods have been widely used for development of the other soft magnetic films, such as CoNiFe and CoFe.

6.3 Design of High-Performance Soft Magnetic Films

6.3.1 *Theoretical Concept of Soft Magnetic Properties*

Because of the requirement of new soft magnetic thin films with a high value of B_s and high resistivity, high-performance soft magnetic thin films are overviewed. Only Ni, Co, Fe, and their alloys and some of their oxides show ferromagnetism at room temperature, and Fe has the highest magnetic moment of a single element. Materials with soft magnetic properties have a low value of H_c and a high μ. H_c, and the grain size are related by the following equation: [9]

$$H_c = \frac{K_1{}^4 D^6}{A^3 M_s} \tag{6.1}$$

where, K_1 is the constant of magnetocrystalline anisotropy, D is the crystal grain size, and A is the exchange stiffness constant. M_s and K_1 mainly depend on the composition of metals in the film. From this equation, decreasing the grain size is an effective means of obtaining soft magnetic properties for films with high magnetocrystalline anisotropy. On the other hand, K_{eff} is given by the following equation:

$$K_{eff} \propto \frac{K}{n} \tag{6.2}$$

where, K is the constant of magnetocrystalline anisotropy, and n is the number of crystal grains. This equation shows that an increase in the number of grains, that is, a decrease in the grain size, is an effective means of obtaining low K; therefore, decreasing the grain size is very effective for obtaining soft magnetic thin films.

6.3.2 *Design of Soft Magnetic Thin Films with High B_s*

Since the 1980s, many researchers have discovered new, soft magnetic materials with high values of B_s . Figure 6.9 shows the composition – B_s and H_c diagrams for CoNiFe bulk ternary alloy published by Bozorth [10]. Fe has the highest magnetic moment of a single element, and $Fe_{60}Co_{40}$ is the magnetic alloy with the highest

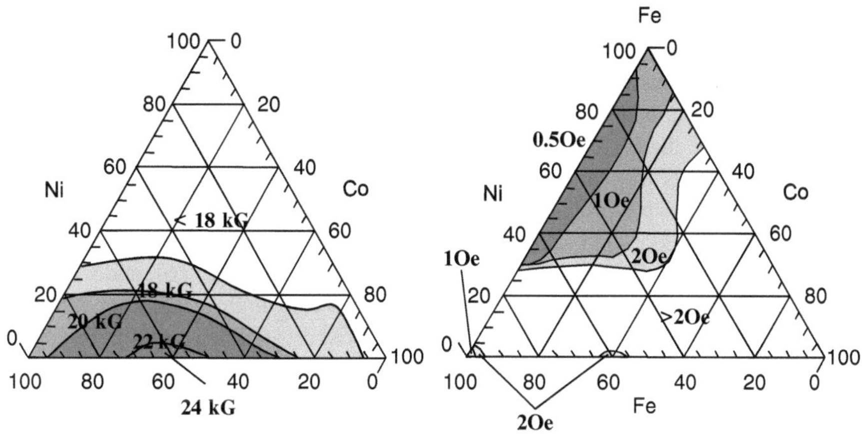

Fig. 6.9 The composition – B_s and H_c diagrams for CoNiFe bulk ternary alloy

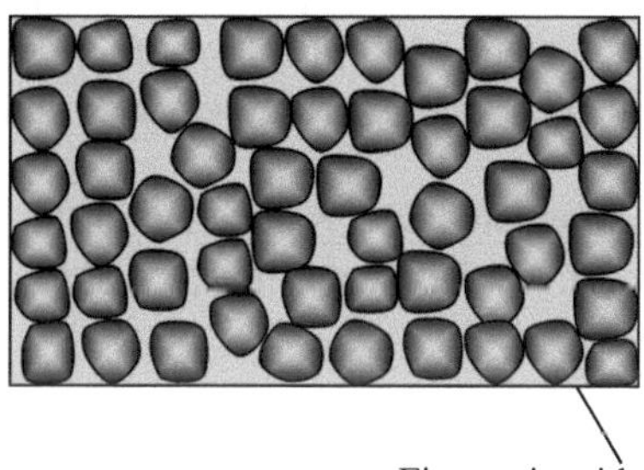

Fig. 6.10 Design of a microstructure for soft magnetic films with a high B_s value. Small and fine grains that have a high B_s, such as Fe, are included in the thin films

known B_s value of 24 kG. However, these materials do not show the soft magnetic properties that only $Ni_{80}Fe_{20}$ and $Ni_{45}Fe_{55}$ films have. Therefore, the CoFe- and Fe-based alloys have been studied with the aim of obtaining new soft magnetic materials. Figure 6.10 shows the design of a microstructure for soft magnetic films with a high B_s value. The ideal high-B_s soft magnetic film has small and fine grains having high B_s value. Various techniques for forming high-B_s soft magnetic thin films are shown below.

The films are in amorphous state under the as-deposited condition; however, if they are annealed above the crystallization temperature, nano Fe or CoFe grains are deposited, and other components inhibit the grain growth. This basic concept is used for the preparation of Fe-based nano-crystalline alloys, such as Fe-Cu-M (M=Nb, Zr, Hf, Ta)-Si-B; (B_s=12–13 kG) [11] and Fe-M (M=Ta, Nb, Hf)-N(C) (B_s=15–16 kG) [12]. These films are mainly prepared by sputtering or evaporation. Rapidly quenched ribbons such as Fe-Cu-M (M=Nb, Zr, Hf, Ta)-Si-B (B_s=12–13 kG) [13] are formed by the same technique. Furthermore, sputtered CoAlO, FeAlN, and

Fe(CoFe)HfN [14, 15] alloys with granular structure are reported to be high-B_s and high-ρ materials with ρ=150–3,500 μΩcm.

The film contains an impurity to inhibit grain growth during film deposition. Therefore, the film is in the nanocrystalline state under as-deposited condition. The additive is used during electrodeposition, whereas during electroless deposition, codeposited atoms from the reducing agent are used as the inhibitor. This basic concept is used for the preparation of electrodeposited CoFe (18–19 kG) [16] and CoNiFe (16–19 kG) [17–21]-based alloys, and electroless-deposited NiFeB (7–10 kG) [22, 23], CoB (14 kG) [24], CoFeB (16–18 kG) [25, 26], and CoNiFeB (15–20 kG) [27]. Electrodeposited CoNiFe with a very high B_s value of 20 kG is also reported [28–33]. This film has a bcc-fcc mixed crystalline structure; therefore, it is suggested that crystal grains with different crystalline structures act as the inhibitors. Furthermore, Fe-M-N (M=Mo, Rh, Zr, Ta, Al) films prepared by sputtering have higher B_s values of nearly 20 kG [34], and sputtered granular oxide soft magnetic thin films, such as Fe-M-O (M=Mg, Al, Hf) [35, 36], Co-M-O (M=Al, Zr, Cr) [37, 38], CoFe-M-O (M=Al, Mg, Hf) [39], formed without annealing have very high B_s value of 16–20 kG. Recently, CoFeN [40, 41] and CoFeAlO [42] were reported to have the highest B_s value of 24 kG. These films have a nanocrystalline structure in the as-deposited state, and grain growth was inhibited by very small amount of metal nitride.

The film contains an intermediate layer as the inhibitor for grain growth during film deposition. Fe/FeC and Fe/CoNbZr, multiplayer films have been produced with such an intermediated layer [43].

6.3.3 Design of Soft Magnetic Thin Films with High B_s and High ρ

Since 1995, new, soft magnetic materials with high B_s and high resistivity have been discovered. Co-based amorphous alloys (CoTaZr, CoNbZr, CoPdHf, etc.) have high B_s values of 11–14 kG and a high ρ of 100–200 μΩcm compared with the values for permalloy. However, the B_s value of these Co-based amorphous alloys is not sufficient to realize high-density recording. Therefore, fine grain-based soft magnetic thin films are required.

Figures 6.11 and 6.12 show microstructure designs for soft magnetic films with high values of B_s and high ρ. In Fig. 6.11, the ideal soft magnetic film, with a high B_s and a high ρ, has small and fine grains with a high B_s value, surrounded and separated by a high-ρ region. Sputtered (Co,Fe,CoFe)-M-O (M=Al, Mg, Hf, Cr, Zr) [35–38] alloys with a granular structure have been reported to be high-B_s and high-ρ materials with ρ values of 150–3,500 μΩcm. Using Al–O and Mg–O for grain boundary, these films have a high B_s and high resistivity because Fe or Co grains are separated by the insulating grain boundary of Al–O and Mg–O. High-ρ soft magnetic thin films prepared by electro- and electroless deposition have also been reported. The high value of ρ was realized by the codeposition of impurities such

Fig. 6.11 Design of a microstructure for soft magnetic thin films with high B_s and high ρ. Fine grain with high B_s is separated by high ρ materials

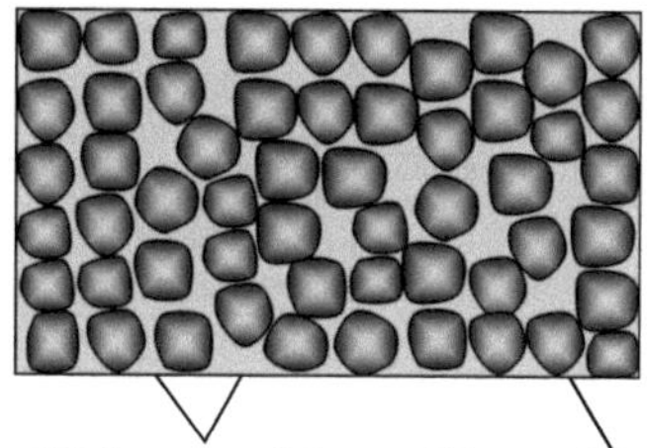

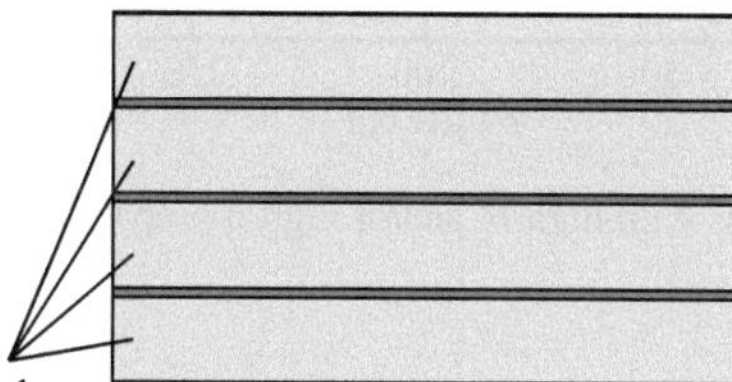

Fig. 6.12 Design of a microstructure for soft magnetic thin films with high B_s and high ρ. Soft magnetic film is separated by intermediate layer with high ρ

as Mo, Cr, and S; however, the details of this mechanism by which a high value of ρ has been achieved have not been reported.

In Fig. 6.12, the film contains an intermediate layer for suppressing the eddy current loss. Insulators such as SiO_2 and Al_2O_3 are widely used for the intermediate layer, and the formation of films such as FeAlN/Al_2O_3 [44], FeTaN/AlN [45], have been reported. Soft magnetic thin films with high resistivity and a low B_s value are used for the intermediate layer instead of an insulator, because of the small decrease in value of B_s. Fe(Fe-C)/CoNbZr [43, 46] and FeRhN/CoZrCr [47] films have also been reported.

6.4 High-Performance Soft Magnetic Alloy Films Prepared by Electrodeposition

Soft magnetic thin films with high B_s and/or high ρ prepared by electrodeposition are shown in Fig. 6.13. Soft magnetic films prepared by electroless deposition are also discussed in this section.

6.4.1 NiFe Alloy and NiFe-Based Alloy Films

Electroplated permalloy films for use as thin-film heads are based on the technology of magnetic wire memory developed in the 1950s. Ever since IBM applied the

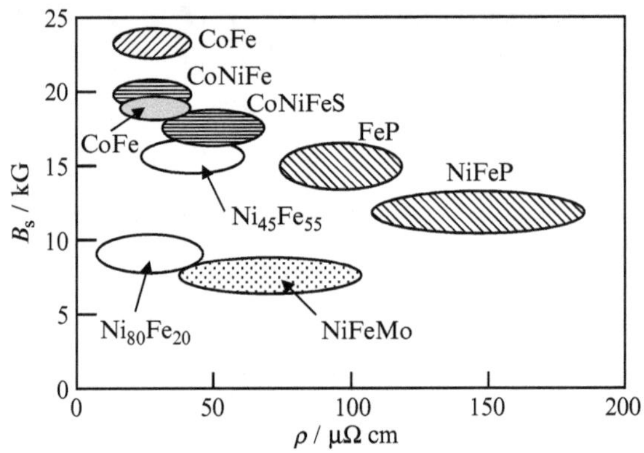

Fig. 6.13 Soft magnetic thin films with high B_s and/or high ρ prepared by electrodeposition

electroplated $Ni_{80}Fe_{20}$ permalloy to the thin-film head, electrodeposited permalloy films are being widely used , even though alternative soft magnetic films are available with high values of B_s [3]. This is because the electrodeposited permalloy film has several advantages, such as a high B_s (=10 kG), high permeability (3,000–5,000), near-zero magnetostriction, and high corrosion resistance. However, it is difficult to prepare the electrodeposited NiFe alloy films because their magnetic properties are very sensitive to the composition. Furthermore, maintaining constant composition is very difficult because of the serious problem of Fe-based plating, known as anomalous co-deposition [48–50]. Romankiw solved this problem by using the paddle plating cell system and the frame plating method [7, 8].

Though $Ni_{80}Fe_{20}$ permalloy has superior properties for magnetic recording head core, the film has low resistivity (20 μΩcm). Therefore, some researchers have reported permalloys with high resistivity, such as Moco-deposition permalloys [51]. Reports also indicate that an increase in the resistivity can be achieved by co-deposition with metalloid-forming elements, such as sulfur, carbon, or phosphorus [52, 53]. The inclusion of impurities is said to result in an increase in resistivity. $Ni_{45}Fe_{55}$ permalloy had been widely used for write head core because the film has high B_s value (14–16 kG). $Ni_{45}Fe_{55}$ permalloy co-deposited Mo or Cr is also reportedly in use [54].

6.4.2 CoNiFe Alloy Films

CoFe alloy has a high B_s value; and $Co_{90}Fe_{10}$ bulk alloy shows both a high B_s (=19 kG) and a low H_c (<2 Oe) [16]. There have been several reports on the preparation of soft magnetic CoFe films and the development of a magnetic recording head using CoFe alloy. However, CoFe alloy has poor corrosion resistance; thus, CoNiFe ternary alloy has been the focus of research because of its high value of B_s and high

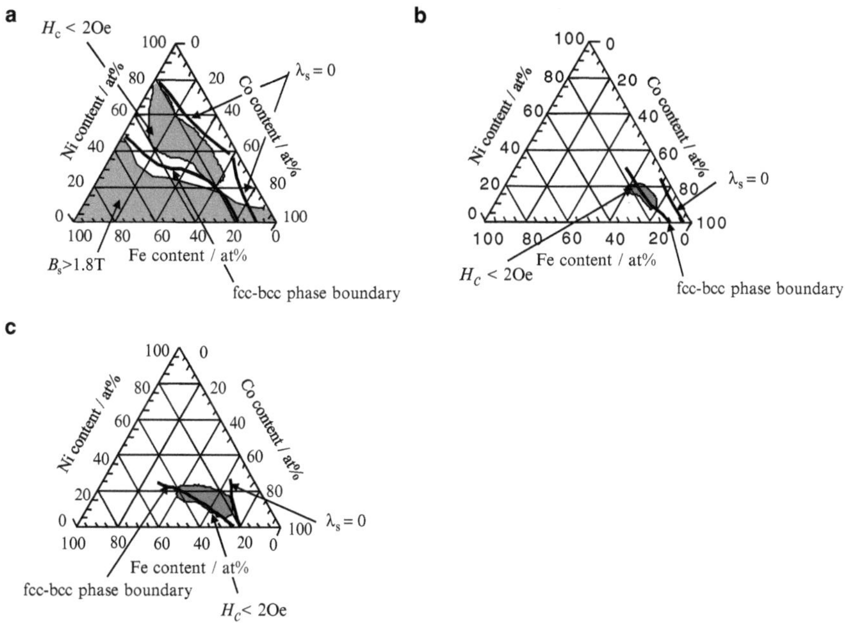

Fig. 6.14 Magnetic properties and fcc-bcc phase boundary of electrodeposited CoNiFe from saccharin bath (**a**), CoNiFeS from thiourea bath (**b**) and high B_s CoNiFe from additive free bath (**c**) [29]. © 1998 IEEE

corrosion resistance. For the composition used, CoNiFe alloys do not have a low H_c, and it is believed that the use of additives is necessary for decreasing H_c. Preparation of CoNiFe films deposited from the saccharin contained bath had been reported in 1994 [18]. There have been several reports on the preparation of CoNiFe soft magnetic thin films using the baths containing additives [17–19]. CoNiFe films prepared from the bath containing thiourea have a high $B_s = 17$ kG and low $H_c = 0.7$ Oe because of small grains cased by sulfur impurities from thiourea (Fig. 6.14) [20, 21]. The first CoNiFe soft magnetic thin films with the highest reported B_s value of 20 kG, and developed using a bath without additives were reported (Fig. 6.14) [28]. These new films had a fine bcc-fcc mixed crystal structure, and a low value of H_c was realized by this crystal structure (Fig. 6.15) [28–31]. This film has bcc-fcc mixed crystalline structure, and therefore, it is suggested that crystal grains with different crystalline structure from each other act as the inhibitor. Moreover, new heads using these high-B_s CoNiFe films show superior write performance compared with heads using conventional magnetic thin films [28, 30, 55, 56]. Therefore, some researchers have reported soft magnetic CoNiFe film with high resistivity. Mo-codeposition CoNiFe are also reported [57, 58]. Moreover, it has been reported that an increase in the resistivity can be achieved by co-deposition with a metalloid-forming element such as sulfur [20, 21] or carbon [59, 60]. It is suggested that the inclusion of impurities can result in an increase in resistivity [61].

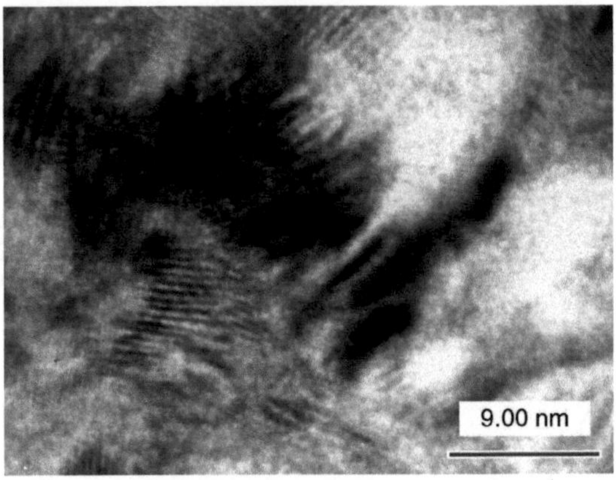

Fig. 6.15 High-resolution transmission electron micrograph of the electrodeposited high B_s CoNiFe thin film [28]

6.4.3 CoFe Alloy Films and its Preparation Technology

$Co_{35}Fe_{55}$ binary alloy is known as a material with B_s of about 24 kG, which is close to the limiting value achievable with ferromagnetic alloys. Several soft magnetic thin films with the highest B_s value of 24 kG have recently been prepared by using a dry sputtering process, e.g., NiFe/CoFeN/NiFe trilayer film [40, 41] and Co–Fe–Al–O granular film [42]. However, thin films containing only Co and Fe do not exhibit soft magnetic properties. Electrodeposited soft magnetic thin films of CoFe binary alloy with the highest B_s of 24 kG are now in strong demand. However, CoFe films with a value of B_s as same as that of bulk alloy cannot be obtained using the conventional electrodeposition method (Fig. 6.16) [62, 63]. To clarify the reason for the low B_s value of the electrodeposited CoFe films compared with that of the bulk alloy, the state of Fe ions was analyzed [62, 63]. As a result, the ratio increased dramatically after electrodeposition in the conventional bath and Fe^{2+} could be oxidized to Fe^{3+} at the potential of the anode during electrodeposition [64]. Thus, it is suggested that Fe^{3+} was produced by the anodic reaction. Moreover, the solubility product constant, K_{sp} of $Fe(OH)_3$ is very low, 66×10^{-38} [65]; therefore, Fe^{3+} must have formed $Fe(OH)_3$ at the surface during electrodeposition because the pH near the surface during electrodeposition increased to a value above 2.3. Thus, $Fe(OH)_3$ is *assumed* to be present in the outer shell of the deposit, and it is suggested that the adsorbed $Fe(OH)_3$ is included into the film, and that the inclusion causes the decrease in B_s value. Confirmation of the *presence* of Fe^{3+} is important for understanding the electrodeposition of Fe and Fe alloy films.

A $Co_{35}Fe_{65}$ film was electrodeposited from a bath containing trimethylamineborane, TMAB to suppress the oxidation of Fe^{2+} in the plating bath. It is suggested that the addition of TMAB completely suppressed the formation of Fe^{3+} in the bath *before* and *after* the electrodeposition. As shown in Table 6.1, the $Co_{35}Fe_{65}$ thin film deposited from the TMAB bath exhibited a B_s value of 23 kG, which was higher than that of the film deposited from the conventional bath. It is suggested that $Fe(OH)_3$ inclusion was suppressed with the decrease in Fe^{3+} concentration.

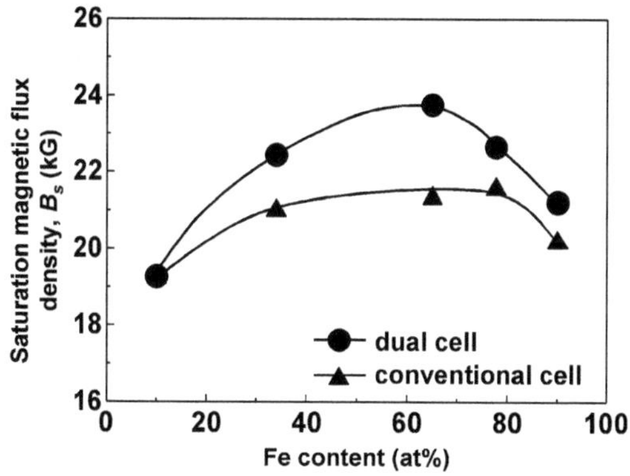

Fig. 6.16 B_s value of electrodeposited Co-Fe thin films as a function of Fe content

Table 6.1 B_s and H_c values of electrodeposited high moment CoFe thin films

	Conventional bath	TMAB bath	Bath used with dual cell system
B_s/kG	20	23	24
H_c/Oe	15	14	15

However, the B_s value of this film was still lower than 24 kG. This result is attributed to the presence of about 0.3% of boron in the film.

An improved electrodeposition cell system was investigated for suppressing the oxidization of Fe^{2+}. A separated compartment dual cell system was used, in which the anode and cathode cells were separated from each other but electrochemically connected using a salt bridge. A 0.1 mol dm^{-3} H_2SO_4 solution was used in the anodic cell. With this cell system, the oxidization of Fe^{2+} could not occur at anode because the anode cell did not contain any metal ions. From the result of iodometric analysis, it is seen that the concentration of Fe^{3+} ion did not increase after electrodeposition. As shown in Table 6.1, the B_s value of the $Co_{35}Fe_{65}$ thin film was about 24 kG, which was equal to that of bulk $Co_{35}Fe_{65}$ alloy. This result indicates that the film did not contain $Fe(OH)_3$ because of the low Fe^{3+} concentration. As shown in Fig. 6.16, the B_s values of the CoFe thin film with various Fe content were equal to that of bulk CoFe alloy. This result also indicates that the film did not contain $Fe(OH)_3$ because of the low Fe^{3+} concentration.

6.4.4 *High-Performance Soft Magnetic Alloy Films Prepared by Electroless Deposition*

Electroless deposition has been widely used for preparing magnetic thin films as described above. Note that this process is advantageous for depositing films onto a

Table 6.2 Magnetic and electrical properties of electroless-deposited soft magnetic thin films

	Electroless-deposited films (at %)			
Properties	$Ni_{70}Fe_{27}B_3$[a]	$Co_{96}B_4$[b]	$Co_{89}Fe_9B_2$[a]	$Co_{77}Ni_{13}Fe_9B_1$[a]
H_c/Oe	0.5	0.5	0.7	1.2
B_s/kG	10	14	16–18	15–17
μ [at 1 MHz]/–	1,000	400	650	600
H_k[as-deposited]/Oe	10	50	30	20
stability of heating/°C	400	300	400	–
ρ/μΩcm	40	–	30	40
$\lambda_s \times 10^{-6}$/–	+5.0	−1.0	+5.4	+2.0
Crystal structure	fcc	hcp	fcc	fcc

[a]Substrate: Cu(50 nm)/Ti(5 nm)/glass and Film thickness: 1.0 μm
[b]Substrate: Cu(50 nm)/Ti(5 nm)/glass and Film thickness: 0.7 μm

fine, selective area of a non-conductive surface. Magnetic and electrical properties of electroless-deposited soft magnetic thin films are shown in Table 6.2. Compared with electroless soft magnetic thin films such as NiFeP, CoNiP and CoP [66, 67] prepared from baths containing $NaHPO_2$ as the reducing agent, the films such as NiFeB [22, 23], CoB [24], CoFeB [25, 26] and CoNiFeB [27] prepared from baths containing dimethylamine-borane (DMAB) have higher B_s values, because of the difference in crystalline structure and codeposited elements between the two different classes of films. It is suggested that the formation of a fine crystalline microstructure or an amorphous-like phase results in a low H_c value, which probably is caused by the co-deposition of boron in the film. Soft magnetic thin films with magnetic properties similar to those of electrodeposited NiFe and CoFe can thus be obtained by electroless deposition using DMAB as the reducing agent. In particular, the CoNiFeB film has good magnetic properties and is suitable for use in a magnetic recording head. Electroless CoNiFeB films with high resistivity were also reported [68]. It is suggested that the inclusion of impurities such as carbon can result in an increase in resistivity.

References

1. Osaka T (2000) Electrodeposition of highly functional thin films for magnetic recording devices of the next century. Electrochim Acta 45:3311–3321
2. Andricacos PC, Robertson N (1988) Future directions in electroplated materials for thin-film recording heads. IBM J Res Develop 42:671–680
3. Jones RE Jr (1980) IBM 3370 film head design and fabrication. IBM Disk Storage Technology, pp 6–9
4. Tsang C et al (1990) Gigabit density recording using dual-element MR inductive heads on thin-film disks. IEEE Trans Magn MAG-26:1689–1693
5. Tanaka Y (2005) Recording performance and system integration of perpendicular magnetic recording. J Magn Magn Mater 287:468–474

6. Osaka T (1998) Development of soft magnetic materials with high B_s by the electroplating method and their application to an MR write head core. J Magn Soc Jpn 22:1182–1188
7. Castellani EE et al (1978) U.S. Patent 4102756
8. Powers JV, Romankiw LT (1972) U.S. Patent 3652442
9. Hoffmann H (1964) Quantitative calculation of magnetic ripple of uniaxial thin permalloy films. J Appl Phys 35:1790–1798
10. Bozorth RM (1951) Ferromagnetism. Van Nostrand, New York, NY, p 160
11. Yoshizawa Y et al (1988) New Fe-based soft magnetic-alloys composed of ultrafine grain-structure. J Appl Phys 64:6044–6046
12. Ishiwata N (1987) Magnetic and structural-properties of dual ion-beam sputtered pure iron films. IEEE Trans Magn MAG-23:2152–2154
13. Todd I et al (2000) Magnetic properties of ultrasoft-nanocomposite FeAlSiBNbCu alloys. J Magn Magn Mater 215:272–275
14. Huijbregtse J et al (1998) High-frequency permeability of soft-magnetic Fe-Hf-O films with high resistivity. J Appl Phys 83:1569–1574
15. Sato T et al (1998) New applications of nanocrystalline Fe(Co–Fe)-Hf-O magnetic films to micromagnetic devices. J Appl Phys 83:6658–6660
16. Liao SH (1987) High moment CoFe thin-films by electrodeposition. IEEE Trans Magn MAG-23:2981–2983
17. Anderson NC, Chesnutt RB (1987) U.S. Pattent 4661216
18. Shinoura O (1994) Soft magnetic properties of electrodeposited CoNiFe films. J Magn Soc Jpn 18:277–280
19. Nakamura A (1996) Preparation and magnetic properties of CoNiFe thin film by electrodeposition. J Surf Finish Soc Jpn 47:934–938
20. Takai M (1997) Electrochemical preparation of soft magnetic CoNiFeS film with high saturation magnetic flux density and high resistivity. J Electrochem Soc 144:L203–L204
21. Takai M (1997) High frequency permeability of electrodeposited CoNiFeS films with high B_s and high ρ. J Magn Soc Jpn 21(S2):443–446
22. Kim DH et al (1996) Electroless Ni-Fe-B alloy plating solution using DMAB as a reducing agent. Plat Surf Finish 83:78–80
23. Takai M (1995) Magnetic properties of electroless-deposited NiFeB and electrodeposited NiFe alloy thin films. IEICE Trans Electron E78-C:1530–1535
24. Osaka T et al (1992) Co-based soft magnetic-films produced by electroless deposition. J Electrochem Soc 139:1311–1314
25. Osaka T et al (1994) Preparation of electroless-deposited CoFeB soft-magnetic films with high saturation magnetic-flux density. Denki Kagaku (presently Electrochemistry) 62:987–988
26. Osaka T et al (1994) Soft magnetic properties of electroless-deposited CoFeB films. J Magn Soc Jpn 18(S1):183–186
27. Yokoshima T et al (2000) Electroless CoNiFeB soft magnetic thin films with high corrosion resistance. J Electroanal Chem 491:197–202
28. Osaka T et al (1998) A soft magnetic CoNiFe film with high saturation magnetic flux density and low coercivity. Nature 392:796–798
29. Osaka T (1998) New soft magnetic CoNiFe plated films with high Bs = 2.0–2.1 T. IEEE Trans Magn 34:1432–1434
30. Ohashi K et al (1998) Newly developed inductive write head with electroplated CoNiFe film. IEEE Trans Magn 34:1462–1464
31. Osaka T et al (1999) Influence of crystalline structure and sulfur inclusion on corrosion properties of electrodeposited CoNiFe soft magnetic films. J Electrochem Soc 146:2092–2096
32. Tabakovic I et al (2000) Organic additives in the electrochemical preparation of soft magnetic CoNiFe films. J Electrochem Soc 147:219–226
33. Liu X et al (2000) Electrodeposition of soft, high moment Co–Fe–Ni thin films. J Appl Phys 87:5410–5412

34. Kim KH et al (2000) The magnetic properties of nanocrystalline Fe-Co(Cr)-Hf-N thin films. J Appl Phys 87:5248–5250
35. Kim SR et al (2000) Soft magnetic properties of as-sputtered Fe–Al–O films. J Magn Magn Mater 215:365–367
36. Ohnuma S (1996) High-frequency magnetic properties in metal-nonmetal granular films. J Appl Phys 79:5130–5135
37. Morikawa T (1998) Soft magnetic properties of Co-Cr-O granular films. J Appl Phys 83:6664–6666
38. Ohnuma S et al (2001) Co-Zr-O nano-granular thin films with improved high frequency soft magnetic properties. IEEE Trans Magn 37:2251–2254
39. Ohnuma S et al (1999) Magnetostriction and soft magnetic properties of $(Co_{1-x}Fe_x)$–Al–O granular films with high electrical resistivity. J Appl Phys 85:4574–4576
40. Wang SX et al (2000) Sandwich films-properties of a new soft magnetic material. Nature 407:150–151
41. Sun NX, Wang SX (2000) Soft high saturation magnetization $(Fe_{0.7}Co_{0.3})_{(1-x)}N_x$ thin films for inductive write heads. IEEE Trans Magn 36:2506–2508
42. Shintaku K et al (2003) High-B_s Fe–Co–Al–O soft magnetic films. J Appl Phys 93:6474–6476
43. Kim EH et al (2001) Permeability enhancement in Fe/CoNbZr multilayers prepared by Ar/H_2 mixed gas sputtering and heat treatment. J Magn Magn Mater 233:L142–L146
44. McNeill KA et al (2000) Effect of lamination period and deposition angle on FeAlN-Al2O3 multilayers. J Appl Phys 87:5837–5839
45. Hong J et al (1999) Magnetic properties and high-frequency responses of high moment FeTaN/AlN laminates for high-data-rate magnetic recording. IEEE Transactions on Magnetics vol 35:2502–2504
46. Choi KK et al (2003) High Frequency Properties of CoZrNb/Fe–C Multilayer Films. Trans Magn Soc Jpn 3:55–58
47. Chen YJ et al (2000) Laminated FeRhN films for high speed writers. IEEE Trans Magn 36:3476–3478
48. Brenner A (1963) Electrodeposition of Alloys. Academic, New York, NY
49. Dahms H, Croll IM (1965) The anomalous codeposition of iron-nickel alloys. J Electrochem Soc 112:771–775
50. Hessami S, Tobias CW (1989) A mathematical-model for anomalous codeposition of nickel-iron on a rotating-disk electrode. J Electrochem Soc 136:3611–3616
51. Shinoura O (1995) Electrodeposition of Ni–Fe–Mo multilayered soft-magnetic films with high specific resistance. Denki Kagaku (presently Electrochemistry) 63:473–478
52. Takai M et al (1998) Electrodeposition of soft magnetic Ni–Fe-based films with high resistivity. J Surf Finish Soc Jpn 49:292–296
53. Takai M (1998) Increasing the resistivity of NiFeP films by means of electrodeposition. J Magn Soc Jpn 22:629–632
54. Hoshino K (1999) Magnetic properties and thermal stability of electroplated NiFeCr and NiFeMo films with high resistivity. IEEE Trans Magn 35:3433–3435
55. Ohashi K et al (1999) Write performance of heads with a 2.1-tesla CoNiFe pole. IEEE Trans Magn 35:2538–2540
56. Nonaka Y et al (2000) Co–Ni–Fe write heads with a 10 μm yoke length for high-speed recording. IEEE Trans Magn 36:2514–2516
57. Sogawa Y (2000) Preparation of electrodeposited high-B_s and high-ρ CoNiFe thin films by Mo addition. J Magn Soc Jpn 24:699–702
58. Yokoshima T (2003) Effect of carbon inclusion on properties of electrodeposited CoNoFeMo thin films. In: Krongelb S et al (eds) Electrochemical society proceedings, vol 2002–27. The Electrochemical Society Inc, pp 365–375
59. Yokoshima T et al (1999) Increasing the resistivity of electrodeposited high B_s CoNiFe thin film. IEEE Trans Magn 35:2499–2501

60. Kaseda M et al (2000) Preparation and characterization of electrodeposited high-B_s CoNiFe thin films with high resistivity and improvement of their corrosion resistance. In: Romankiw LT et al (eds) Electrochemical society proceedings, vol 99–34. The Electrochemical Society Inc, pp 263–272
61. Osaka T (2001) Effects of impurities on resistivity of electrodeposited high-B_s CoNiFe-based soft magnetic thin films. IEEE Trans Magn 37:1761–1763
62. Osaka T et al (2003) A high moment CoFe soft magnetic thin film prepared by electrodeposition. Electrochem Solid State Lett 6:C53–C55
63. Yokoshima T et al (2004) Preparation of high-B_s, Co-Fe soft magnetic thin films by electrodeposition. IEEE Trans Magn 40:2332–2334
64. Milazzo G, Caroli S (1978) Tables of standard electrode potentials. Wiley, New York, NY, p 320
65. Chemical Society of Japan (1984) Chemical handbook-basic volume, 3rd edn. Maruzen, Tokyo p. II–179
66. Kim DH (1994) Preparation of soft magnetic films by electroless Ni-Fe-P plating. J Surf Finish Soc Jpn 45:203–206
67. Kim DH et al (1995) Soft-magnetic films by electroless Ni–Co–P plating. J Electrochem Soc 142:3763–3767
68. Sobue M et al (2002) Increase of the resistivity of electroless-deposited high-B_s CoNiFeB thin films. IEEE Trans Magn 38:2228–2230

Chapter 7
Magnetic Thin Films for Perpendicular Magnetic Recording Systems

Atsushi Sugiyama, Takuma Hachisu, and Tetsuya Osaka

7.1 Introduction

In the advanced information society of today, information storage technology, which helps to store a mass of electronic data and offers high-speed random access to the data, is indispensable. Against this background, hard disk drives (HDD), which are magnetic recording devices, have gained in importance because of their advantages in capacity, speed, reliability, and production cost. These days, the uses of HDD extend not only to personal computers and network servers but also to consumer electronics products such as personal video recorders, portable music players, car navigation systems, video games, video cameras, and personal digital assistances.

IBM introduced the first HDD of the IBM 350 Disk File in 1956, which was the storage unit of the IBM RAMAC 305 computer [1]. It was composed of fifty 61 cm aluminum disks coated with iron oxide paint, and provided an areal density of 0.002 Mbit/inch^2; in the last half century, the density has been increasing to beat the band.

Such an increase in the density was achieved through epochal developments in the magnetic thin film materials used in magnetic recording media and heads. After a magnetic thin film had been made by using electroless deposition in place of coating, a rapid rise in the density was achieved. Product PATTY@ by Nippon Telegraph and Telephone (NTT), which used electroless Co-Ni-P film reached a density of 9 Mbit/inch^2 [2]. NEC Corp. had continuously developed higher level of 24.5 Mbit/inch^2. In a disk-manufacturing process, an electroless Ni-P film with amorphous structure was also used for the underlayer plated on the aluminum alloy substrate. A higher system of perpendicular magnetic recording media was being developed by electroless Co-Ni-Re-P and Co-Ni-P alloy films [3, 4].

A. Sugiyama
Waseda Institute for Advanced Study, Waseda University, 1-6-1 Nishiwaseda, Shinjuku-ku, Tokyo, 169-8050, Japan

T. Hachisu and T. Osaka (✉)
School of Advanced Science and Engineering, Waseda University, 3-4-1 Okubo, Shinjuku-ku, Tokyo 169-8555, Japan
e-mail: osakatets@waseda.jp

T. Osaka et al. (eds.), *Electrochemical Nanotechnologies*,
Nanostructure Science and Technology, DOI 10.1007/978-1-4419-1424-8_7,

In this section, our recent study on a challenge of new materials for the next generation of magnetic recording approaching an ultra high areal recording density (beyond 1 Tbit/inch^2) is overviewed.

7.2 Soft Magnetic Underlayer by Electroless-Deposition for Perpendicular Magnetic Recording

A double-layered perpendicular magnetic recording medium [5–7], which is composed of a hard magnetic recording layer with perpendicular magnetic anisotropy and a soft magnetic underlayer (SUL), is a promising system for realizing an ultra high areal recording density. Therefore, we should pay attention not only to the recording layer but also the SUL in order to develop high performance perpendicular magnetic recording media. Concerning the SULs, a large value of $B_s\delta$, where δ is the thickness of the SUL, is required for making the most of the writing ability of the magnetic recording head, and various sputter-deposited SULs with the thickness of more than a 100 nm have been reported [8–11]. However, we face the following critical issues in developing the SULs for practical use. First, mass productivity is a rather low when such a thick SUL is sputter-deposited. Second, peculiar layered structures are needed to suppress the "spike noise" caused by magnetic domain boundaries in the SUL [12–14].

We have investigated the possibility of the application of Co–Ni–Fe-based alloy thin films electroless-deposited on a disk substrate to the SUL, as a trial to solve the above-mentioned issues. We succeeded in electroless-depositing the SULs of a Co–Ni–Fe–P or Co–Ni–Fe–B alloy without marked macroscopic magnetic domain boundaries [15–17]. Figures 7.1a and b show magnetic domain images observed through the *Kerr effect* for the Co–Ni–Fe–P SULs, which were prepared without and with the applied magnetic field of ca. 800 Oe during electroless-deposition, respectively.

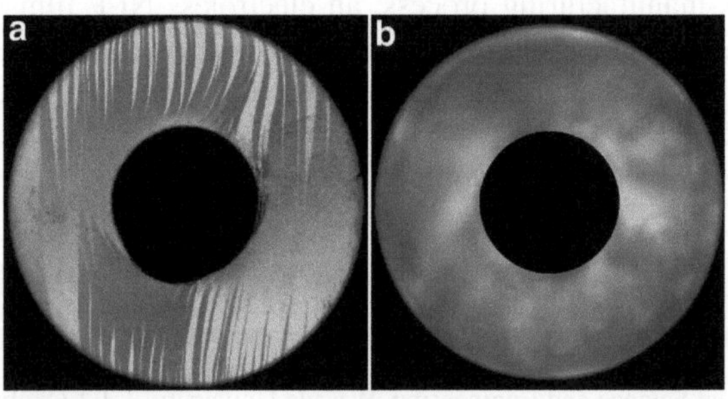

Fig. 7.1 Magnetic domain images were observed through the Kerr effect of Co–Ni–Fe–P soft magnetic underlayer on a disk substrate electroless-deposited (**a**) without and (**b**) with an external applied magnetic field. The thickness of each sample is 1 μm [15]. © 2004 IEEE

The magnetic field was applied in the direction parallel to the surface of disk substrate while the disk substrate was rotated. The Co–Ni–Fe–P SUL deposited without the magnetic field tends to show clear and complicated magnetic domain boundaries, whereas, the SUL deposited with the magnetic field has no marked domain boundaries. The magnetic domain boundaries of the former are readily moved by an imposition of an external magnetic field, and the domain boundaries cause an problem of spike noise. On the other hand, no domain boundaries are induced in the latter even when the external magnetic field was imposed.

Figures 7.2a and b show in-plane *M–H* hysteresis loops of the Co–Ni–Fe–P SULs in Figs. 7.1a and b, respectively. The SUL deposited without the magnetic field had a good soft magnetic property, and it exhibited magnetic anisotropy with the easy axis in the film plane. On the other hand, we obtained an interesting result that the Co–Ni–Fe–P SUL deposited with the magnetic field exhibited almost isotropic magnetic property in the film plane, but slightly perpendicular magnetic anisotropy, which is likely to be induced by fine stripe magnetic domains. The stripe magnetic domains of ca. 1 μm in width were actually observed by precise magnetic force microscopy in this kind of SUL [18].We consider that the complicated fine stripe domains make macroscopic magnetic domain boundaries inconspicuous. This phenomenon is similar in the case of the so-called anomalous magnetic thin films, e.g., permalloy thin films with a large Ni composition more than 85 wt.% [19]. From read and write experiments, it was confirmed that the double-layered perpendicular magnetic recording medium with the SUL shown in Fig. 7.1b generated *no* spike noise, whereas, that with the SUL shown in Fig. 7.1a showed, explicitly, the spike noise.

However, in a magnitude of root-mean-square noise amplitude, i.e. DC noise, the SUL without the generation of spike noise takes poor level. Figure 7.3 shows a variation in DC noise voltage of the Co-Ni-Fe-B SUL with in-plane coercivity. DC noise was increased with increasing the value of coercivity [18]. Since the coercivity

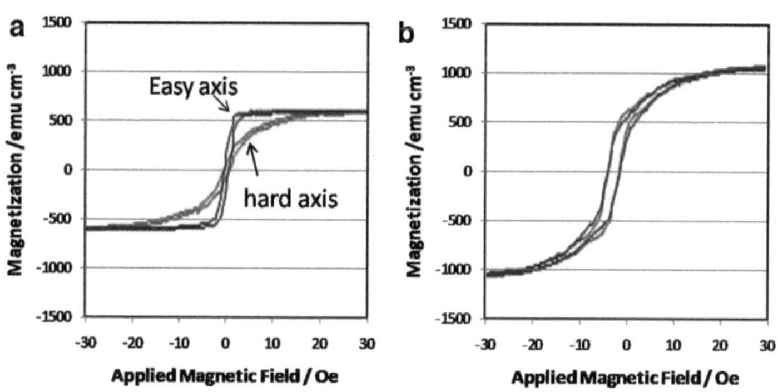

Fig. 7.2 *M–H* hysteresis loops of the samples shown in Fig. 7.1(a) Co–Ni–Fe–P soft magnetic underlayer electroless-deposited, (**a**) without and (**b**) with an external applied magnetic field [15]. © 2004 IEEE

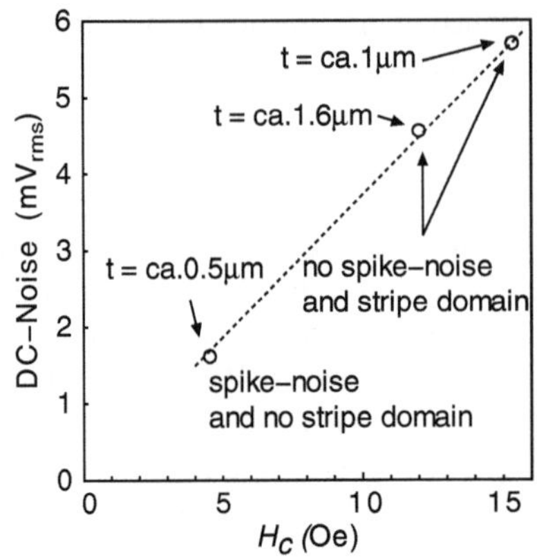

Fig. 7.3 DC noise voltage of Co–Ni–Fe–B films as a function of in-plane coercivity [18]. © 2005 IEEE

of the SUL without the generation of spike noises inevitably takes a large value by the perpendicular magnetic leakage of striped magnetic domains, then DC noise of the SUL without spike noise cannot be reduced more than that of SUL with the generation of spike noise.

In order to solve this problem, we proposed that the dual-layered SUL consists of two layers having different magnetic domains [20]. The schematic of the double-layered SUL is shown in Fig. 7.4. The lower layer is the film with the stripe domain layer, and the upper layer is the film with the both of small coercivities and no stripe domain. The upper layer works as a return-path layer, i.e., the perpendicular magnetic leakage generated from the lower layer is refluxed into the SUL through the upper layer. The optimized thickness of the upper layer is the height of the magnetic leakage on the surface, and Favieres reported that the height becomes equal to the width of the striped magnetic domains [21]. Figure 7.5 shows the noise spectra for these SULs, which were measured by using a spin-stand tester with a merged GMR head. The double-layered SUL has allowed the noise to be reduced to the same level as in the SUL without the striped magnetic domain. Thus, we have successfully developed a double-layered medium with the SUL fabricated by electroless-deposition, which generates no spike noise. This study has just opened up a new way to manufacture SULs with high mass productivity.

7.3 Control of Characteristics of Magnetic Recording Media by Metal Clusters Formed by Electrochemical Process

In a double-layered perpendicular magnetic medium, an intermediate layer between a recording layer and an SUL plays an important role as a seed to control the crystal growth of the recording layer and to reduce the magnetic domain size of the recording layer. Moreover, the intermediate layer should be as thin as possible to minimize the

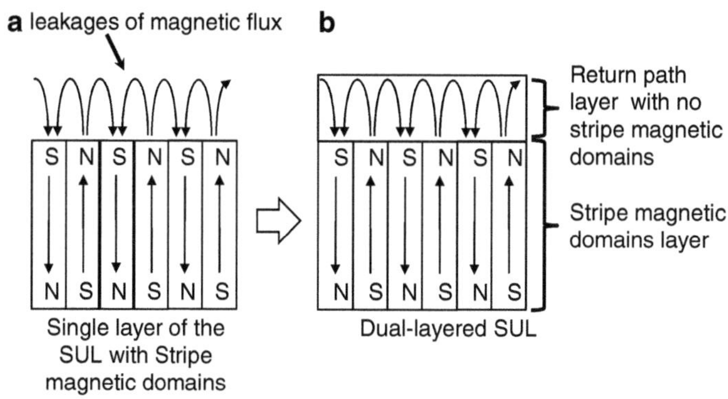

Fig. 7.4 Schematic of the dual-layered SUL

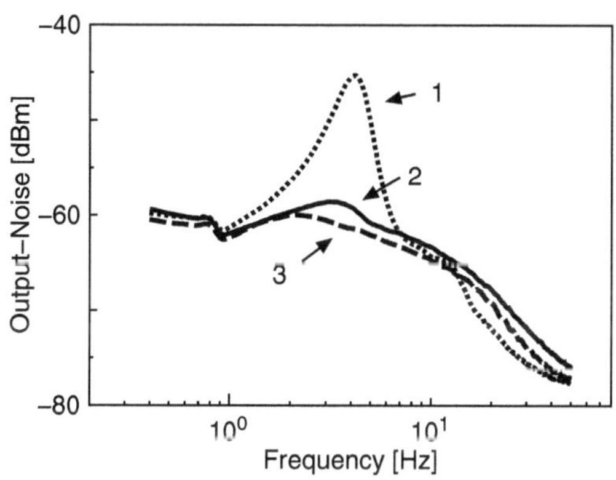

Fig. 7.5 Noise spectra for electrolessly Co-Ni-Fe SULs. 1: SUL without the generation of spike noise (stripped magnetic domain emerged film), 2: Dual layered SUL, 3. SUL with the generation of striped magnetic domain (low H_0 film) [20]

magnetic spacing loss between the recording layer and the SUL for high recording efficiency. The intermediate layers meeting these demands have been developed by sputtering for various perpendicular magnetic recording media.

For instance, in the case of the medium with Co/Pd multilayered perpendicular magnetization film [22, 23], Pd-based intermediate layers such as Pd/ITO [24], Pt–B/Pd/MgO [25], Pd–Si–N [26], and Pd/Si [27–29] were reported to successfully improve magnetic properties of the Co/Pd multilayered films. In these intermediate layers, fine Pd grains provide the suitable nucleation site for the magnetically isolated Co/Pd crystal grains, leading to weak intergranular exchange coupling and good read and write characteristics. On the other hand, we have focused on an electrochemical substitution deposition technique as a novel method to prepare the metal seed for the double-layered media. Electrochemical process is useful to modify a film surface and to form uniformly nanoscopic structure on the surfaces. Since the electrochemical substitution reaction takes place only at the interface between the surface and the electrolyte solution, a deposited metal is expected to form a very thin film or island-like clusters rather than a thick continuous film, which could be applicable to the seed for the double-layered media. In this section, the Pd formation on a Co-based Co–Zr–Nb SUL as the seed for the sputter-deposited Co/Pd perpendicular magnetization film using the substitution reaction caused by the difference in redox potential between Pd and Co is mainly described [30, 31].

Figures 7.6a and b show high resolution SEM images of the Co–Zr–Nb SUL surface without any treatment and that treated with $PdCl_2$ solution, respectively. An island-like structure with the mean diameter of ca. 9 nm was formed on the SUL with a high density, where the mean distance between adjacent clusters was ca. 18 nm. From X-ray photospectroscopy and diffractometry, the island-like structure was identified as the Pd metal with fcc structure [30], which is hereafter called "Pd cluster seeds".

Figures 7.7a and b show MFM images at an ac-demagnetized states of the Co/Pd multilayered films with a sputter-deposited continuous Pd seedlayer (film I) and with the Pd cluster seeds (film II). The maze-like domain pattern with large magnetic domains was observed in the former, whereas, smaller magnetic domains with a diameter of ca. 100 nm were observed in the latter [31]. These results indicated that the Pd cluster seeds formed by the electrochemical process were quite effective to suppress the intergranular exchange coupling, leading to the reduction of magnetic domain size. To elucidate the effect of the Pd cluster seeds, TEM observations were carried out. Figures 7.7c and d show the plan-view TEM bright field images of films I and II. Well-defined grain boundaries were observed in film II, providing the Co/Pd grains with the mean diameter of 15.8 nm, while the boundaries were hardly seen in film I [31]. From this result, it is found that the electrochemically formed Pd cluster seeds worked as nucleation sites for the Co/Pd multilayered film, and enhanced the columnar grain growth of the Co/Pd multilayered film with wide grain boundaries. Apparently, in the Co/Pd multilayered film, the grain boundaries

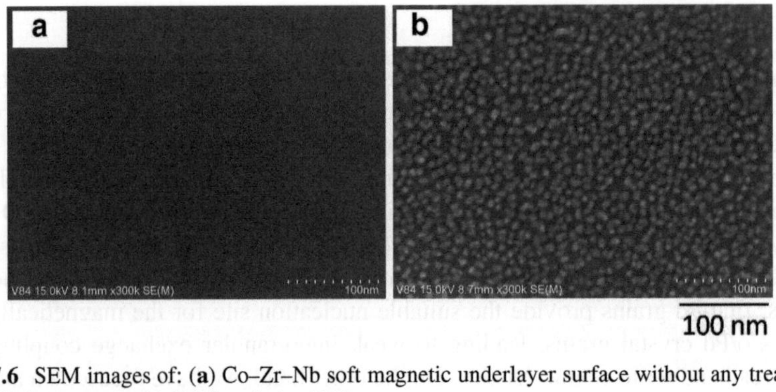

Fig. 7.6 SEM images of: (**a**) Co–Zr–Nb soft magnetic underlayer surface without any treatment and (**b**) that treated with $PdCl_2$ solution [31]

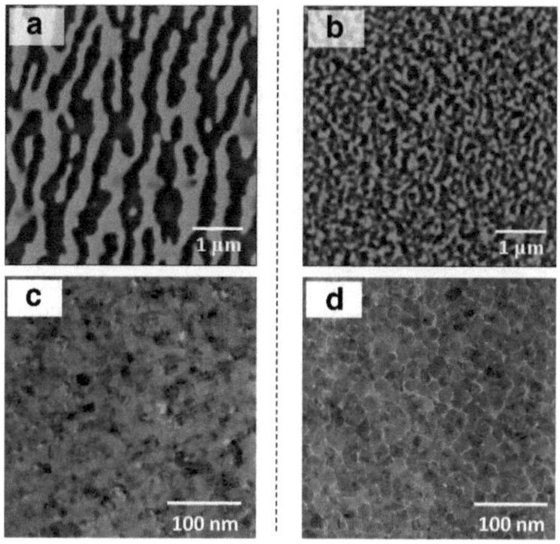

Fig. 7.7 MFM images at the ac-demagnetized states of the Co/Pd multilayered films: (**a**) with the sputter-deposited 10 nm-thick Pd seedlayer (film I) and (**b**) with the Pd cluster seeds (film II). Plan-view TEM bright field images of (**c**) film I and (**d**) film II [31]

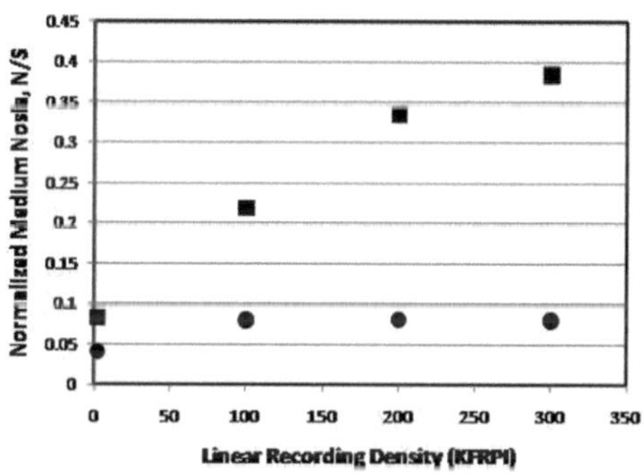

Fig. 7.8 Dependence of normalized medium noise on the linear recording density for the recording media consisting of film I (*squares*) and film II (*circles*) [31]

of voids with low material density contributed to the reduction of intergranular exchange coupling in the lateral direction, resulting in a fine magnetic domain structure as shown in Fig. 7.7b.

Figure 7.8 shows the dependence of normalized medium noise (N/S_0) on the linear recording density for the recording media consisting of films I and II, where the medium noise (N) was normalized to the reproduced voltage at the low recording density of 2 kFRPI (S_0). Medium II with the Pd cluster seeds exhibited lower medium noise than medium I with the sputter-deposited Pd seed layer. This result suggested that the transition noise caused by the fluctuation of the magnetization transition region between recorded bits was suppressed by employing the Pd cluster seeds, which bring about weak intergranular exchange coupling and fine magnetic domains [31].

Our study clarified that the size and nucleation density of the Pd cluster seeds were controlled by applying a pretreatment with a $SnCl_2$ solution, which mostly produced the finer Pd cluster seeds with a higher density and further decreased the intergranular exchange coupling, magnetic domain size, and medium noise. Moreover, the Pd cluster seeds formed on the electroless-deposited Co–Ni–Fe based SUL, mentioned in the previous section, were also effective for improving the magnetic properties of Co/Pd multilayered films [32]. This is a promising way to improve not only the magnetic properties but also the mass productivity of double-layered perpendicular magnetic recording media. Through these studies, we have demonstrated that the combination of wet and dry processes is useful in fabricating perpendicular magnetic recording media.

7.4 Development of Permanent Magnet Thin Films with Very High Perpendicular Magnetic Anisotropy

From the viewpoint of the materials of recording layer, thin films exhibiting high perpendicular magnetic anisotropy are indispensable for increasing areal recording density against superparamagnetism. Various attempts have been initiated to achieve the goal of ultra high areal density beyond 1 Tbit/inch^2, and, in particular, a thin film with high perpendicular magnetic anisotropy with an anisotropy constant, K_u, greater than 1.2×10^7 erg/cm^3 is required to maintain the stability of magnetization in exceedingly small crystal grains less than 5 nm in diameter against thermal agitation [33, 34]. The $SmCo_5$ alloy, a representative permanent magnet material, is a promising candidate for the materials because of its extremely high uniaxial magnetocrystalline anisotropy, whose K_u value is 1.1×10^8 erg/cm^3 or greater in the form of bulk alloy [34]. However, an $SmCo_5$ thin film exhibiting distinct perpendicular magnetic anisotropy had not been prepared, although the film with in-plane anisotropy is relatively easy to prepare, for instance, by using a Cr underlayer [35–39]. Recently, we succeeded in developing sputter-deposited $SmCo_5$ thin films with very high perpendicular magnetic anisotropy [40–45]; it will be stated in detail in this section.

The introduction of appropriate underlayer is a key to controlling the magnetic properties. First, various materials were examined as an underlayer of Sm–Co thin films in order to impart perpendicular magnetic anisotropy. As a result, it was found that the Sm–Co thin films deposited on a glass substrate coated with a (111)-oriented fcc-Cu underlayer with a large thickness of 100 nm or more exhibited distinct perpendicular magnetic anisotropy. The Sm–Co thin films with the Cu underlayer contained the hexagonal $SmCo_5$ phase with (001) orientation, revealed by an X-ray diffractometry. Thus, the Cu underlayer was found to be useful for affording perpendicular magnetic anisotropy to $SmCo_5$ thin films [40]. This was the first time when the $SmCo_5$ thin films exhibiting perpendicular magnetic anisotropy were developed. Takei et al. almost simultaneously reported a similar result [46].

Figure 7.9a shows *M–H* hysteresis loops for a typical sample with the Cu underlayer (sample A). The substrate temperature was set at an elevated temperature of 345°C for the deposition of all layers. The values of H_c and squareness ratio (SQR) measured in the perpendicular direction are listed in Table 7.1 together with the values of K_u and $\Delta\theta_{50}$ of the $SmCo_5$(002) reflection in an X-ray diffractometry, defined as the values of full width at half maximum obtained from the rocking curves. Both H_c and SQR in the perpendicular direction were greater than those in the film plane, indicating that perpendicular magnetic anisotropy was clearly generated. The K_u value of 1.7×10^7 erg/cm^3 is greater by a factor of 10 than a conventional material for recording layer, such as a Co–Cr–Pt-based alloy. Figure 7.9b shows *M–H* hysteresis loops for sample B, which has the same layer configuration with sample A deposited under a different condition; namely, the substrate temperature was set at a room temperature of 20°C for the deposition of the underlayer, followed by the deposition of the Sm–Co layer at 325°. The values of H_c, SQR, and K_u were greater than those of sample A. The enhancement of perpendicular magnetic anisotropy was revealed to be attributable to the suppression of the surface roughness of Cu underlayer [42, 43]. The surface roughness was attempted to decrease further by introducing

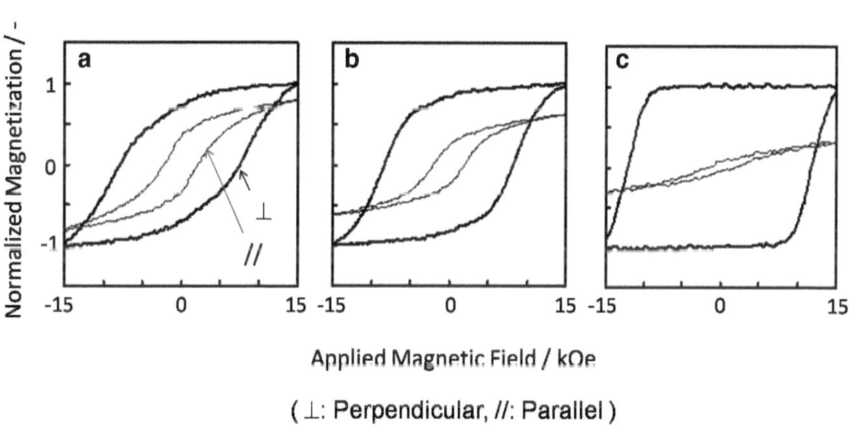

Fig. 7.9 *M–H* hysteresis loops of $SmCo_5$ thin films with perpendicular magnetic anisotropy. (**a**) Sample A, (**b**) sample B, and (**c**) sample C correspond to those in Table 7.1. Thick and thin lines represent the loops measured in the directions perpendicular and parallel to the film plane, respectively [45]

Table 7.1 Magnetic properties and values of full width at half maximum of $SmCo_5$(002) reflection ($\Delta\theta_{50}$) for samples A, B and C shown in Fig. 7.9

Sample	Underlayer	H_c (kOe)	SQR	K_u ($\times 10^7$ erg/cm^3)	$\Delta\theta_{50}$ $SmCo_5$(002)
A	Cu	7.7	0.72	1.7	Undetectable
B	Cu	8.2	0.85	2.2	10.1
C	Cu/Ti	12.0	1	4.0	3.1

a Ti layer prior to depositing the Cu underlayer. The Ti layer was expected to make smoother the Cu layer deposited upon it, because the melting point and the surface energy of Ti are higher than those of Cu. Indeed, the surface roughness of underlayer for sample C became smaller than that for sample B [42]. The $\Delta\theta_{50}$ values for samples B and C were 10.1° and 3.1°, respectively. This indicates that the higher degree of preferred orientation of the *c*-axis (easy axis) of hexagonal $SmCo_5$ was achieved with samples having a smoother underlayer surface. The *M–H* hysteresis loop for sample C is shown in Fig. 7.9c. The magnetic properties of sample C were significantly improved; the values of H_c and SQR became 12.0 kOe and unity, respectively. The K_u value reached an extremely high value of 4.0×10^7 erg/cm^3. We estimate the minimal stable grain diameter of sample C to be 3.7 nm from the K_u value, which is a value highly advantageous for an ultra high-density magnetic recording medium.

In summary, we developed, by using a conventional sputtering process, the $SmCo_5$ thin film with very high perpendicular magnetic anisotropy. This result contributes to the progress in applications of the $SmCo_5$ thin films to various magnetic devices, especially to magnetic recording media with ultra high areal recording density.

References

1. Noyes T, Dickson WE (1957) The random-access memory accounting. machine II, the magnetic-disk, random-access memory. IBM J Res Dev 1:72–75
2. Hattori S et al (1982) Continuous thin film media for 3.2 Gbyte multi-device disk storage. Rev Elec Comm Lab NTT 30:24–35
3. Matsubara H et al (1989) Recording characteristics of electroless-plated CoNiReP/NiWP double layers. J Magn Soc Jpn 13:153–156
4. Osaka T et al (1990) High density perpendicular recording media by an electroless plating method. ITEJ Tec Rep VIR90-29:1–9
5. Iwasaki S, Nakamura Y (1977) An analysis for the magnetization mode for high density magnetic recording. IEEE Trans Magn 13:1272–1277
6. Iwasaki S, Ouchi K (1978) Co–Cr recording films with perpendicular magnetic anisotropy. IEEE Trans Magn 14:849–851
7. Iwasaki S (1980) Perpendicular magnetic recording. IEEE Trans Magn 16:71–76
8. Tanahashi K et al (2002) Low-noise FeTaC underlayer for double-layered perpendicular recording media. J Magn Magn Mater 242–245:325–327
9. Soo EW, Zhou TJ, Wang JP (2002) FeCoC film as the soft magnetic underlayer for perpendicular media. J Appl Phys 91:8019–8021
10. Kong S, Okamoto T, Nakagawa S (2003) Improvement of soft magnetic properties of Fe–Co–B underlayer with large saturation magnetization by Ni–Fe–O seedlayers. J Appl Phys 93:6778–6780
11. Uchida M, Suzuki T, Ouchi K (2001) Preparation of Fe–Pt perpendicular double-layered media with high electric resistivity backlayer. J Magn Magn Mater 235:143–147
12. Nakamura F, Hikosaka T, Tanaka Y (2001) Low noise multi-layered FeAlSi soft magnetic films for perpendicular magnetic recording media. J Magn Magn Mater 235:64–67
13. Ando T, Nishihara T (2001) Exchange-coupled CoZrNb/CoSm underlayer for perpendicular recording media. IEEE Trans Magn 37:1228–1233

14. Jung HS, Doyle WD (2001) FeTaN/IrMn exchange-coupled multilayer films as soft underlayers for perpendicular media. IEEE Trans Magn 37:2294–2297
15. Asahi T (2004) Novel soft magnetic underlayer for double-layered perpendicular magnetic recording media: Electroless-deposited films of CoNiFe-based alloy. IEEE Trans Magn 40:2356–2358
16. Osaka T (2005) Electroless-deposited soft magnetic underlayer on silicon disk substrate for double-layered perpendicular magnetic recording media. J Magn Magn Mater 287:292–297
17. Sugiyama A (2005) Electroless deposition of soft magnetic underlayer for a double-layered perpendicular recording media. IEICE Tech Rep MR2004-48:15–20
18. Sugiyama A (2005) Magnetic microstructure and noise property of electrolessly deposited CoNiFeB soft magnetic underlayer. IEEE Trans Magn 41:3163–3165
19. Saito N, Fujiwara H, Sugita Y (1964) A new type of magnetic domain structure in negative magnetostriction Ni–Fe films. J Phys Soc Jpn 19:1116–1125
20. Sugiyama A (2005) Application of electrolessly dual-layered CoNiFeB film to softmagnetic underlayer for a double-layered perpendicular recording media. IEICE Tech Rep MR2005-19:53–58
21. Favieres C et al (2002) Continuous change of surface magnetization direction from perpendicular to planar in soft magnetic CoP multilayers. J Appl Phys 91:9995–10001
22. Carcia PF, Meinhaldt AD, Suna A (1985) Perpendicular magnetic anisotropy in Pd/Co thin film layered structures. Appl Phys Lett 47:178–180
23. Lairson BM, Perez J, Baldwin C (1994) Application of Pd/Co multilayers for perpendicular magnetic recording. Appl Phys Lett 64:2891–2893
24. Peng W et al (2002) CoB/Pd multilayers with indium tin oxide seedlayer on NiFe underlayer for perpendicular magnetic recording. J Appl Phys 91:8070–8072
25. Nakagawa H et al (2003) CoB/Pd multilayers with PtB/Pd/MgO intermediate layers for perpendicular magnetic recording. IEEE Trans Magn 39:2311–2313
26. Matsunuma S et al (2002) Co/Pd multilayer media with Pd inorganic granular seed layer for perpendicular recording. J Appl Phys 91:8073–8075
27. Kawaji J et al (2003) Effect of the preparation conditions of a Pd/Si dual seedlayer on the magnetic properties of Co/Pd multilayered perpendicular magnetic recording media. Trans Magn Soc Jpn 13:1–7
28. Kawaji J et al (2004) Optimization of the thickness ratio of Pd/Si intermediate layer in a Co/Pd multilayer perpendicular magnetic recording media. Trans Magn Soc Jpn 4:78–84
29. Kawaji J et al (2004) Micrositructure and magnetic properties of a Co/Pd multilayer on a controlled Pd/Si seed layer for double-layered perpendicular magnetic recording media. J Appl Phys 95:8023–8029
30. Kawaji J et al (2004) Formation of Pd nanocluster seeds by electrochemical process for control of magnetic properties in Co/Pd multilayered film. IEEE Trans Magn 40:2473–2475
31. Tanaka M et al (2005) Microstructure of a Co/Pd multilayered perpendicular recording medium with Pd seeds prepared by electrochemical process. J Magn Magn Mater 287:188–192
32. Kawaji J (2006) Electrochemical formation of intermediate layer for Co/Pd multilayered media. J Magn Magn Mater 303:e128–e132
33. Wood R (2000) The feasibility of magnetic recording at 1 terabit per square inch. IEEE Trans Magn 36:36–42
34. Strnat KJ (1988) Rare earth-cobalt permanent magnets. In: Wohlfarth EP, Buschow KHJ (Eds.) Ferromagnetic materials, vol 4. North-Holland, Amsterdam, p 131
35. Rani R et al (1997) SmCo based sputtered films with $CaCu_5$ and $TbCu_7$ structures. J Appl Phys 81:5634–5636
36. Mizukami M, Abe T, Nishihara T (1997) Magnetic stability and noise characteristics of high He CoSm/Cr thin films. IEEE Trans Magn 33:2977–2979
37. Liu Y et al (1998) Phase formation and magnetic properties of Co-rare earth magnetic films. J Appl Phys 83:6244–6246
38. Cadieu FJ et al (1999) Fully in plane aligned SmCo based films prepared by pulsed laser deposition. J Appl Phys 85:5895–5897

39. Neu V, Shaheen SA (1999) Sputtered Sm–Co films: microstructure and magnetic properties. J Appl Phys 86:7006–7009
40. Sayama J et al (2004) Newly developed $SmCo_5$ thin film with perpendicular magnetic anisotropy. J Phys D Appl Phys 37:L1–L4
41. Sayama J et al (2004) Thin films of $SmCo_5$ with very high perpendicular magnetic anisotropy. Appl Phys Lett 85:5640–5642
42. Sayama J et al (2005) Magnetic properties and microstructure of $SmCo_5$ thin film with perpendicular magnetic anisotropy. J Magn Magn Mater 287:239–244
43. Sayama L et al (2005) $SmCo_5$-based thin films with high magnetic anisotropy for perpendicular magnetic recording. IEEE Trans Magn 41:3133–3135
44. Sayama J et al (2006) Origin of perpendicular magnetic anisotropy of $SmCo_5$ thin films with Cu underlayer. J Magn Magn Mater 301:271–278
45. Sayama J et al (2006) Magnetic properties and domain structure of the SmCo5 perpendicular magnetization films prepared by using a UHV sputtering system. J Magn Soc Jpn 30:423–428
46. Takei S, Morisako A, Matsumoto M (2004) Fabrication of Sm–Co films with perpendicular magnetic anisotropy. J Magn Magn Mater 272–276:1703–1705

Chapter 8
Cusp-Field Single-Pole-Type (CF-SPT) Head for Perpendicular Recording

Kiyoshi Yamakawa

8.1 Introduction of CF-SPT Head

Maintaining the thermal stability of small magnetic grains is important for perpendicular magnetic recording; for achieving higher recording densities, it is inperative to continue to increase the strength and sharpness of the magnetic recording field of single-pole-type (SPT) heads because the anisotropy energy of the recording media becomes large, thereby threatening the thermal stability. Development of soft magnetic pole material with a high saturation magnetic flux density, B_s, is the first requirement. Furthermore, improvement of the head structure is important because little room is left for the B_s increasing to its practical limit of 2.45 T.

The first single-pole head was developed as an auxiliary-pole-driven-type head [1]. This head underscored the importance of the head-energizing method for realizing a strong and sharp recording field in which the coil created the strongest field at the top in the main pole located at the air-bearing surface (ABS). This concept was inherited by the thin film SPT head [2] depicted in Fig. 8.1a. In the head, the coil of the helical structure is wound closely around the main pole throat to excite the pole tip directly with the strongest coil field. As depicted in Fig. 8.1b, field calculations reveal that the highest head field was obtained when the coil recession height, h, from the ABS was zero.

The cusp-field single-pole type (CF-SPT) head [3] was proposed recently as a modification of the head described above to meet the requirements for practical use. As illustrated in Fig. 8.2, the head structure has a novel coil and a dual return yoke. The coil consists of a pair of windings in mutually opposite current directions, generating a cusped field. The main pole is placed in an off-center arrangement of the axisymmetric coil field to be energized strongly. The recession height of the outermost conductor of each winding is set at zero. For this coil configuration, it is easy to increase the number of coil turns so that the coil is useful at sufficiently low

K. Yamakawa (✉)
Akita Research Institute Of Advanced Technology (AIT), 4-21 Sanuki, Araya, Akita 010-1623, Japan
e-mail: yamakawa@rdc.pref.akita.jp

T. Osaka et al. (eds.), *Electrochemical Nanotechnologies*, Nanostructure Science and Technology, DOI 10.1007/978-1-4419-1424-8_8,

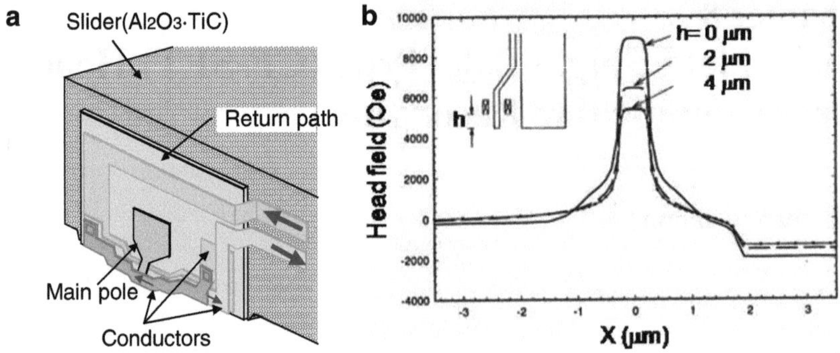

Fig. 8.1 (**a**) Schematic illustration of thin film head with a helical coil at the main pole tip. (**b**) Dependence of field distribution on coil recession height, *h* [2]. © 1999 IEEE

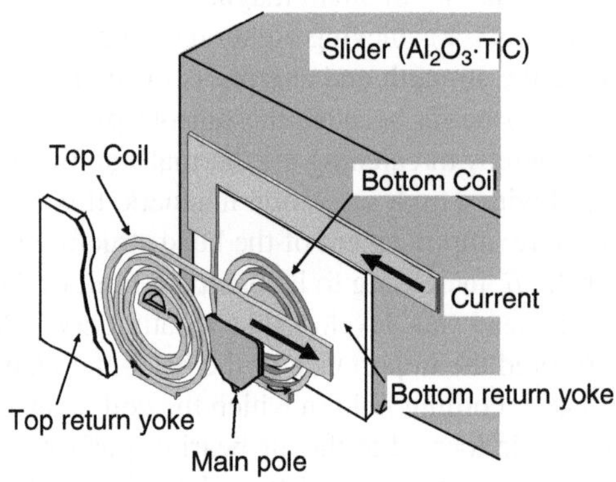

Fig. 8.2 Schematic illustration of CF-SPT head

current for practical use. The return yoke consists of two planar magnetic films. The main pole must be centered between these two films. This configuration has resolved the problem preventing its practical use; previously used conventional single pole heads having a one-side return yoke are vulnerable to stray fields. The CF-SPT head is anticipated for use in future high-density recording, although SPT heads modified from ring type heads for longitudinal recording have been used in first-generation commercial perpendicular recording systems.

In figures, use italic for variables.

8.2 Feasibility of Head Fabrication

The CF-SPT head has a simply stacked structure comprising planar films of magnetic, conductive and insulating materials, except for the connection part between the two coil windings. Therefore, the fabrication process of this head uses simpler techniques than those used for conventional heads. The CF-SPT head is fabricated using wet processes. The electrodeposition technique has been widely used for manufacturing conventional recording heads because it presents distinct advantages such as a high aspect ratio deposition, precise definition of pattern width, and a high deposition rate. To increase the track density, the critical dimension of the pole tip has been reduced by adopting advanced lithography techniques, reaching 40 nm, which corresponds to more than 500 Gbit/in^2 [4]. Electrodeposition will continue to be an important technology for manufacturing heads in the future also.

The head fabrication procedure involving wet processes is described below. As depicted in Fig. 8.3, a bottom-return yoke is first formed using electrodeposition. A soft magnetic material, e.g. CoNiFe, is electrodeposited on a sputter-deposited seed layer such as NiFe/Ti (Fig. 8.3a), which will be described later in detail together with the main pole fabrication. A SiO_2 insulating layer is then sputter-deposited on it. In the next step, a bottom coil is fabricated using a damascene process (Fig. 8.3b), which will also be described later in detail. After deposition of the insulating layer, the main pole material is electrodeposited on a sputter-deposited seed layer (Fig. 8.3c). Next, a Cu stud is made to connect the bottom and the top coils at their center and an insulating layer is deposited to planarize the surface. Then, a top coil is formed in the same manner as that of the bottom coil (Fig. 8.3d), which is followed by deposition of an insulating layer. At the final step, a top-return yoke is formed by electrodeposition (Fig. 8.3e), then, an insulating material as a protective layer is sputter-deposited. After completion of these wafer processes, the head wafer is cut into individual pieces and the ABS is formed by polishing (Fig. 8.3f).

In the test fabrication, the soft magnetic material of CoNiFe [5], with B_s of 2 T as developed by the Osaka group of Waseda University, was used for both the main pole and the return yoke. The CoNiFe film, with a target composition of $Co_{63}Ni_{13}Fe_{24}$, was electrodeposited using a paddle plating system, as depicted in Fig. 8.4, with the plating bath of the composition presented in Table 8.1. In addition, a frame plating method in a magnetic field to induce magnetic anisotropy was applied to form the main pole and the return yoke. The deposited film exhibited

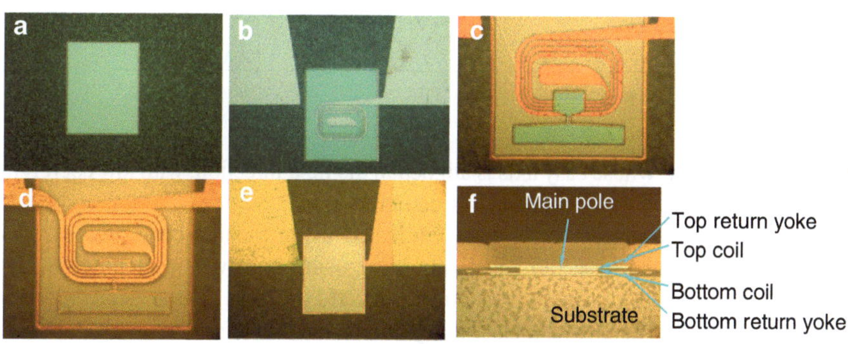

Fig. 8.3 Main process steps of CF-SPT head fabrication; (**a**) bottom return yoke, (**b**) bottom coil, (**c**) main pole, (**d**) top coil, (**e**) top return yoke, and (**f**) ABS

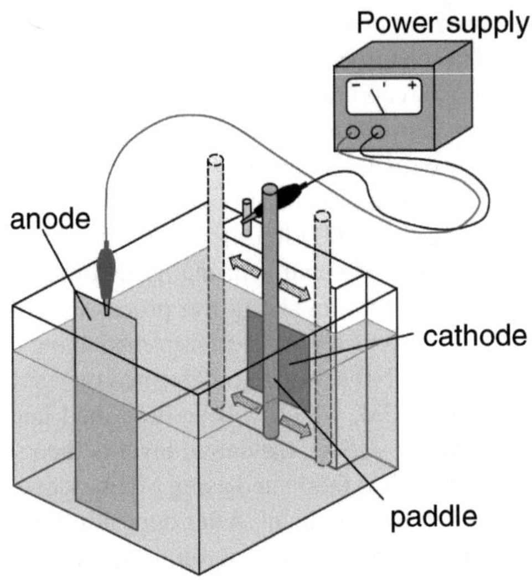

Fig. 8.4 Paddle plating system

Table 8.1 Plating bath composition for CoNiFe soft magnetic material

Chemicals	Concentration (mol/dm^3)
H_3BO_3	0.4
NH4Cl	0.28
$CoSO_4{\cdot}7H_2O$	0.064
$NiSO_4{\cdot}6H_2O$	0.2
$FeSO_4{\cdot}7H_2O$	0.011
SDS	10 ppm
pH	2.8
Current density	20 mA/cm^2
Agitation speed	144 cpm
Bath temperature	18–21°C
Counter electrode	Co plate
Magnetic field	100 Oe
Working electrode	Cu/Ti/glass (1 inch square)

uniaxial anisotropy, whose easy axis is aligned in the field direction, with low coercivity for widely varying film thicknesses, as depicted in Fig. 8.5.

The top and the bottom coils were fabricated using a damascene process. Coil trenches were formed in a SiO_2 insulating layer using reactive ion etching. A seed layer for electrodeposition such as a Cu/Ti stacked layer was sputter-deposited onto the etched surface. Then, the Cu electrically conductive material was electrodeposited on the seed layer using generic copper sulfate solution. Finally, lapping was performed

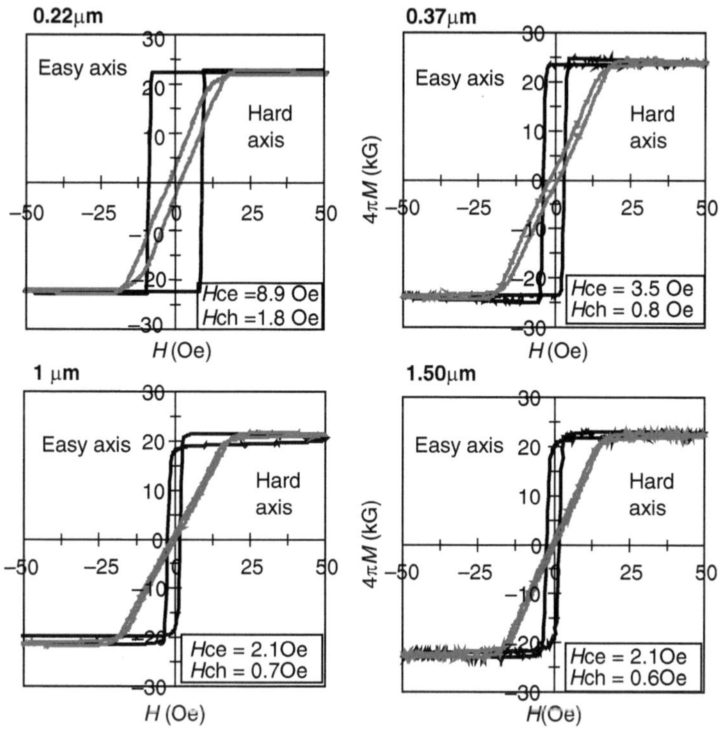

Fig. 8.5 Hysteresis loops of CoNiFe films in both easy and hard axes for various film thicknesses of 0.22–1.50 μm

to remove the excess Cu layer material. Coil pitch is an important parameter that influences the head efficiency because it governs the effective coil recession. The coil pitch is expected to be reduced. In very narrow pitches required for future high-density recording heads, forming the seed layer in the trenches with high aspect ratio will become problematic. For this issue, the new process technique developed for interconnects in ULSI [6] is useful for seed layer deposition. In the process, the NiB seed layer electroless-deposited on the self-assembled-monolayer (SAM) can be formed in high-aspect-trench patterns with excellent uniformity and conformability. Figure 8.6 shows the cross section of the Cu-filled trench pattern formed with the NiB/SAM seed layer. From this result, a narrow pitch of less than 1 μm is expected.

8.3 Performance of CF-SPT Head

In the CF-SPT head, the main pole is energized using a part of the cusped field generated by the opposing coil conductors. Field calculations using finite element method (FEM) revealed that the fields created at the main pole by the cusp coil and

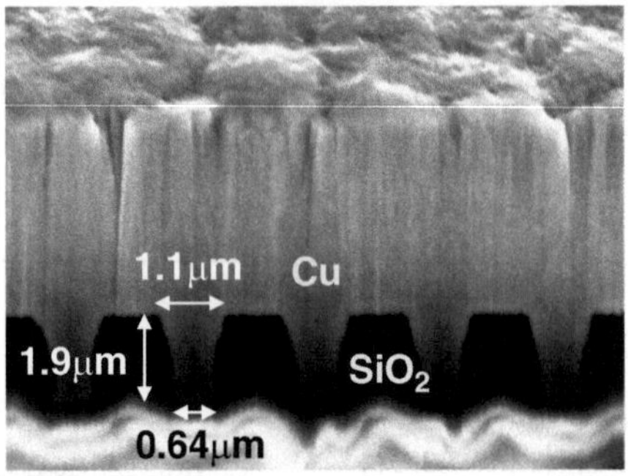

Fig. 8.6 Cu-filled coil trench using a NiB/SAM seed layer

the helical coil wrapped around the pole tip like that depicted in Fig. 8.2, are almost identical [7]. This similarity reflects that the cusp coil is equivalent in terms of main pole excitation to a helical coil with half the number of physical coil turns of the cusp coil. Therefore, the highest field applicable to the main pole tip located between the coil conductors and the field rapidly decreases with distance from the conductors [8]. This coil field scheme provides a stronger head field than that of the conventional SPT head with a recessed coil, as depicted in Fig. 8.7.

In addition to the head field strength, the coil field distribution confined between the opposing conductors uses a different pole-energizing method from the conventional head. The main pole tip region is magnetized preferentially. Therefore, only a small amount of flux generation is required, and it is localized at the main pole tip region. In contrast, for the conventional head, whole magnetic yokes are magnetized because of a closed magnetic circuit and a widely distributed coil field. Consequently, the CF-SPT head exhibits little field underneath the return yoke [7] that would affect the wide-area track erasure causing data loss by erasure of recorded information over a large area. In addition, the head inductance is small, helping to operate in the high data rate. An example of measured inductance as a function of frequency for the heads with effective two and three turn coils is depicted in Fig. 8.8 [9]. The resonance frequencies of the heads are much higher than that for the conventional ring-type head with seven turn coil, as shown in the figure. Therefore, superior performance of the CF-SPT head is expected.

Early SPT heads had a fatal flaw by which the head occasionally erased recorded signals without a writing operation when the head was exposed to an external stray field [10]. That phenomenon was a difficult issue for practical head implementation because the stray field easily magnetizes the main pole. For the head of a conventional yoke structure, the externally applied field, as depicted in Fig. 8.9, significantly degrades the signal output. In contrast, because the CF-SPT has a return yoke structure that sandwiches the main pole, the main pole can be shielded against a stray field. Consequently, the CF-SPT head has good stray field immunity, as depicted in Fig. 8.9.

As the head track width decreases, remanent magnetization of the main pole tip is likely to occur because of stress-induced anisotropy and increased shape anisotropy in the normal direction to the ABS. The field from this remanent magnetization can

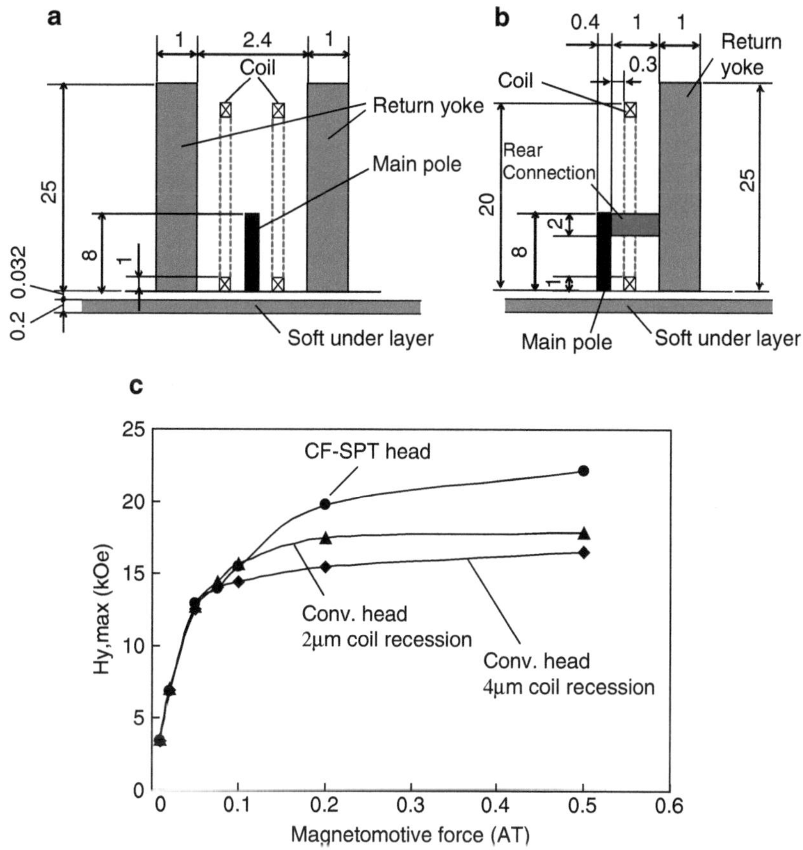

Fig. 8.7 Head model of (**a**) CF-SPT head and (**b**) conventional SPT head. (**c**) Comparison of head field for the CF-SPT head and conventional heads with various coil recessions

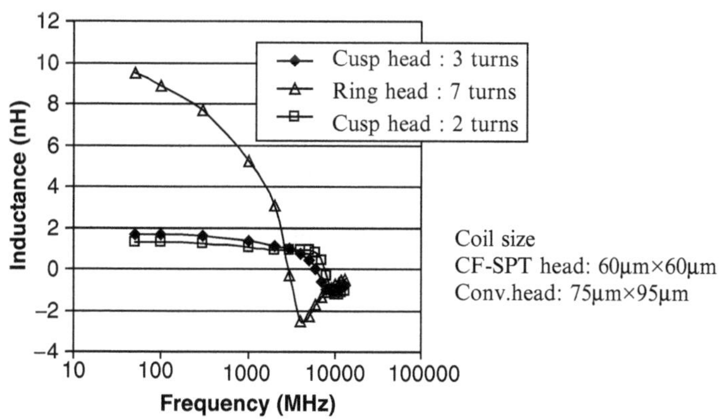

Fig. 8.8 Head inductance as a function of frequency for CF-SPT heads with various effective coil turns and a conventional head [9]. © 2003 IEEE

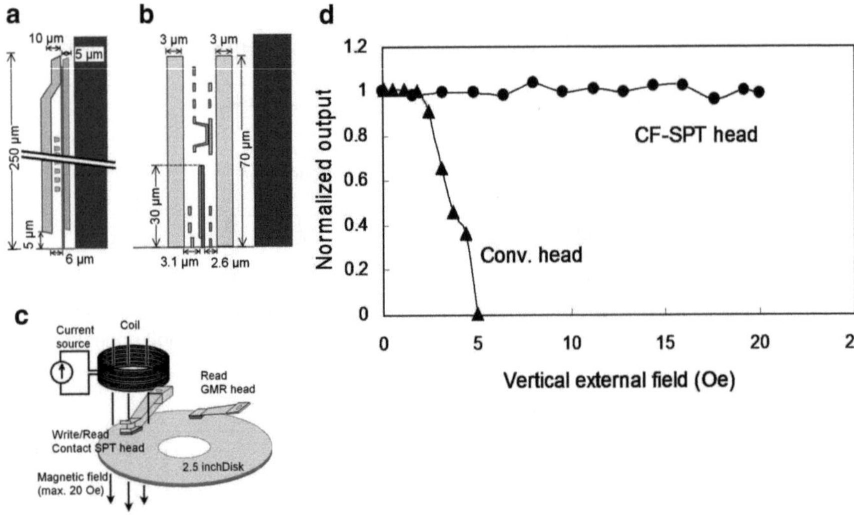

Fig. 8.9 Schematic illustration of (**a**) conventional SPT head and (**b**) CF-SPT head. (**c**) Experimental setup for stray field immunity. (**d**) Influence of the stray field on signal decay for CF-SPT head and conventional head

cause another type of head-induced erasure or pole erasure of recorded signals, even in an environment with no stray field [10]. The pole erasure can be eliminated through several means to align the magnetization in the track width direction. These solutions are a multilayer film structure [11], high anisotropy material [12], controlling stress-induced anisotropy [13], and the proper aspect ratio of the pole tip [14]. In addition, optimization of the main pole shape is a countermeasure for avoiding pole erasure. As depicted in Fig. 8.10, the main poles with different flare angles display different domain structures. The conventional flare angle of 45° exhibits an irregular domain structure containing a wall at the center running in the direction parallel to the flux propagation. In contrast, for 14°, a regular closed domain structure is realized by which the magnetization in the center domains lies in the track-width direction, as depicted in Fig. 8.10a. These domain structures show different signal decay over time. As depicted in Fig. 8.10b, the output signal decreases significantly for a 45° flare angle, although no signal deterioration is observed for the 14° flare angle. This difference suggests that the irregular domain structure causes the remanence of the pole tip. Consequently, domain control by optimizing the pole shape is an effective means to eliminate the pole erasure.

8.4 Advanced CF-SPT Head

Introduction of controlled inter-granular exchange coupling and hard/soft stacked layer design with inter-layer exchange coupling has been considered for granular media to alleviate writability problems for increased media anisotropy.

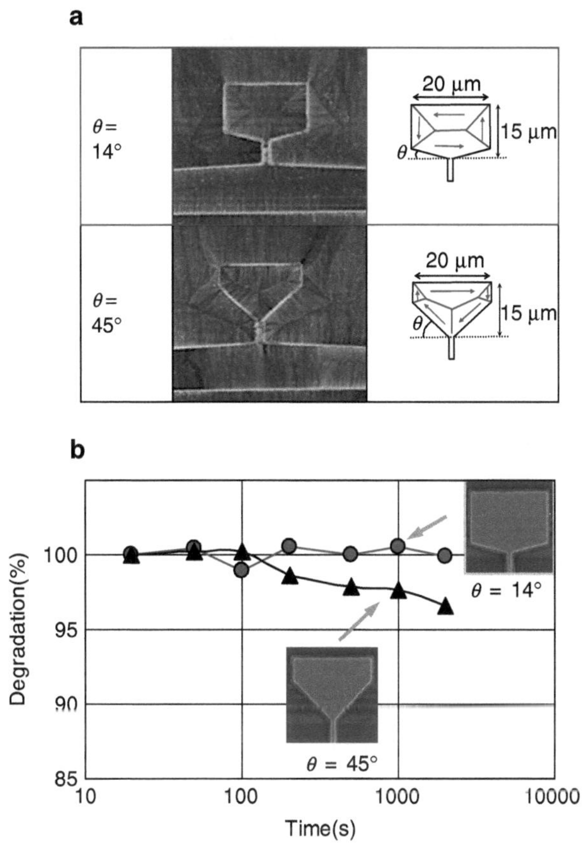

Fig. 8.10 (**a**) Effect of flare angle on domain structure of main pole. (**b**) Signal decay with time for different flare angles

Moreover, discrete track media, which enable the use of larger track-width writing heads, thereby providing a stronger head field than that for continuous media and bit-patterned media that have large switching volume, have been proposed as promising candidates for future high-density recording media. Nevertheless, a still larger head field will be necessary because the head track width will be continuously narrowed for an increased recording density. Furthermore, a higher field gradient is required because reduction of transition noise of granular media including discrete track media and an increased write-timing margin for bit-patterned media are responsible for the head field gradient and for the media property.

A shield structure has been proposed to improve the field gradient where the shield yoke is placed near the main pole [15, 16]. An example of a shielded SPT head based on a CF-SPT head [17] is depicted in Fig. 8.11a. The shield yoke absorbs the magnetic flux emanating from the side of the main-pole and prevents it from

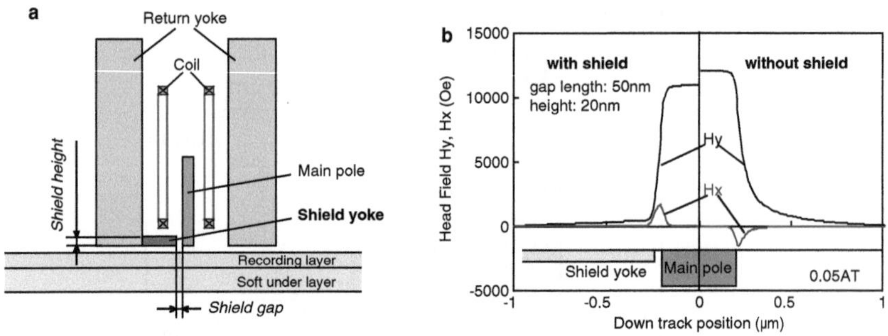

Fig. 8.11 (**a**) Structure of shielded CF-SPT head. (**b**) Comparison of head field between shielded and normal CF-SPT head

flowing into the medium. Consequently, the field gradient observed at the medium position is improved at the cost of the deterioration of recording flux, or head field strength, as depicted in Fig. 8.11b. Field strength loss occurs because the shield is small for an optimized shield design and the head exhibits no negative under-shoot observed for a shielded head modified from a conventional ring type head.

Shield design optimization for the CF-SPT head shows that critical dimensions such as shield gap length and shield height are expected to be in the order of a few tens of nanometers, as shown in Fig. 8.12. It is difficult to realize such a small shield structure using normal head fabrication techniques. However, the fabrication of a shielded CF-SPT head by a unique process has been demonstrated [17]. This process is as follows: a soft magnetic material is deposited on the head surface followed by focused ion bead (FIB) milling to form the shield gap, as depicted in Fig. 8.13. Improvement in writing resolution was actually obtained for the fabricated shield head.

On the analogy of a pole piece of an electromagnet, a main pole with a tapered structure is expected to produce a large field in association with the medium soft underlayer, thereby surpassing the material's limitation. The results of analytical calculations, depicted in Fig. 8.14, show that a tapered pole with a pyramidal shape generates a larger field for a larger base core size, W_2 and Tm_2, which engenders the possibility of a field beyond the saturation magnetization of the pole material [18–20]. In the figure, the stepped type of main pole is also shown to exhibit a large field, but the field is low compared to the tapered pole, especially for a large base core size. The reason for the strong fields of these heads, a so-called multi-surface pole head [19], is that the field from the charge appearing on the tapered surfaces is superimposed on the field from the charge on the pole tip surface. This mechanism suggests that the head field depends on the area of the tapered surface. Figure 8.15 compares the fields for the three different pole structures: tapered structures in both the down track and cross track directions, tapered structure only in the down track direction and no tapered structure [21]. The all-round tapered pole exhibits a much stronger field than the others. The tapered and the stepped pole

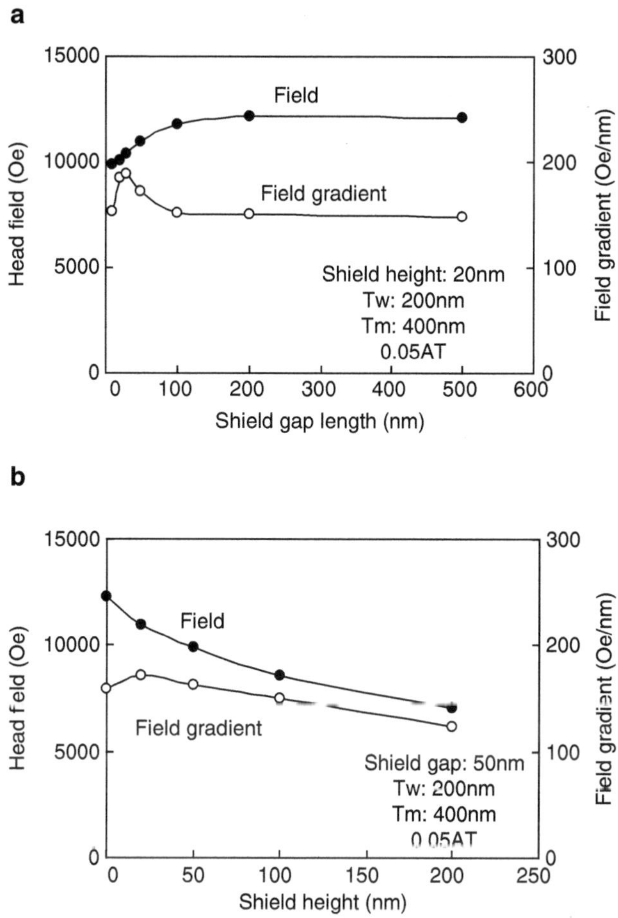

Fig. 8.12 Effect of (**a**) shield gap length and (**b**) shield height on head field strength and gradient

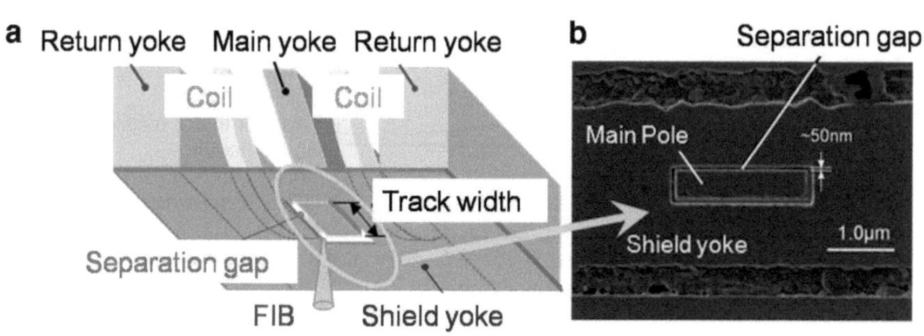

Fig. 8.13 (**a**) CF-SPT head with shield structure formed by FIB. (**b**) ABS view of shielded CF-SPT head

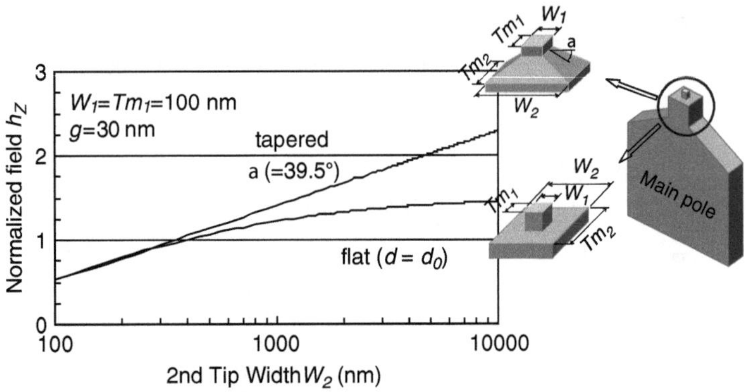

Fig. 8.14 Head field as a function of base core size, W_2 and Tm_2, for tapered and stepped main poles. Head field is normalized by saturation magnetization

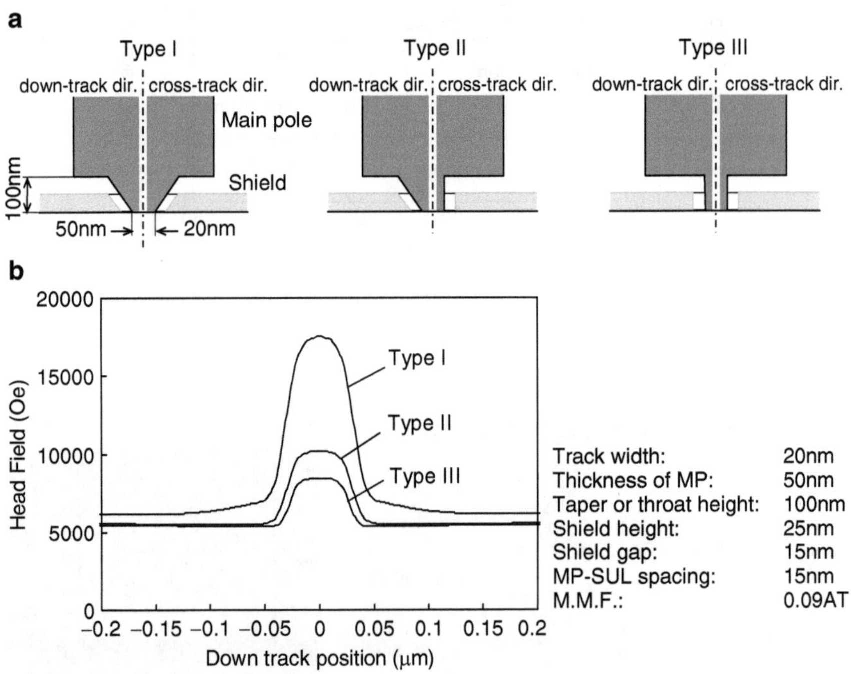

Fig. 8.15 (**a**) Variation of pole tip structure: *Type I* – tapered structure in both down and cross track directions; *Type II* – tapered only in the down track direction; *Type III* – no tapered structure. (**b**) Comparison of head fields for tapered main poles of various types

configurations fundamentally produce a large fringing field, especially at large magneto-motive forces, because of the field from the charge on the recessed surfaces. For that reason, the shield structure described previously must be combined with the main pole to obtain a sharp and strong field.

With a combination of the tapered pole structure and the shield structure, an advanced CF-SPT head has been proposed for 1 Tbit/in^2 recording [22]. As depicted in Fig. 8.16, the head has shields not only in the down track direction but also in the cross track direction to reduce the fringing field at adjacent tracks for achieving a high track density. The head exhibits high field strength and sharpness. Therefore, the CF-SPT head is a promising candidate as a writing head for future high-density recording. This advanced head is challenging in terms of its manufacture because of its structural complexity and nanoscale dimensions. It is expected to be a *breakthrough* development in resolving issues with the help of advanced nanoscale electro-deposition and electroless-deposition techniques.

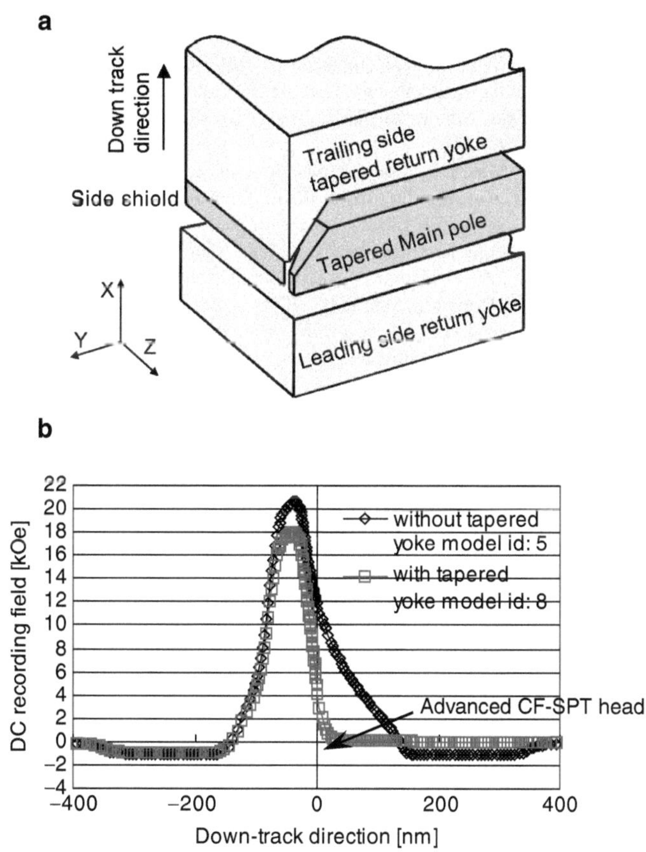

Fig. 8.16 Structure, cut at the track center (**a**), and field distribution (**b**) of an advanced CF-SPT head [22]. © 2003 IEEE

References

1. Iwasaki S, Nakamura Y (1978) The magnetic field distribution of a perpendicular recording head. IEEE Trans Magn 14:436–438
2. Muraoka H et al (1999) Low inductance and high efficiency single-pole writing head for perpendicular double layer recording media. IEEE Trans Magn 35:643–648
3. Ise K et al (2000) High writing-sensitivity single-pole head with cusp-field coils. IEEE Trans Magn 36:2520–2523
4. Brankovic SR et al (2006) Pulse Electrodeposition of 2.4 T $Co_{37}Fe_{63}$ alloys at nanoscale for magnetic recording application. IEEE Trans Magn 42:132–139
5. Osaka T et al (1998) A soft magnetic CoNiFe film with high saturation magnetic flux density and low coercivity. Nature 392:796–798
6. Yoshino M et al (2005) All-wet fabrication process for ULSI interconnects technologies. Electrochim Acta 51:916–920
7. Yamakawa K et al (2002) A new single-pole head structure for high writability. IEEE Trans Magn 38:163–168
8. Yamakawa K et al (2007) FEM model analysis of single-pole-type heads with different coil structures. IEICE Trans Electron E90-C:1555–1560
9. George P et al (2003) High-frequency inductance measurements and performance projections made for cusp-field single-pole heads. IEEE Trans Magn 39:1949–1954
10. Payne W, Cain A, Bauldwinson M, Hempstead R (1996) Challenges in the practical implementation of perpendicular magnetic recording. IEEE Trans Magn 32:97–102
11. Nakamoto K et al (2004) Single-pole/TMR heads for 140-Gb/in^2 perpendicular recording. IEEE Trans Magn 40:290–294
12. Hirata K et al (2005) Material properties and domain structure influence on pole erasure occurrence in perpendicular recording heads. IEEE Trans Magn 41:2902–2904
13. Hirata K, Roppongi T, Noguchi K (2005) A study of pole material properties for pole erasure suppression in perpendicular recording heads. J Magn Man Mater 287:352–356
14. Gao KZ, Bertram HN (2002) 3-D micromagnetic simulation of write field rise time in perpendicular recording. IEEE Trans Magn 38:2063–2065
15. Mallary ML (1987) Vertical magnetic recording arrangement. US Patent #4 656 546
16. Mallary ML, Das SC (1992) Reissued #33 949
17. Ise K, Yamakawa K, Honda N (2003) High-field gradient cusp field single-pole writing head with front return yoke. IEEE Trans Magn 39:2374–2376
18. Takahashi S, Yamakawa K, Ouchi K (2001) 2 steps type of single pole head for ultra narrow track. Tech Rep IEICE MR2001-1:1-8
19. Takahashi S, Yamakawa K, Ouchi K (2002) Single-pole type head with multicharged surfaces for ultrahigh density recording. J Appl Phys 91:6839–6841
20. Gao KZ, Bertram HN (2002) Write field analysis and write pole design in perpendicular recording. IEEE Trans Magn 38:3521–3527
21. Ise K et al (2006) New shielded single-pole head with planar structure. IEEE Trans Magn 42:2422–2424
22. Kanai Y et al (2003) Recording field analysis of narrow-track SPT head with side shields, tapered main pole, and tapered return path for 1 Tb/in^2. IEEE Trans Magn 39:1955–1960

Chapter 9
Perpendicular Magnetic Recording Medium for a Density Beyond 1 Tera Bit/inch^2

Kazuhiro Ouchi and Naoki Honda

9.1 Necessity of Patterned Media

Perpendicular magnetic recording was proposed by Professor S. Iwasaki in 1977 [1] as a scheme superior to that of longitudinal recording in terms of high density recording performances. The new HDD (hard disk drive) system of perpendicular recording was commercialized in 2005. The area recording density started at 133 Gbit/inch^2 [2], which far surpassed the achieved density of the conventional HDD of longitudinal recording. In 2006, successful demonstrations of the highest density at around 350–420 Gbit/inch^2 were announced, one after another, by HDD manufacturers [3]; no other new information storage technology superior to magnetic recording has been proposed as yet. Thus, perpendicular recording is expected to dominate over the existing information storage technology in the near future.

Perpendicular magnetic recording (PMR) has the great advantages of a single pole high writeability of recording in the gap between the head and the medium soft under layer, a high recording resolution of anti-parallel magnetization transition with no demagnetizing field, and a high thermal stability with a rather thick recording layer, when compared with the longitudinal magnetic recording (LMR) used so far . These advantages in PMR and the lately diagnosed limitation of thermal stability of the LMR media accelerated the commercialization of PMR at around a density of over 100 Gbits/inch^2,where the PMR media have a large-enough margin for the limit of thermal stability. Construction of the commercialized PMR system is based on the original principle of PMR, in which the combination of a single pole head and a composite medium with a soft magnetic back layer was essential. Presumably, however, as long as granular type media are used, even the PMR system would face thermal

K. Ouchi (✉) and N. Honda
Akita Research Institute of Advanced Technology (AIT), 4-21 Sanuki, Araya, Akita 010-1623, Japan
e-mail: ouchi@rdc.pref.akita.jp

N. Honda
Faculty of Engineering, Tohoku Institute of Technology, 35-1 Yagiyama, Kasumicho, Taihakuku, Sendai, Miyagi 982-8577, Japan

T. Osaka et al. (eds.), *Electrochemical Nanotechnologies,*
Nanostructure Science and Technology, DOI 10.1007/978-1-4419-1424-8_9,

instability of the media or the restriction of writing by single pole heads when a high density over 1 Tera bits/inch^2 is designed. The former issue can be answered by employing very high anisotropy energy materials such as Fe–Pt, Sm–Co, Fe–Nd–B, etc. But it means an extremely high switching field of such media; thus, the latter issue of head writeability would, in the final outcome, become very serious.

In order to solve the dilemma of the PMR system design, heat-assisted magnetic recording (HAMR) [4], discrete track or bit-patterned media have been proposed [5]. HAMR is inevitably based on perpendicular magnetic recording that makes magnetization switching easy with the aid of thermal energy provided by some probe light beams irradiated bit by bit locally or over a wider area. However, there is neither a proper light beam source of 10 nm spot nor practical media with the required thermal properties. Furthermore, a medium of lower coercivity due to an increased temperature may, in principle, show a slower switching speed. Thus, a high speed and a high density may hardly coexist by HAMR.

A discrete track medium has been successfully developed in order to form a very narrow recording track without any adjacent track interference even for a fairly wide track head [6, 7]. However, it has been reported that the discrete media can improve the track margin by around 10% when compared with the conventional media [8]. Furthermore, the technologies to fabricate a few ten nanometer grooves on the disk media have almost the same difficulties as those for a so-called patterned medium. Therefore, it may be necessary to think that bit-patterned array media, rather than the discrete track media, should take over the granular recording layer, as the second-generation perpendicular recording media; hence, introduction of HAMR should preferentially be taken up after the development of patterned media.

The fundamental idea of bit-patterned media was first proposed by I. Nakatani et al. in 1991 [9] followed by a demonstration of the fabrication of 65 Gbits/inch^2 columnar dots by S. Y. Chow et al. in 1994 [10]. However, its superiority in the area density when compared to conventional recording media has not yet been proved. Around the same period, Y. Nakamura [11] proposed a "Tera bit spinic storage" that stores a bit per grain, indicating preferred perpendicular anisotropy in terms of magnetostatic interactions between bits, but with no suggestions for fabrication and design of the media. In 1997, R. L. White et al. [12] refocused attention on patterned media in terms of thermal stability after the suggestion of the thermal stability limit of conventional longitudinal magnetic recording at around 100 Gbit/inch^2 by S. H. Charap [13]. Early studies on patterned media [14, 15], with the exception of the study by Y. Nakamura, had not focused on the media with perpendicular anisotropy. Although many papers on the fabrication of magnetic dot arrays have been published [16, 17], few have focused either on the thermal stability design, the influence of magneto-static interaction between dots, or the recording conditions such as writability of heads. Especially, write conditions without a bit interference with neighbor dots, which may determine the dot array configuration, such as dot separation, shape and magnetic properties have not been discussed. Fundamental read-write simulations were first performed by G. H. Hughes [18], aiming at recording on a 100 Gbits/inch^2 patterned bits array, which had a large margin of the energy ratio of more than 500 for thermal stability of recorded bits. Hence, thermal stability issues were not taken up for discussion. However, since an area density of more than 300 Gbits/inch^2 has become possible with conventional perpendicular media of a granular type, the target area density should be increased to more than 1 Tera bits/inch^2, where the bit size is roughly within 25×25 nm square or less.

9.2 Materials for Patterned PMR Medium

The medium design for a patterned-medium should be reconsidered because the fine granular structure used in the recent PMR system will no longer be required even though the same perpendicular recording scheme will be used. Since the patterned-dot is used for recording a bit unit, the individual dot itself should show very ideal single domain behavior without any magnetic defects for a high-quality signal bit. Moreover, the dot array should be precisely fabricated in terms of the arrangement, the dot pitch, and the dot size. Hence, the original thin film for the patterning of either a single crystal, amorphous or multilayered, should show very high homogeneity without any imperfection. Even where a self-organized dot array is concerned, the individual dots must show exactly the same shape and magnetic properties. The candidate materials have to be chosen from the view point of conditions to simultaneously satisfy the long-term thermal stability of magnetization, a large saturation moment, and the magnetization switching feasibility. For practical use, chemical stability, namely, a high corrosion resistance and mechanical strength are also very essential.

Figure 9.1 shows the relationship between saturation moment M_s and anisotropy energy K_u for various candidate magnetic materials. The larger anisotropy energy

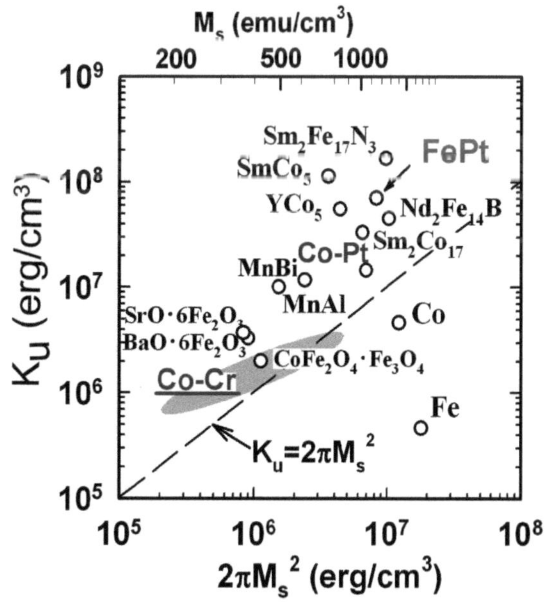

Fig. 9.1 Various candidate materials for perpendicular patterned media on the map of M_s vs. K_u [37]. (2003@IEEE)

provides a smaller critical particle diameter for superparamagnetic behavior independently from the saturation moment M_s. Hence, smaller single domain dots can be realized for materials of a large anisotropy energy constant, K_u, such as Sm–Fe, Sm–Co, Fe–Pt, Co–Pt and so on. Among them, Co–Pt or Fe–Pt alloy would be much better for the patterned-media because of not only a large K_u with high saturation moment M_s but also a high corrosion resistance as a thin film state, differing from Sm–Co, Nd–Fe–B, etc.

It has been reported, for example, that a dot pattern array of Co–Pt films, whose deposited film state showed a maze-pattern-like domain structure implying a single crystal-like structure, has been tentatively fabricated by a focused ion beam (FIB) technique [19].

9.3 Dot Shape and Arrangement

Dot shape of the bit-patterned media could be generally a circular plate, a columnar type or a square type. The shape depends principally upon the fabrication process and the method. A columnar type of Ni dot array is typically obtained by using micro-photolithography [10]. A column-like dot array has been realized by a self-organized chemical synthesis method for Fe–Pt nanometer-scale particles [20, 21]. This type can also be fabricated by an electro-deposition method in the hole of the anodic oxidized layer of Al [22]. The size of the alumite hole can be easily controlled by the conditions of anodic oxidation [23]. A further improved technique, which is based on Si wafer holes etched by electron beam lithography with electroless deposited Co–Ni–P or electrodeposited CoPt, FePt, has been developed recently [24, 25]. The square dot is the most popular shape as in the original patterned medium [9]. This shape has been fabricated by electron beam lithography. Two types of arrangements have been thought of for the bit-patterned media. One is a honeycomb-like arrangement with hexagonally close packed dots obtained for plated dots or self-organized dot patterns. This hexagonal dot array has been found to have the most efficient arrangement to achieve the highest area density in terms of reproducing S/N ratio [26]. However, as far as the writing process is concerned, the hexagonal array is inconvenient to discriminate the neighbor tracks with proper timing, while the influence of the write head on the adjacent tracks is the same. Thus, in the case of e-beam lithography or FIB fabrication, the square-shaped dot might be preferred for either the cost or the fabrication feasibility. The transition width can be small enough for the square-shaped dots when compared with the other shapes in the down track direction.

9.4 Demagnetizing Field Effects of Magnetic Dots Array

Patterned-media have non-magnetic boundaries between the dots so as to discriminately switch and detect the magnetization of each individual dot as an information bit. The localized demagnetizing field around the dots is a very important factor to determine the

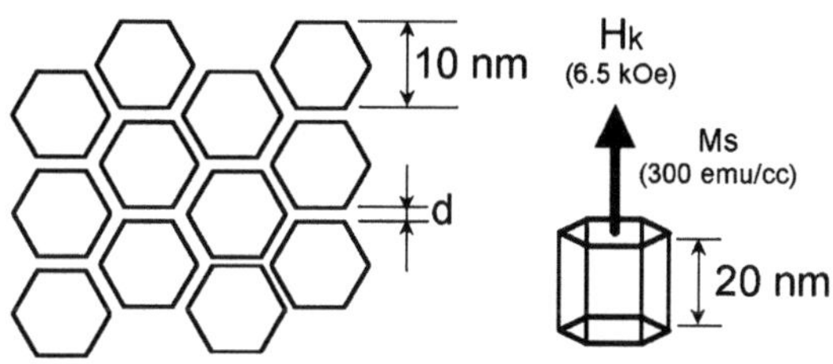

Fig. 9.2 Simulated model for evaluation of the demagnetizing effect on *M–H* loops assumed for a highly packed dots array [27]. © 2002 IEEE

write margin as well as the writing feasibility. Therefore, micromagnetic demagnetizing field or magnetostatic interaction effect on the *M–H* loop is considered first [27].

The media were assumed to consist of a monolayer of hexagonal dots with saturation magnetization of 300 emu/cc, a perpendicular magnetic anisotropy field H_k of 6.5 kOe, a hexagon size of 10 nm, and a height of 20 nm as shown in Fig. 9.2. This model has been used for the design of a perpendicular granular type medium [27]. However, the results can be applicable to the fundamental demagnetization effect on the *M–H* loops of the patterned-media if we assume that the exchange interaction between the grains, that is, the patterned dots, is neglected. It is to be noted that the M_s and the H_k are smaller than those required for a patterned-medium, suggesting a slightly underestimated interaction compared to the practical case.

The hysteresis loops of the dot arrays with the net demagnetizing field are simulated as shown in Fig. 9.3. Figure 9.3a shows the net *M–H* loops of the array indicating that the net smaller saturation magnetization is observed for a larger separation *d*. The loop inclination, α, does not seem to change so much with the separation *d* changes. This means that the change in the net *M–H* loop can be understood by the packing fraction change with the d variations. On the contrary, if we pay attention to magnetization reversal of the individual dot, each dot should have the same M_s as that of the original material for any value of *d*, while the loop inclination α_g changes drastically with the increase in the dot separation, *d*, as shown in Fig. 9.3b. Consequently, the magnetization of each dot switch obeys the grain *M–H* loop but not the net *M–H* loop.

A dramatic increase in the *M–H* loop slope, α_g, is seen with increase in the grain separation, *d*, while H_c and α change a little. According to the increase in α_g, the required maximum magnetic field, namely, the saturation field, H_s, to reverse the dot magnetization in the hardest case becomes small. This can be explained by the reduced magneto-static interaction between the dots as the separation *d* increases. This means that even for a rectangular-shaped *M–H* loop in a state of isolated dots, the *M–H* loop for the dot array is sheared by the dot by dot magnetostatic interaction, indicating the dispersion of the switching field. The slope α_g is the major factor to determine the writing feasibility of the patterned media. Therefore, a very important design factor is paying attention to the separation *d* between dots. It is also implied that the higher M_s of dot material induces much larger magnetostatic interaction while it increases the stored magnetic energy of the dot.

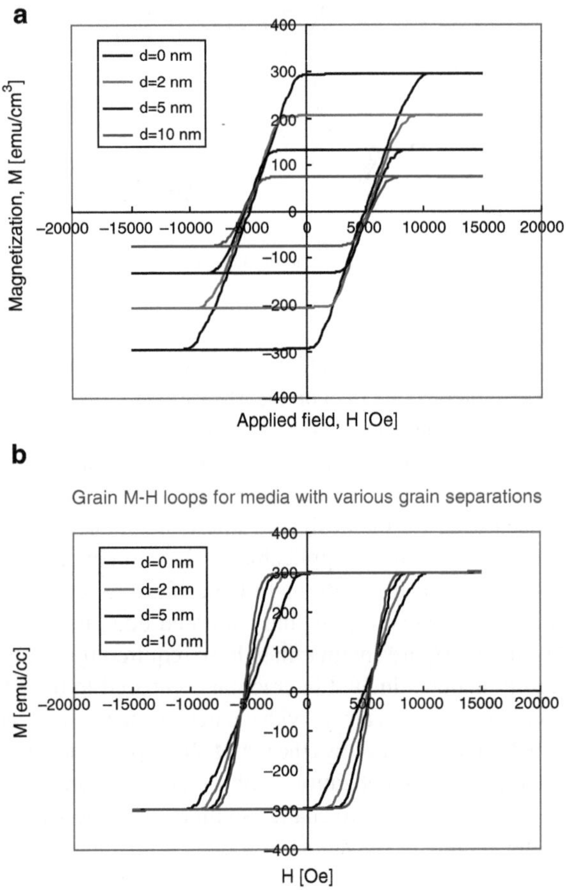

Fig. 9.3 Hysteresis loops for various values of separation d of hexagonal dots array (a) net *M–H* loops with net demagnetizing field, (b) the *M–H* loops in average of individual dot [27]. © 2002 IEEE

9.5 Thermal Stability and Write Feasibility

Let us consider the condition for thermal stability of the patterned-media with perpendicular anisotropy, based on a perpendicular *M–H* loop [28]. The net *M–H* loop for a dot array is the statistic result of small Stoner–Wohlfarth model like square *M–H* loops of each magnetic dot. The thermally stable condition for a magnetic dot array is just that of the condition for a dot, which is the easiest to reverse among the whole dots. Using the beginning field of the reversal, namely, the nucleation field of the magnetic dot array, H_n, the condition is expressed as,

$$E_m / K_B T = \frac{1}{2} H_n M_s V / K_B T \geq 60 \tag{9.1}$$

where, E_m, M_s, V, k_B and T represent magnetic energy, saturation magnetization of dot, dot volume, Boltzmann constant, and temperature in K, respectively. The nucleation field H_n is defined as the value at the time of zero second in the present study. Equation (9.1) assumes the error rate of less than 10^{-7} in 30 years for the weakest dot according to the thermal decay theory of magnetic particles proposed by M. P. Sharrock [29]. The energy ratio E_m/k_BT was set at 70 for $T = 300$ K so that it can maintain the value to be greater than 60 even at +70°C.

The value of H_n is always smaller than the coercivity, H_c, for a dot array as shown in Fig. 9.4. The fact comes from (1) dispersion in H_c of individual dot and/or (2) the magnetostatic interaction between dots. In case (1), the estimation of the magnetic energy by (9.1) is reasonable. However, in case (2), the estimated value may include error because the magnetic energy should be reduced by $(1-H_d/H_c)^2$, where H_d is the demagnetizing field from the neighboring dot magnetization. Nevertheless, the error is estimated as less than −10% for patterned-media with substantial separation between dots, because $H_d/H_c < 1/10$. The other case of the error may come from incoherent reversal of the magnetization of the dots. The extreme incoherent reversal for the exchange-coupled composite (ECC) medium [30] would cause a 50% error in the estimation of the magnetic energy. However, the error is estimated to be less than 10% even if the dot is stacked on a halved anisotropy softer dot. When the M_s is assumed as 600 emu/cc, the smallest nucleation field $-H_n = 15$ kOe is sufficient to keep thermal stability of magnetization for square dots with 7×7 nm and 11 nm height. The saturation field H_0 is estimated as more than 21 kOe. Therefore, the required head field strength is more than 21 kOe on track position and less than 15 kOe on the position of neighbor track of 25 nm distance from the center of pole surface. A field gradient of 240 Oe/nm is, accordingly, sufficient to write on the aimed dot without any interruption against the other dot. This type of head field distribution could be realized only by the multisurface pole head [31].

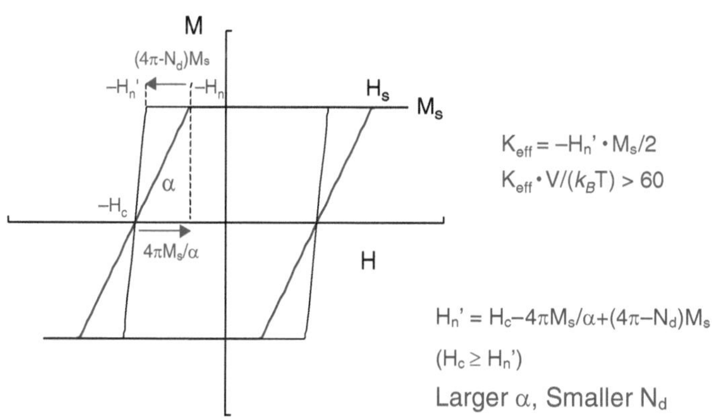

Fig. 9.4 Model *M–H* loop and definition of H_n, H_c, K_{eff} [28]. © 2002 IEEE

Finally, typical solution to keep thermal stability of the magnetization, together with write feasibility by a perpendicular single pole head is to set the square dot, placed on a soft magnetic underlayer (SUL), in a size of 7.5×7.5×10 nm with $M_s = 1{,}000$ emu/cm³. Then, magnetic field strength to saturate the dot, H_s, is prospected as 18 kOe with the nucleation field, H_n, of 11 kOe, where H_k of 15 kOe is required for the dot material under the consideration of magnetostatic interaction between dots. The H_k value of 15 kOe corresponds to an anisotropy energy constant, K_u, of 7.5×10^6 erg/cc for $M_s = 1{,}000$ emu/cm³.

These magnetic properties are easily obtainable with presently well-known high K_u materials such as FePt [32] and Co–Pt alloy [33] as described in the previous section. Consequently, it is very much possible to design a medium for a density of 1 Tera bits/inch² in terms of thermal stability, recording feasibility and applicable materials without any corrosion problems.

9.6 Simulation Analysis on Basic Recording Conditions

The authors first simulated the recording process for a model of bit-patterned media of 1 Tera bits/inch² as shown in Fig. 9.5. The single pole head main pole has 25×25 nm² square surface area. Karlqvist-type field distribution is assumed and the perpendicular field H_y exhibited a half-height width of 42.5 nm with the maximum field of 12.5 kOe in both down and cross track directions at a spacing of 25 nm between the head pole surface and the soft magnetic underlayer (SUL). Various dot models from M1 to M13, shown in Table 9.1, were examined in order to investigate the dot shape effect in terms of recording feasibility and recording characteristics. Each magnetic dot is constituted with 2.5 nm size cubic elements. The elements are exchange-coupled with the nearest neighbor elements with each other. Therefore,

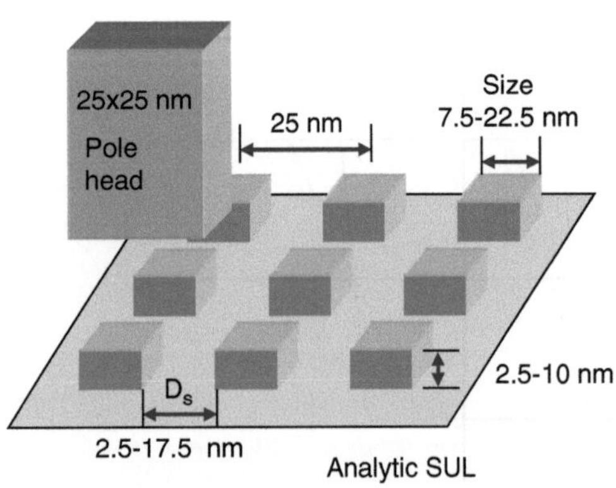

Fig. 9.5 Model scheme used for simulation of recording process on a patterned perpendicular medium with a soft magnetic under layer by a single pole head

Table 9.1 Simulated patterned dot array media of various shapes, sizes, magnetic properties

Media	a [nm]	b [nm]	t [nm]	Volume [nm^3]	H_k [kOe]	H_{nr} [kOe]	H_{cr} [kOe]	H_{sr} [kOe]
M-1	7.5	7.5	10	563	15	12.0	14.8	18.0
M-2	12.5	12.5	5	781	15	84	9.9	12.0
M-3	15	15	2.5	563	22	11.4	13.2	15.0
M-4	17.5	17.5	2.5	766	19	84	9.9	11.7
M-5	75	12.5	10	938	10	72	8.4	10.5
M-6	7.5	15	5	563	19	11.4	13.8	16.5
M-7	7.5	17.5	5	656	18	10.2	12.4	15.0
M-8	75	20	5	750	17	9.6	11.0	13.5
M-9	7.5	22.5	5	844	16	8.1	9.7	12.3
M-10	12.5	15	5	938	15	7.8	9.5	11.1
M-11	12.5	20	5	1,250	13	5.1	6.3	8.7
M-12	12.5	20	2.5	625	21	10.2	12.0	14.1
M-13	12.5	22.5	2.5	703	20	9.0	10.8	12.6

$KV/k_BT > 70$
25 nm period array
Element: $2.5\times2.5\times2.5$ nm^3, $M_s=1{,}000$ emu/cm^3, $A\sim1\times10^{-6}$ erg/cm, $\sigma H_k=15\%$, $\sigma\theta=2°$, $t_{IL}=5$ nm, Analytic SUL

an in coherent rotation mode of magnetization reversal in the dot can be treated by this model. Each element has a saturation magnetization of $M_s = 1{,}000$ emu/cm^3, an exchange stiffness of 0.98×10^{-6} erg/cm, dispersion in H_k and perpendicular orientation of 15%, and 1.5°, respectively. A non-magnetic intermediate layer of 5 or 1 nm thick is assumed on the analytic SUL.

Series of M1–M4 are square dot arrays of various sizes of edge and dot thickness. Media of M5–M9 have the same edge size of D_1 but different sizes of D_2 with the same thickness of 5 nm except M5 of 10 nm for comparison. Series of M10–M13 are for investigation of the elongation effect with an increased D_1. The dot size variety is from 7 nm to 22.5 nm, and the thickness is from 2.5 nm to 10 nm.

For these models, thermal stability conditions were first discussed as the absolute prerequisite condition. Consequently, the nucleation field of the each dot arrays H_n has been determined to satisfy the stability conditions of $E_m/kT > 70$ at room temperature, where the required perpendicular anisotropy indicated by anisotropy field H_k has been, beforehand, decided by the simulation of the remanence curve for the objective variety of the dots. Every patterned-medium has a soft magnetic backlayer of infinite permeability. It is noted that the remanence coercivity H_{cr} is substantially smaller than the averaged anisotropy field H_k as seen in Table 9.1. This is caused by the shape anisotropy of the dots and assumed dispersion of 1.5° in orientation. Hence, every dot has slightly different magnetic properties in comparison with the others. This implies occasional error of writing in some critical conditions. Every dot pattern is, needless to say, designed to accommodate 1 Tera bits/inch^2 in terms of thermal stability.

9.7 Recording with Karlqvist Head Field

The model media has three tracks of dots and the center track is just the recorded track in the simulation. All the tracks are initially DC erased before recording. Only the center track is written with all 1's signal. The on-track write error rates for 25 recorded bits with a linear density of 1,016 kFCI (flux reversal per inch, namely, 25 nm bit length) were evaluated by shifting the reverse timing (position) of the recording head field in the down track direction keeping the on-track condition. Evaluation was performed for the four media (M1–M4) of different sizes but similar volume. The remanence nucleation field H_{nr} value for thermal stability is also not so much different among them. The maximum magnetic field required for enough write-ability was at around from 9.6 to 13.7 kOe.

The write shift margin or the write window was 17.5 nm for dots less than 12.5 nm. However, error-free recording was not possible for a large dot of 17.5 nm where the H_s value was larger than the maximum head field of 12.5 kOe. Generally speaking, the ideal shift margin should be 25 nm for a dot array with a period of 25 nm like this case. The write shift margin, W_w, is roughly estimated using $(H_{sr}-H_{nr})$ and perpendicular write field gradient, $\mathrm{d}H_y/\mathrm{d}x$, as $W_w = 25-(H_{sr}-H_{cr})/(\mathrm{d}H_y/\mathrm{d}x)$. As $(H_{sr}-H_{rc}) = 3.6$ kOe and $\mathrm{d}H_y(\mathrm{eff.})/\mathrm{d}x = 370$ Oe/nm for M-2, where effective field, $H_y(\mathrm{eff.})$, compensating for the angular dependence of the switching field of S–W particle, [35] was used instead of H_y, W_w was estimated as small as 15.3 nm for the simulation value of 17.5 nm. The difference between the estimated and simulated values would be caused by the reduced saturation field in the recording process. As decrease in the field gradient by averaging over the dot width was estimated at about only 10%, the observed reduced W_w for larger dots would be caused by another reason such as, for instance, the incoherent reversal of magnetization in large dots with a small thickness of 2.5 nm.

The reverse rates at an adjacent track, when the head was shifted in the cross track direction, have been also simulated. The error occurrence in this case was different from that in down track writing where the field gradient determines the discriminability of bit writing. The full shift margin in the cross track direction was less than 1/3 of that in the down track direction. The small shift margin in the cross track direction is mainly attributed to the broad field distribution in the cross track direction. Therefore, it is essential to use a write head with a high field gradient in the down track direction and a narrow field distribution in the cross track direction. One of the solutions would be to use shield type heads.

The shift margins were investigated for media in Table 9.1 including dots with elongated shapes in the down track direction as summarized in Fig. 9.6.

It was found that less decreased or even increased shift margin in the down track direction was obtained for the dots with the elongated shape compared with the square dots (solid line indicated by $D_1 = D_2$). The dot length in down track direction, D_2, could be elongated to over 20 nm for dots with the size in the cross track direction, D_1, of less than 12.5 nm. At the same time, the shift margin in cross track direction is larger than that of the square dots. In this case, the dot volume can be increased so that the required H_n for thermal stability could be reduced. In order to

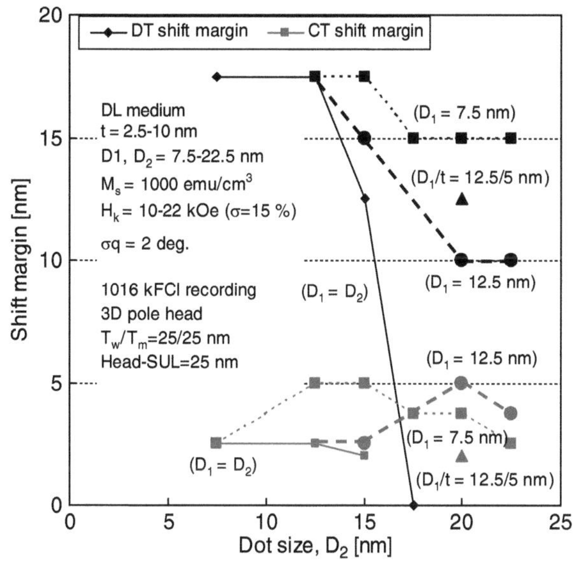

Fig. 9.6 Down track (DT) timing shift margin and cross track (CT) head shift margin for shapes of various D_1, D_2

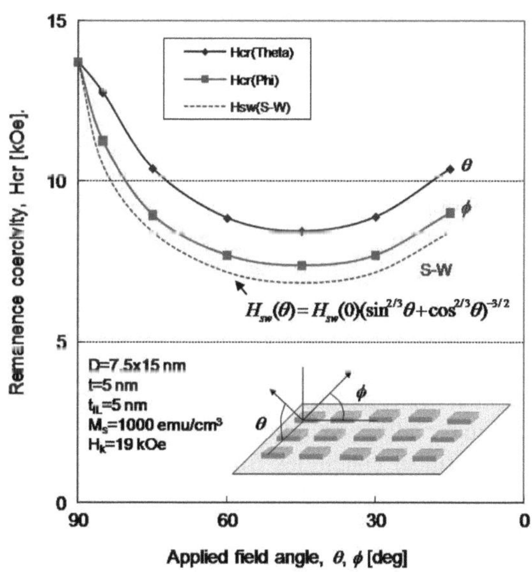

Fig. 9.7 Difference in angular dependence of the switching field, H_{rc}, in the two typical applied field angle directions of the anisotropic shape dot medium, M-6

understand the effect of elongation of the dot, angular dependence of the remanent coercivity for the M-6 medium was calculated for both in the cross track and the down track directions comparing with a so-called Stoner–Wohlfarth model as shown in Fig. 9.7. The angular dependences in ϕ direction (DT) are similar to that of S–W model. However, the variation of the remanence coercivity is larger for ϕ

direction (DT) than that for theta direction (CT), meaning that CT direction is harder to reverse the dot magnetization than that of DT. The fact makes the dot to show a larger shift margin in the cross track direction compared with square dots. It could be concluded that the elongated dot shapes are useful for obtaining patterned-media with larger shift margins in the cross track direction than square dots in terms of the head design.

9.8 Recording with Multi Surface Pole Head

Recording simulation has been carried out using the head field distribution obtained by 3 D-FEM analysis for a multisurface pole head proposed by S. Takahashi et al [36]. The head has a front side shield to make the head field narrower in the cross track direction than conventional single pole heads. The head core and the shield material are assumed to have a saturation flux of 2.4 T. Needless to say, combination with a soft magnetic backlayer of the medium is essentially assumed. The bump core size is 14×45 nm and the head-medium magnetic spacing is assumed as 6 nm. The half-height width of the field in the cross track direction was 34 nm, which was smaller than the field for Karlqvist one by 8.5 nm. This multisurface pole head can provide the very strong maximum magnetic field by the flux concentration effect and/or back- side core field superposition effect of the pole shape even for a very narrow track width. This is a large discriminatory property in the cross track direction compared with the conventional single pole head.

The shift margin is wider than that obtained for Karlqvist type head field in both directions. The elongation effect of dots to increase the shift margin in the cross track direction has also been verified for the multisurface head with side shields.

Although our simulation does not include the dynamic reversal of magnetization of the dot, the result means that bit-patterned media of variety of size and shape can realize, principally, an area density of at least 1 Tera bits/inch^2. Furthermore, it is important that considering the shift margins in both cross and down track directions, the size and pitch of the patterned-dot are very much essential for the design of array. In order to make a large margin of healthy recording scheme, the spacing of the dot has some critical conditions from the view point of write feasibility implied by larger *M–H* loop slope of the dot arrays through the reduction of interaction field from the neighboring dots.

9.9 Experiments on Patterned Media

There are many reports on fabrication of patterned-media. However, few papers have discussed the dot-spacing effect in terms of the *M–H* loop shape. From the view point of importance of write feasibility of media, we have investigated experimentally effect of the dot spacing on magnetization reversal. A typical machined pattern

Fig. 9.8 Co_{80}–Pt_{20} patterned dots array fabricated by focused ion beam technique. Pitch of the dots is 200 nm and the dot size is 70 nm [38]

observed by magnetic force microscopy is shown in Fig. 9.8. The source material is Co–Pt as has been mentioned in Sects. 9.1 and 9.4 [19]. The original Co–Pt film showed a continuous structure, which has a maze like domain structure and very low coercivity compared with the anisotropy field, suggesting that wall motion reversal of magnetization is dominant. The 1 μm square was patterned as an array of 16 dots of 70 nm in edge size. The dot shape should be square but seems an edge-rounded shape due to the resolution limit of the used MFM (magnetic force microscopy). The designed size of the dots is at around 70 nm with a dot-by-dot spacing of 200 nm. The bright dots, for instance, represent the reversed domains from the original direction of magnetization of the dark dots by applying external field.

Multiple patterned array areas were formed on the same Co_{80}–Pt_{20} perpendicular anisotropy film to evaluate the DC remanence curve by counting the number of reversed dots with changing applied magnetic field. The Co–Pt film thickness is 15 nm.

Figure 9.9 shows remanent magnetization curves of dot arrays with the same size of 70 nm and various spacing of 20–200 nm. In case of the composite-patterned dots array with a soft magnetic backlayer of Co–Zr–Nb, when the spacing changes from 200 nm to 20 nm, the inclination of the curve decreased as shown in Fig. 9.9a, while the remanence coercivity changed a little. According to these facts, the nucleation field H_n decreases and the saturation field H_s increases when the dot spacing becomes small. The tendency coincides with that of the simulation results shown in Fig. 9.3. Therefore, recording feasibility becomes worse for the high packing of dot arrays of small spacing, while thermal stability of magnetization of the dots also decreases although situation at the recording scheme by a head must differ from that in this experiment of a uniform applied field. On the other hand, the dot array of 200 nm spacing still exhibits a finite inclination and not like S–W model even though, the spacing seems large enough to isolate magnetostatically each dot.

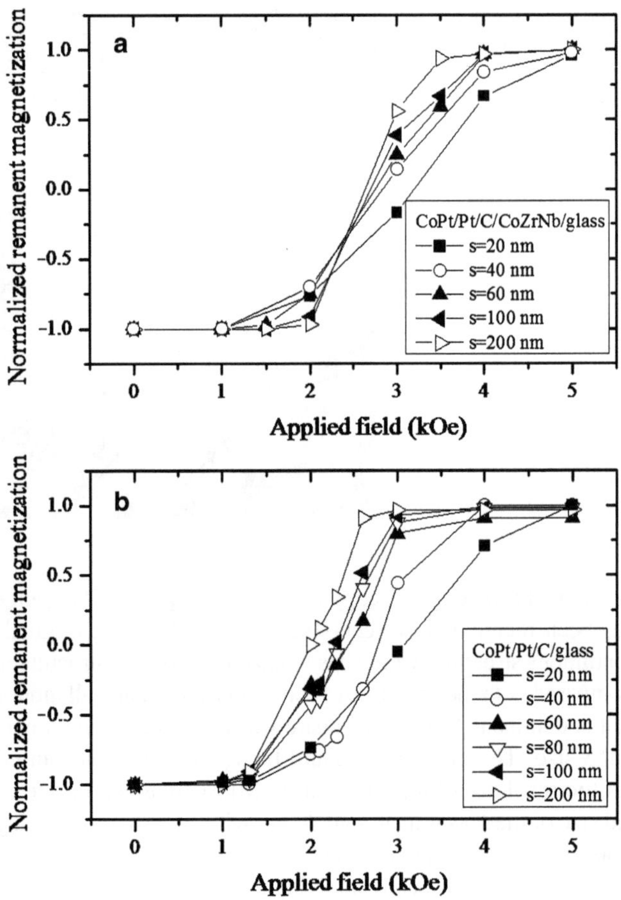

Fig. 9.9 Remanent magnetization curves for (**a**) with SMBL and (**b**) without SMBL [38]

Hence, it was thought that anisotropy distribution for the dots might still be broad presumably by the non-uniformity or fluctuation of machining process by the FIB technique.

On the other hand, in case of single layer of Co–Pt without a soft magnetic back layer, showed different remanent coercivity H_{cr} when the spacing changes, while H_n does not changes but H_s dramatically increases for a larger spacing. According to the calculation considering the surface charges of dots, the magnetostatic interaction from the neighbors is 80 Oe, and 430 Oe for the dot with a soft magnetic backlayer, and without the layer, respectively, for a dot spacing of 200 nm. Consequently, it is suggested that the slope of the curve is larger in the former case than the latter case. However, the experimental results showed similar inclination for both the cases. This discrepancy of the results may come from the too-thick intermediate layer between Co–Pt and Co–Zr–Nb layers. When we aim at 1 Tera bits/inch^2, the intermediate layer should be thinner than present. The other reasons for imperfection in fabrication and so on would also possibly exist. Hence, further study in this direction is very much necessary in the future. In any case, the present experimental study implies that magnetostatic interaction is the key factor to design a proper dot array for the development of a high density recording media.

9.10 Future Issues for 1 Tera bits/inch^2 Density

Patterned dots should be an ideal information-recording medium in terms of thermal stability and write feasibility. This article has discussed writing possibility and thermal stability simultaneously for the patterned media of various dot shapes and sizes, considering interaction effects on the magnetization reversal of the dot. Finally, it was concluded that 1 Tera bits/inch^2 could be realized by using the proper size and shape of dots combined with an advanced head design.

The main phenomena to govern the recording performance of the media are magnetostatic interaction and head medium separation. Both should be *reduced* to achieve a higher density in perpendicular scheme of recording. It is also concluded that *perpendicular* recording seems the best to make easy writing on thermally stable high anisotropy dot media. The most important breakthrough in the future should be nano order fabrication technology since the candidate materials with a high thermal stability are very realistic. The head structure is also an important issue to be improved so as to have higher field strength and gradient. A planar type head proposed recently would have great advantages in this regard.

Because we have not yet found out a hopeful information storage system with a low cost and a high capacity except HDD systems, magnetic recording technology should be advanced more than 20 years at least to answer the demands in the information society in the future. If the manufacturing method is enhanced, patterned-media will be the most hopeful media to realize a density higher than 1 Tera bits/inch^2.

References

1. Iwasaki S, Nakamura Y (1997) An analysis of the magnetization mode for high density magnetic recording. IEEE Trans Magn 13:1272–1277
2. Press Release (Dec. 14, 2004) Toshba 133G
3. Press Release (Aug. 2006) Web 300G-420G
4. Rottmayer RE et al (2006) Heat assisted magnetic recording. IEEE Trans Magn 42:2417–2421
5. Lambert SE et al (1987) Recording characteristics of submicron discrete magnetic tracks. IEEE Trans Magn MAG-23(5):3690–3692
6. Press Release (Oct. 03, 2006) TDK
7. Soeno Y et al (2005) Performance evaluation of discrete track perpendicular media for high recording density. IEEE Trans Magn 41:3220–3222
8. Greaves S, Kanai Y, Muraoka H (2006) Trailing shield head recording in discrete track media. IEEE Trans Magn 42:2408–2410

9. Nakatani I et al (1991) Japan patent 1888363, publication JP03-022211A
10. Chou SY et al (1994) Single-domain magnetic pillar array of 35 nm diameter and 65 Gbits/in.^{2} density for ultrahigh density quantum magnetic storage. J Appl Phys 76:6673–6675
11. Nakamura Y (1994) A challenge to terabit perpendicular spinic storage. J Magn Soc Jpn 18(S1):161–170
12. White Robert L et al (1997) Patterned media: a viable route to 50 Gbits/in^2 and up for magnetic recording? IEEE Trans Magn 33:990–995
13. Charp SH, Lu P, He Y (1997) Thermal stability of recorded information at high densities. IEEE Trans Magn 33:978–983
14. Terris BD et al (1999) Ion-beam patterning of magnetic films using stencil masks. Appl Phys Lett 75:403–405
15. Ross CA et al (1999) Fabrication of patterned media for high density magnetic storage. J Vac Sci Technol B 17:3168–3176
16. Rottner CT, Best ME, Terris BD (2001) Patterniing of granular magnetic media with a focuses ion beam to produce single-domain islands at >140 Gbits/in^2. IEEE Trans Magn 37:1649–1651
17. Aoyama T, Sato I, Ishio S (2003) Fabrication and magnetic properties of patterned magnetic recording media. Oyo Butsuri 72:298–302
18. Hughes GF (2000) Patterned media write designs. IEEE Trans Magn 36:521–526
19. Kondo Y et al (2006) Magnetic properties of magnetic dot arrays with a soft magnetic underlayer. J Magn Soc Jpn 30:112–115
20. Sun S et al (2000) Monodisperse FePt nanoparticles and ferromagnetic FePt nanocrystal superlattices. Science 287:1989–1992
21. Wang S et al (2003) Magnetic properties of self-organized L10 FePtAg nanoparticle arrays. J Magn Magn Mater 266:49–56
22. Gapin AI et al (2006) CoPt patterned media in anodized aluminum oxide templates. J Appl Phys 99:08G902
23. Arai K, Ohoka Y, Wakui Y (1988) Preparation and magnetic properties of anodic oxide magnetic films. IEICE Trans Electron J71-C:994–1000
24. Kawaji J et al (2005) Area selective formation of magnetic nanodot arrays on Si wafer by electroless deposition. J Magn Magn Mater 287:245–249
25. Huang YH et al (2002) CoPt and FePt nanowires by electrodeposition. J Appl Phys 91:6869–6871
26. Nutter PW et al (2005) Effect of island distribution on error rate performance in patterned media. IEEE Trans Magn 41:3214–3216
27. Honda N et al (2002) Role of *M–H* loop slope of media for recording properties in perpendicular magnetic recording. IEEE Trans Magn 38:2030–2032
28. Honda N, Ouchi K, Iwasaki S (2002) Design consideration of ultrahigh–density perpendicular magnetic recording media. IEEE Trans Magn 38:1615–1621
29. Sharrock MP (1990) Time-dependent magnetic phenomena and particle-size effects in recording media. IEEE Trans Magn 26:193–197
30. Victora RH, Shen X (2005) Composite media for perpendicular magnetic recording. IEEE Trans Magn 41:537–542
31. Takahashi S et al (2005) Magnetic recording head for patterned medium with 1 Tbit/inch^2. Abs int'l symp. creation of magnetic recording materials with nano-interfacial technologies, Waseda University, Tokyo, PS08:27
32. Suzuki T, Honda N, Ouchi K (1997) Preparation on magnetic properties of sputter-deposited Fe–Pt thin films with perpendicular anisotropy. J Magn Soc Jpn 21-S2:177–180
33. Shimatsu T et al (2004) High perpendicular magnetic anisotropy of CoPtCr/Ru films for granular-type perpendicular media. IEEE Trans Magn 40:2483–2485
34. Honda N (2005) Design of patterned media for 1 Tbit/in^2 recording.Tech Rep IEICE MR2005-15:51–56

35. Stoner EC, Wohlfarth EP (1948) A mechanism of magnetic hysteresis in heterogeneous alloys. Phil Trans Roy Soc 240:599–644
36. Takahashi S, Yamakawa K, Ouchi K (2003) Design of multisurface single pole head for high-density recording. J App Phy 93:6546–6548
37. Suzuki T et al (2003) Design and recording properties of Fe–Pt perpendicular media. IEEE Trans Magn 39:691–696
38. Kondo Y et al (2006) Magnetic properties of magnetic dot arrays with a soft magnetic underlayer. J Magn Soc Jpn 30:112–115

Part III
Nanotechnology for Bio-Chip Applications

Chapter 10
Micro pH Sensors and Biosensors Based on Electrochemical Field Effect Transistors

Junji Sasano, Daisuke Niwa, and Tetsuya Osaka

10.1 Fabrication of On-Chip FET pH Sensor

10.1.1 Introduction

A study on ion-sensing using field effect transistor (FET) was begun by Bergveld in the 1970s [1–3]. The ion-sensitive (IS) FET is now widely used as a miniaturized pH sensor, commercialized by some companies. First, the principle and structure of the ISFET are introduced in this section. A basic design of ISFET is shown in Fig. 10.1a. ISFET has silicon substrate with field-effect structures such as electrolyte/IS layer/(insulator)/semiconductor structures; the space charge region in the semiconductor is modulated depending on the gate voltage (V_g), same as a typical metal-oxide-semiconductor (MOS) FET. A typical bias V_g versus drain-source current (I_{ds}) characteristic of the device that has silicon nitride/silicon dioxide/silicon is shown in Fig. 10.1b. This characteristic is quite similar to the MOSFET. A prominent difference between ISFET and MOSFET is that the gate voltage for the operation of the device is applied by an electrochemical reference electrode through the electrolyte in contact with the gate insulator. The threshold voltage (V_{th}) could shift according to the value of the pH of the solution. In the MOSFET, the V_{th} would shift

J. Sasano
Consolidated Research Institute for Advanced Science and Medical Care, Waseda University, 513 Wasedatsurumaki-cho, Shinjuku-ku, Tokyo 162-0041, Japan; Department of Production Systems Engineering, Toyohashi University of Technology

D. Niwa
Consolidated Research Institute for Advanced Science and Medical Care, Waseda University, 513 Wasedatsurumaki-cho, Shinjuku-ku, Tokyo 162-0041, Japan; Nano Bionics R&D Center, ROHM Co., Ltd.

T. Osaka (✉)
Consolidated Research Institute for Advanced Science and Medical Care, Waseda University, 513 Wasedatsurumaki-cho, Shinjuku-ku, Tokyo 162-0041, Japan; School of Science and Engineering, Waseda University, 3-4-1 Okubo, Shinjuku-ku, Tokyo 169-8555, Japan
e-mail: osakatets@waseda.jp

T. Osaka et al. (eds.), *Electrochemical Nanotechnologies,*
Nanostructure Science and Technology, DOI 10.1007/978-1-4419-1424-8_10,

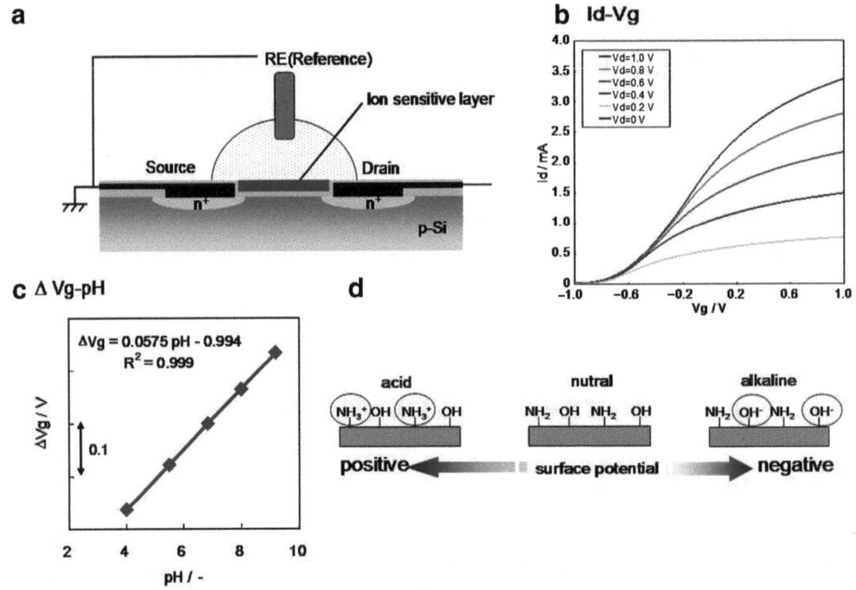

Fig. 10.1 (**a**) Basic design of ISFET. (**b**) I_{ds}–V_g characteristics of the device. (**c**) pH responsibility of the device. (**d**) Change in the surface structure and surface potential of silicon nitride layer in each pH solution

depending on the change in the space charge region in the MOS capacitor structure by the application of V_g. On the other hand, the V_{th} in ISFET would shift according to the change in the surface potential in the electrolyte/IS layer interface. Therefore, the IS layers and their interfaces in ISFET play an important role in the performance of pH responsibility. It is well-known that the silicon nitride surface shows a good pH response in solution. The silicon nitride layer is often formed by plasma-enhanced chemical vapor deposition (PECVD), which is generally formed at the thickness of 100–500 nm. The V_g vs. $I_{ds,}$ characteristics of the silicon nitride-based ISFET indicate a good pH responsibility of 58 mV/decade that shows Nernstian response (Fig. 10.1c). The shift of the V_{th} depends on the changes of surface potential at electrolyte/silicon nitride interface. On the silicon nitride surface immersed in aqueous solution, both amphoteric Si–OH sites and basic Si–NH_2 sites (Fig. 10.1d) are produced by hydrolysis. These sites directly interact with the solution to either bind or release hydrogen ions, leading to bear a certain surface charge on the nitride surface that was opposed to an ionic charge in the solution. This formed a double-layer capacitance across which the potential drop occurs. Therefore, the threshold voltage shifted accompanied by the pH change in solution.

Based on this principle, ISFET is used gradually for the detector as various sensing materials. However, since 1990s, such active research has declined due to various problems, such as the stability of the device and the molecular modification on the electrode surface. On the other hand, the research of FET has begun to steal the limelight again by the progress of the technologies of the semiconductor device fabrication and the surface modification. Especially, the development of IS layers and new device processing has been studied for the improvement of the performance of ISFET itself. Currently, Ta_2O_5 [4] is often used as an insulating layer. However, to construct the high-performance FET sensor, it should be necessary to improve the device architecture.

In this section, new type FET-based sensors are described focusing mainly on the research activities in nanotechnology at Waseda University.

10.1.2 Concept of On-chip pH Sensors Using FETs Modified with Self-Assembled Monolayers

The aim of our study is to fabricate an extremely high-performance ion and biosensing device. In our work, we have studied the fabrication of FET-based ion and biosensor using self-assembled monolayers (SAMs) [5–9]. Our concept and details of device architecture are described in the following lines.

In order to realize highly sensitive biosensing system, *precise fabrication* of the electrode parts for molecular recognition is a significant issue. For this purpose, development of new detection devices with high sensitivity is strongly demanded. It is especially desired that the electrode surfaces have the supramolecular structure that mimics cell systems. In order to fabricate such an electrode, application of the template for the ordered-arrangement of the molecules is effective. Organic monolayers have the ability to self-assemble onto the surfaces [10, 11]; the monolayer-modified electrode is suitable as the template for orderly immobilization of biomolecules. On the other hand, it is preferable that the detection system can detect the signal immediately and very sensitively. A FET type electrode can detect the response of surface reactions as an electric signal, with capability for on-chip integration. Therefore, we have studied the formation of electrodes functionalized by the modification of organic monolayer on silicon wafer surfaces and the development of the detection system utilizing a semiconductor device such as a FET.

Figure 10.2 shows the basic design of the on-chip integrated biosensing devices including reference devices. For the on-chip sensing, reference device as well as sensing device is necessary. In general, a glass-based Ag/AgCl electrode is used as a reference electrode. However, the glass electrode is hard to miniaturize and is easy to break. The use of such a glass electrode acts as a high barrier in the miniaturization of the sensor chip. Therefore, the development of a small and solid-based reference electrode is desired for the realization of the on-chip sensors. Utilizing the functionality of an organic monolayer is expected to be one of the solutions to the problem. In order to fabricate the sensor and the reference devices, it is required that the organic monolayers having different functional groups are area-selectively immobilized on each gate electrode. At the sensing electrode, amino-functionalized monolayer, which is an active site, is suitable for immobilization of biomaterials, enzyme, etc. as well as ion response. For the reference electrode, an alkyl, or perfluoro-alkyl functionalized monolayer having inactive functional group is effective for preventing any undesired adsorptions and ionic reactions at the surface. Hence,

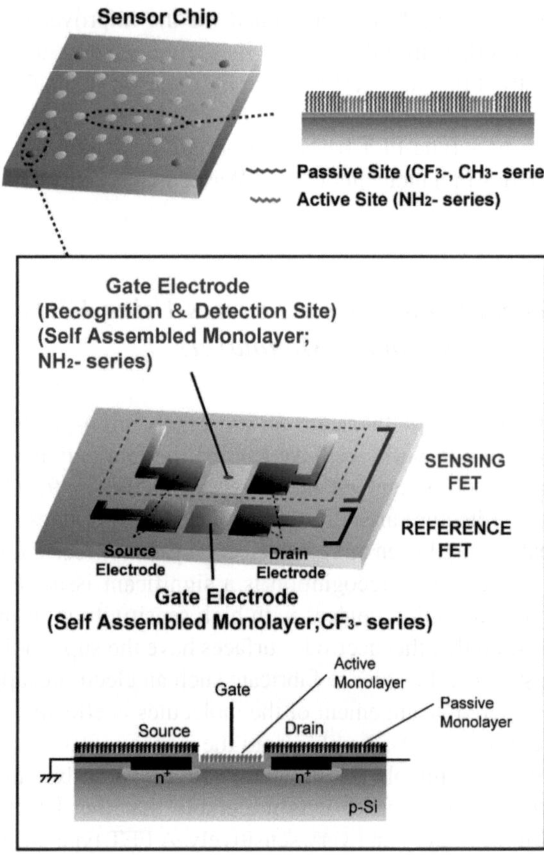

Fig. 10.2 Schematic design of on-chip FET sensing device

the formation of various functionalized monolayers and their patterning are one of the *key* processes for fabricating on-chip biosensing devices.

It is assumed that such an on-chip sensing device has the ability accurately detect the single molecule. Especially, formation of very thin monolayer on the electrode enables immediate detection of the signal that originates in the bioreaction. Moreover, it seems that effective arrangements of molecules, during the bioreaction, onto the electrodes are acquired by using the monolayer templates. The multi-detection of the molecule is expected to be achieved because this device is easy to miniaturize and integrate. On the other hand, FETs are required to have a high chemical durability because they are exposed to various types of solutions during the sensing, the surface cleaning, and the molecular modification processes.

In our work, we have investigated the fabrication of the FET devices with various functionalized SiO_2-gate modified with organosilane monolayers. The pH sensitivity and chemical durability of the devices are evaluated as the basic characteristic of the devices.

10.1.3 Formation of Organosilane Monolayer on Silicon Surface

Organosilane SAMs have been widely applied to control physical and chemical properties of the surfaces of glass, quartz, SiO_2/Si wafer, and silica particle [10]. Many researchers have studied the formation process of the organosilane monolayers and synthesis of silanization reaction in hydrocarbon solvent, such as toluene, bicyclohexyl, hexadecane, etc. [11]. In some cases of the liquid-phase modification, microdefects often exist at the modified surface because the SAM formation is thought to involve self-assembly of monomer or small oligomer units on the surface in a noncovalent manner to give well-ordered monolayer domains or islands on the surface surrounded by bare substrate [12–15]. Such a silane formation on silica surface shows that island formation occurs leading to a multilayer [16]. Hence, in order to achieve complete modification, precise control of the modification procedure is required. On the other hand, more recently, the monolayer modification process by quite a simple method of using a gas-phase silanization reaction, that is a CVD method, is proposed by Sugimura et.al. for application on ultrahigh resolution patterning resists and patterned monolayer templates [17–23]. This method has proved to have the capability to form homogenous, defect-free monolayer coating onto the surfaces [17–20], which is believed to be suitable for our objective.

Si(100) wafers covered with thermally grown silicon oxide were used for the work. The silicon oxide film was formed at 950°C at N_2 atmosphere. Three types of organosilanes, that is, octadecyltrimethoxysilane (ODMS), (heptadeca-fluoro-1,1,2,2-tetrahydro-decyl)trimethoxysilane (FAS), and 3-aminopropyltriethoxysilane (APS), were used as precursors. The wafers were placed together with an organosilane (ODMS or FAS) bial, and then heated at a constant temperature of 110°C. In the case of APS, the wafers were immersed in toluene solvent including 1 vol% APS liquid at 60°C as the formation of the APS monolayer was found to proceed easily in the liquid phase, rather than in the gas phase. It has been reported that complete monolayers are used for these modification procedures. The thickness of the organosilane monolayers formed was estimated to be 20 Å (ODMS), 13 Å (FAS), and 6 Å (APS), respectively. Water-contact angles of these monolayer-covered SiO_2/Si substrates were 105° (ODMS), 120° (FAS), 60° (APS), respectively. These values correspond to previous reports [17–19, 24, 25]. Figure 10.3 shows contact mode AFM images of the modified surfaces measured under a near-contact condition at low tip-pressure. The RMS and R_a values for each modified surface are indicated to be similar to those of the bare silicon oxide surface. Therefore, it is suggested that the modified surfaces are flat and uniformly formed at the monolayer level.

Chemical properties and coverage of the modified surface were characterized by XPS. Figure 10.4 shows the carbon (1s) narrow spectra of the modified surfaces. The coverage of modified surface was calculated by using the integrated peak areas of the carbon (1s) and silicon (2p) XPS narrow scans. Table 10.1 lists the ratio of organic adsorbates per all reaction sites of ideal quartz surface, and the areal density of the adsorbate to maximum areal density [5, 26, 27]. The maximum

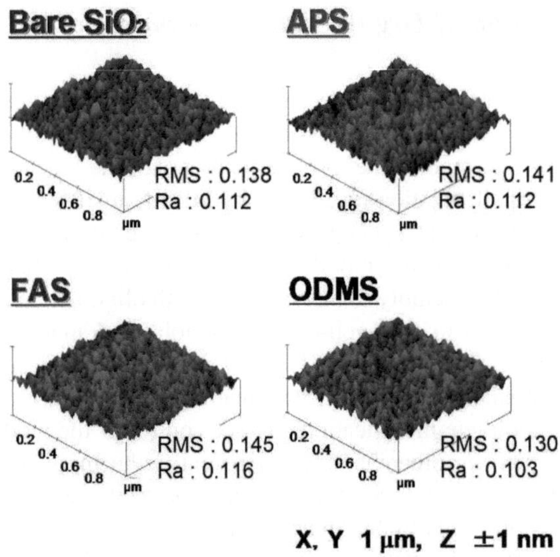

Fig. 10.3 Contact mode AFM images of organosilane modified SiO_2 surfaces

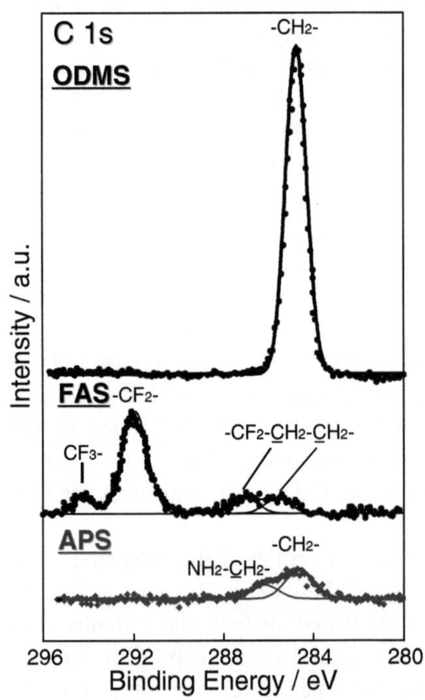

Fig. 10.4 X-ray photoelectron spectra in the C1s region of organosilane modified SiO_2 surface

Table 10.1 Coverage for each monolayer and molecular length of each moeity

	C/Si	Adsorbate areal density/SiO_2 areal density	Adsorbate areal density/maximum areal density	Molecular length/Å	
				Experimental	Calculation
ODMS	1.10	0.85	1.02	26.0	25.5
FAS	0.39	0.58	1.02	15.9	15.6
APS	0.22	0.81	1.01	6.8	6.7

areal densities were adapted to the value of the cross-sectional area of each molecule, which are 18.0 Å^2 (ODMS), 27.5 Å^2 (FAS), and 20.0 Å^2, respectively. The areal density of ideal quartz surface is (15.7 Å^2)$^{-1}$. The areal density of the adsorbate to the maximum areal density became the effective coverage. The areal density of each modified surface indicates nearly 1.0. Moreover, we estimated the molecular length of each moiety from the integrated peak areas of the carbon and silicon. These experimental data show expected values compared with the calculated ones (Table 10.1). Based on these XPS characterizations, it is considered that each modified surface is closely packed and is composed of a single moiety species. In addition, as described above, the modified surfaces were indicated to be flat and homogeneous from the AFM investigation. Therefore, each modified surface was expected to be formed as a monolayer.

10.1.4 Device Fabrication

A basic design of the organic monolayer-modified SiO_2 gate FET device is shown in Fig. 10.2. We selected a SiO_2 as a material for gate and protective layers on source-drain electrodes, in order to form the SAMs onto the device surface. As for the protective and gate layers, high-density, quartz-like SiO_2 such as a thermally grown SiO_2 is desired for accomplishing an ideal molecular modification (perfect coverage), and for preventing a leakage and drift during the device operation in aqueous solution. A SiO_2 as the protective layer is often formed by CVD; however, the CVD–SiO_2 layer generally shows lower density than the thermally grown one. Hence, the improvement of the property of CVD–SiO_2 layer is a critical issue for fabricating the organic monolayer-modified FET with high chemical durability. Thus, we preliminarily examined various treatments to improve the structural property of the CVD layer. In this experiment, the layer was formed by a PECVD with tetraethoxysilane (TEOS)/O_2 gas at 400°C. The property of the layer was evaluated from the value of an ellipsometric refractive index and of the etching rate with 1.0 wt% aqueous HF solution. Each value was listed in Table 10.2. When these values of the CVD–SiO_2 were compared to that of a thermally grown SiO_2 formed by dry oxidation at 950°C, the CVD layer showed a low refractive index and a high etching rate. A decrease in the refractive index and an increase in the etching rate are brought about by the lowering of the film density. It is thought that the CVD layer formed under the above condition is porous. Then, we carried out a postannealing

Table 10.2 Properties of the thermally grown and the CVD SiO_2 layers

	Refractive index	Etching rate with 1.0 wt% HF (Å/min)
Thermally grown SiO_2	1.462	60
CVD–SiO_2 (as deposited)	1.449	368
CVD–SiO_2 (after 800°C annealing)	1.460	65

process for the structural improvement of the CVD layer. By annealing the CVD layer in an O_2 atmosphere at over 800°C, the layer shows stable properties compared with the thermally grown one, as shown in Table 10.2. From these results, we decided to employ the postannealing process for the CVD–SiO_2 layers to stabilize its properties.

Figure 10.5 illustrates the process step for fabrication of the monolayer-modified FETs. Field and gate SiO_2 layer was formed on p-type Si(100) surface. P^+ ions were implanted for forming source and drain channels. In order to fabricate the device with a high temperature process, we selected a $TiSi_2$ as a contact metal. Ti and Pt were sequentially evaporated to source and drain regions, and the substrate was annealed at 800°C to form the $TiSi_2$ electrodes (Fig. 10.5a). Then, the whole surface was covered with the SiO_2 using TEOS–PECVD. After the postannealing for the structural improvement of the CVD–SiO_2 layer described above (Fig. 10.5b), the CVD–SiO_2 layer on the thermally grown gate oxide was positioned selectively removed by a reactive ion etching. Modification of organosilane molecules and their patterning process were carried out (Fig. 10.5c). The FAS and APS were formed by the method described. A position-selective formation of monolayers, having different functional groups, onto the substrate was performed by using photo-lithography process. The FAS monolayer modified substrate was covered with conventional photo-resist, and then patterning was carried out with an ultraviolet (UV) lamp.

This patterned substrate was exposed to O_2 plasma for removal of the monolayer (Fig. 10.5d). The resist pattern was used as the mask for plasma ashing. After this process, photo-resist was removed and the amino silane monolayer was formed on the exposed clean oxide layer (Fig. 10.5e).

10.1.5 Device Characteristics

The fabricated FET devices that are modified with organosilane monolayers are shown in Fig. 10.6. The gate length and width of the devices are 10 μm and 1 mm, respectively. The separation between contacts and SiO_2 gate is 10 μm.

Figure 10.7 shows I_{ds}–V_{ds} curves of the amino-monolayer modified FET in pH 6.86 buffer solution. The characteristics indicated typical FET response, which were stable during a long-time immersion of upto 24 h. Also, the I_{ds}–V_{ds} profile of the device was reproducible even if the monolayer removing by O_2 plasma ashing and the

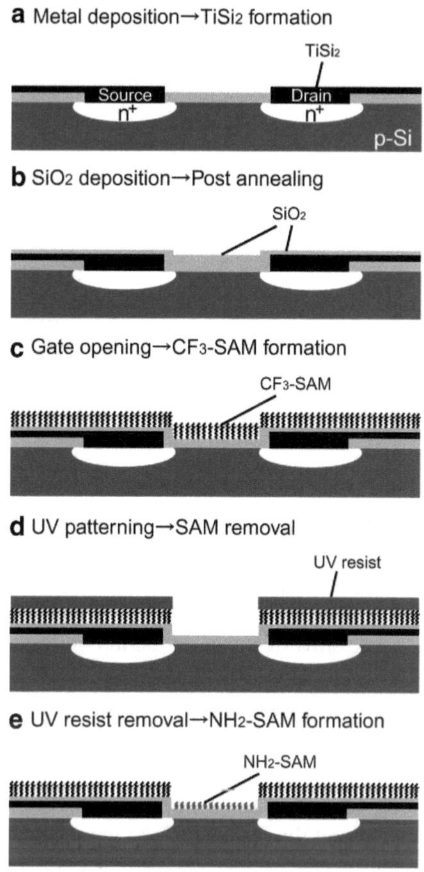

Fig. 10.5 Process steps for fabrication of the monolayer modified FETs

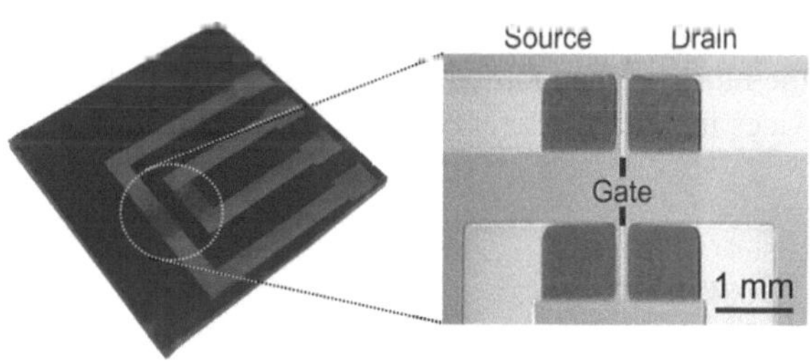

Fig. 10.6 Photograph and optical microscopy image of the fabricated monolayer modified FETs

modification process involving the treatment of strong acid were repeated many times using the process described above. Therefore, this device was quite stable in aqueous solution, and had a high chemical durability. The I_{ds}–V_g curves of the amino-modified FET in various pH solutions are shown in Fig. 10.8a. The V_{th} shifts depending on the change in the pH value in the solution. As seen in Fig. 10.8b, the V_g linearly changes

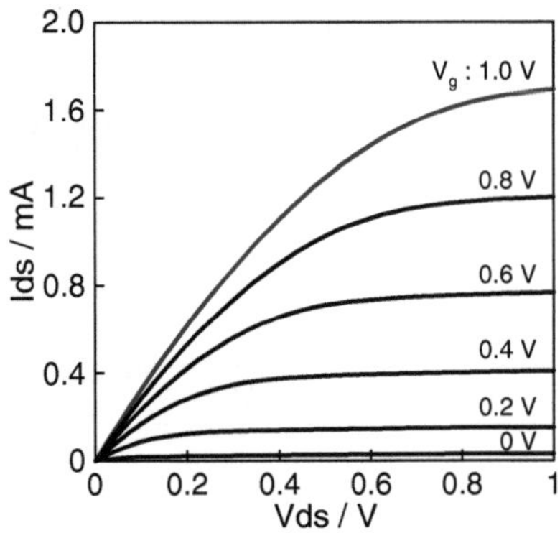

Fig. 10.7 I_{ds}–V_{ds} characteristics of the amino-monolayer modified FET

depending on pH at the sensitivity of 58 mV/pH. This pH sensitivity is quite similar to that of an ISFET such as the Si_3N_4 gate FET [1, 4, 28], which shows a good Nernstian response. On the contrary, in the case of the FAS monolayer modified FET, the potential remains constant, regardless of the pH values in the solution.

The shift of the V_{th} depends on the changes of surface potential at liquid/monolayer interface. For the APS-modified surface, both unreacted Si–OH and O–Si–$(CH_2)_3$–NH_2 sites exist at the surface. Thus, it is thought that the APS-modified FET shows pH response in solution. On the other hand, the FAS-modified surface is a well-ordered and highly hydrophobic surface because the perfluoro alkyl moiety is long-chain and hydrophobic. Such the surface is expected to block the solution, and can inhibit the reaction of the unreacted Si–OH sites with ionic species in the solution, resulting in no pH response in the solution. From these investigations, the pH response of these surfaces is indicated to significantly affect by a variation in functional group. The APS-modified FET has the capability to be used as an ISFET, and the FAS-modified FET is applicable to a reference electrode. It is expected that the fabrication of the monolayer-modified FETs is effective for fabrication of on-chip ion-sensing devices including reference electrodes.

10.2 FET Biosensor

10.2.1 Introduction

Biosensing systems, such as enzyme, immuno sensors and DNA micro arrays, are widely used in the field of medical care and medicine manufacturing [29–31]. In particular, recent progress in genome engineering requires high performance integrated micromulti-biosensing system, which can be utilized for recognition of individual biomolecule and analysis of bioreaction at single molecular level. In the field of the

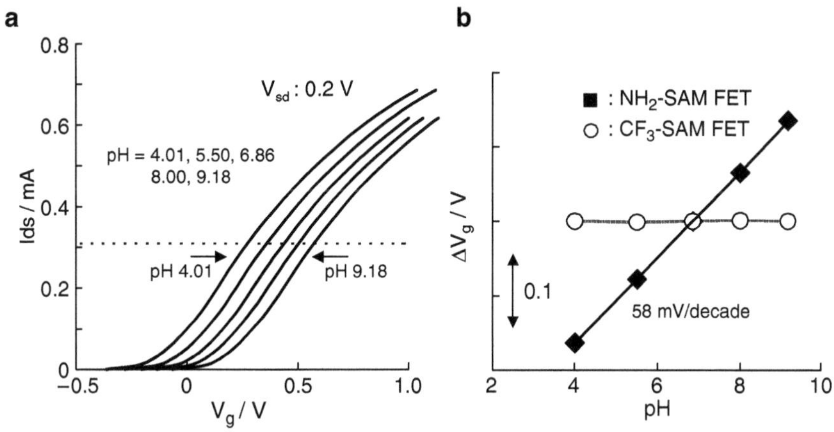

Fig. 10.8 (**a**) I_{ds}–V_g characteristics of the monolayer modified devices in each pH solution. (**b**) pH-potential profiles of each monolayer modified FET

advanced medical care, simple and high accuracy detection systems are essentially required for the application of a tailor-made medical diagnosis. The mainstream of the present biosensor is *fluorescent detection*. This technique has contributed to several analyses for biomolecule reaction including genome sequences. However, since typical biosensing systems based on fluorescent detection, such as DNA microarrays consist of lasers and complex optical systems, the instruments tend to become quite large scale and expensive. Also, modification of fluorescent dye to the target biomaterials is necessary for the fluorescent detection, leading to a high-cost and complicated protocol. Therefore, new detection systems designed for simplicity and high performance are demanded for future advanced medical care.

Here, we have proposed a detection system utilizing a semiconductor device especially to FET as a sensing system that enables to achieve both simplicity and high sensitivity. Semiconductor device manufacturing technologies enable to integrate and miniaturize the device. In addition, it is easy to miniaturize and simplify the system for the device control because the system can operate by using an electrical circuit instead of an optical one. The detection using FET does not require any label materials and mediators since the surface-potential change caused by an interfacial reaction between the solution and the recognition surface can be detected by using FET as an electric signal, directly. Such a detection system by using electric signal is expected to be applicable to biosensors equipped with small size and high performance.

10.2.2 *Attachment of Biomolecules onto the Recognition Region*

Position-selective immobilization of biomolecules has attracted much interest in recent years for performing biological recognition and fabricating miniaturized, array-based assay devices [32–34]; for example, DNA and protein microarray chips are widely used to probe gene sequence and protein level in cells. The miniaturized feature of the device makes it possible to achieve a higher integration density of the arrays of probe molecules, thus allowing the assay to be performed at a high accuracy. The immobilization sites of biomolecules formed precisely in the micro/nano-scopic scale can be applied for conducting highly accurate biomolecular analyses, such as genotyping of single-nucleotide polymorphisms, analysis of proteins for structure and functionality, and recognition of a single molecule. The surfaces for bioanalyses are required to have an ordered, supramolecular structure that mimics cell systems. In order to form such surfaces, the application of templates for the ordered arrangement of the molecules is effective. Surfaces modified with SAMs are suitable for this purpose, and the immobilization of biomolecules on bare gold and SiO_2 surfaces has been reported [11, 35–37]. It is generally recognized that the modified surfaces are preferable to immobilization of specific molecular species. For patterning of the molecules, methods including microcontact printing, ink-jet delivery, dip-pen nanolithography, and nano-manipulation using scanning probe microscopy have been proposed. On the other hand, we investigated the process for fabricating molecular templates at micro and nano scales using UV and electron-beam lithography [5, 38, 39]. These processes aid the controlled formation of SAMs having various functional groups under self-assembling conditions, with a precise position selectivity to form active and passive sites for immobilization of biomolecules.

Figures 10.9b and c show representative fluorescence microscope images of the surface patterned with organosilane monolayers after immobilization of the fluorescence-labeled oligonucleotides. In these images, bright dot-patterns indicating the existence of oligonucleotides are clearly seen. These regions correspond to the APS-patterned regions formed on the surface. In contrast, when patterned surfaces without modification by aminosilane were used, no bright regions were observed in fluorescence microscope images, indicating thereby that the oligonucleotides were *not* immobilized on the surface. These results show that the oligonucleotides were position-selectively immobilized only on APS-modified sites formed on the patterned-monolayer surface at micro and nanometer scales, and ODMS-modified surfaces acted as a layer for preventing the non-specific attachment of oligonucleotides. A similar controlled immobilization was also achieved when a template surface cross-linked with biotin molecules was used. Moreover, the specific reaction between biotin and streptavidin was detected by using the monolayer template surface (Fig. 10.9d, e). These results suggest the usefulness of patterned monolayer templates for *immobilizing* biomolecules and for *promoting* interaction between biomolecules with a high selectivity.

10.2.3 FET Biosensor Using Enzymatic Reaction

As one of the applications of the monolayer-modified FETs to biosensing, we have performed urea detection by using an enzymatic reaction. An enzymatic hydrolysis

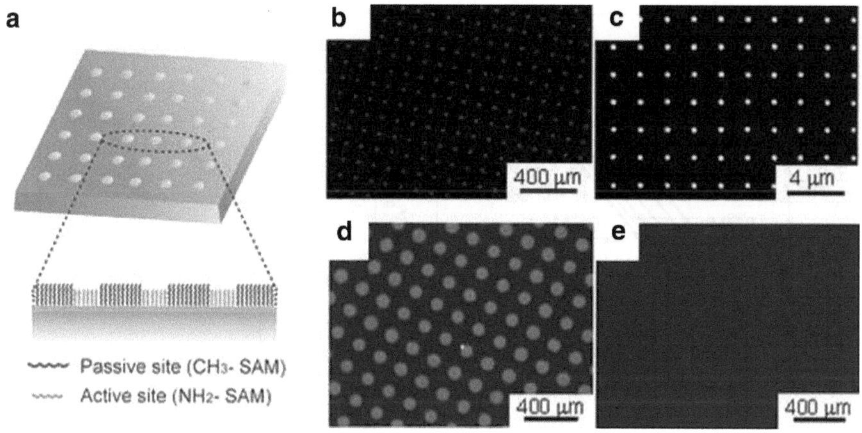

Fig. 10.9 (**a**) Schematic illustration of the surface patterned with different monolayers. (**b, c**) Representative fluorescent microscope images of the APS/ODS patterned surface covalently immobilized with oligonucleotide (**d, e**) fluorescence images of the patterned surface modified (**d**) with and (**e**) without biotin after the reaction

reaction of urea by urease, which shifts the pH toward higher values depending on the quantity of urea, was focused upon [40].

The quantity of urea was estimated from the shift of the gate voltage of an enzyme-immobilized APS FET. Note that the gate voltage of the APS-modified FET shifts depending on the change of pH at the interface between a solution and the gate surface. Figure 10.10a shows I_{ds}–V_g curves of the enzyme-modified FET in the solutions containing different urea concentrations. The I_{ds}–V_g profile shifts to the positive direction (high pH direction) according to the concentration of urea in the solution. The calibration curve is shown in Fig. 10.10b. The gate voltage linearly changes toward urea concentration with a high sensitivity of 64 mV/decade in the range of 10^{-9}–10^{-6} M. However, calibration curve shows the tendency to saturate in the range of over 10^{-6} M. Such a difference in the response is thought to be related to the amount of enzyme immobilized onto the APS surface. Since the APS monolayer surface is quite flat, the amount of enzyme immobilized onto the surface could be relatively low. Therefore, in the case of the urea concentration of over 10^{-6} M, it seems that an ureolysis ability of the enzyme has been exceeded, resulting in the decrease in the voltage shift. From the result, it is indicated that the FET device with enzyme immobilized onto the APS monolayer has a high capability for the detection of trace concentration of urea.

10.2.4 DNA Sensor Using FET

The detection of hybridization reaction of DNA was studied in our active under-investigation research.(Fig. 10.11). If DNA is attached to the surface, the surface

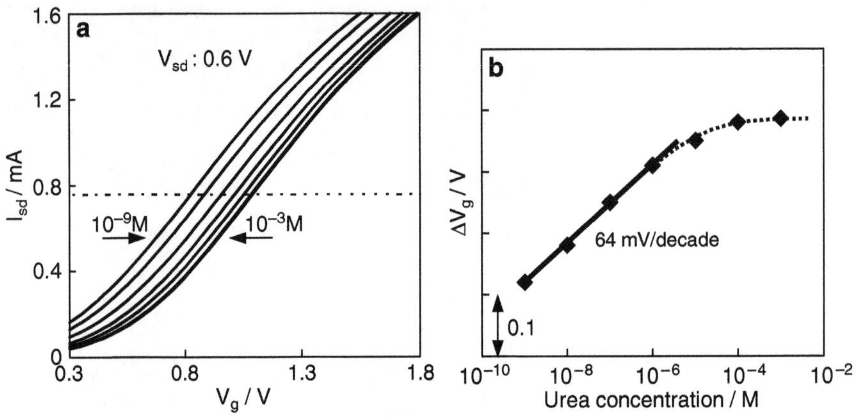

Fig. 10.10 (**a**) I_{ds}–V_g curves of enzyme immobilized FET in the solution containing different concentration of urea. (**b**) Calibration curve of the enzyme immobilized device for urea detection

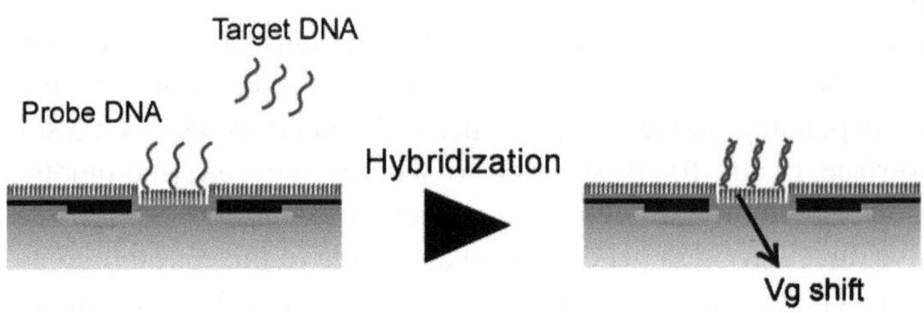

Fig. 10.11 Scheme of hybridization reaction between the immobilized DNA (probe DNA) and Target DNA

potential shifts in the negative direction since DNAs have a negative charge originated from deprotonated phosphate esters in aqueous solution. Moreover by hybridizing DNA at the surface, the potential shifts to more negative. Based on the property of DNA itself, detection of DNA-SNPs was carried out. The V_g of the probe DNA-immobilized FET shifted due to the negative charge while DNA, on the surface, hybridized with complementary DNA. The results revealed that the V_g shifted by 53 mV in the case of using 20-mer DNA. It was also noted that no shift was observed when using noncomplementary DNA. In addition, the voltage shifts depending on the number of mismatches observed in DNA. Therefore, the DNA-immobilized FET is expected to have a high potential for the detection of SNPs with high sensitivity and selectivity.

10.3 Conclusion

The FET device has a high potential for the detection of ions and biomaterials. As described in some examples, the application of the FET as a sensor is expected to be useful for the next generation high-performance, on-chip sensing system. In addition, since the FET sensor enables the miniaturization of the sensor chip itself, it is especially expected to apply the advanced medical care and tailor-made medical diagnosis. Moreover, the combination between nanotechnology and biotechnology will accelerate the fusion of various industries such as the semiconductor industry and bioventures.

References

1. Matsuo T, Esashi M (1981) Methods of ISFET fabrication. Sens Actuators 1:77–96
2. Bergveld P (1972) Development, operation, and application of the ion-sensitive field-effect transistor as a tool for electrophysiology. IEEE Trans Biomed Eng BME-19:342–351
3. Matsuo T, Wise KD (1974) An integrated field-effect electrode for biopotential recording. IEEE Trans Biomed Eng BME-21:485–487
4. Bousse L, Mostarshed S, Van der Schoot B, de Rooij NF (1994) Comparison of the hysteresis of Ta_2O_5 and Si_3N_4 pH-sensing insulators. Sens Actuators B 17:157–164
5. Niwa D, Yamada Y, Homma T, Osaka T (2004) Formation of molecular templates for fabricating on-chip biosensing devices. J Phys Chem B 108:3240–3245
6. Niwa D, Homma T, Osaka T (2004) Fabrication of organic monolayer modified ion sensitive field effect transistors with high chemical durability. Jpn J Appl Phys 43:L105–L107
7. Niwa D, Omichi K, Motohashi N, Homma T, Osaka T (2006) Organosilane self-assembled monolayer-modified field effect transistors for on-chip ion and biomolecule sensing. Sens Actuator B 108:721–726
8. Kuroiwa S, Wang J, Satake D, Nomura S, Osaka T (2009) Effect of surface morphology of reference field effect transistor modified by octadecyltrimethoxysilane on ionic responses. J Electrochem Soc 156:J67–J72
9. Wang J, Ito K, Nakanishi T, Kuroiwa S, Osaka T (2009) Tb^{3+}-enhanced potentiometric detection of single nucleotide polymorphism by field effect transistors. Chem Lett 38:376–377
10. Ulman A (1991) An introduction to ultra thin organic films from langmuir-blodgett to self-asembly. Academic, San Diego, CA
11. Wasserman SR, Tao Y-T, Whitesides GM (1989) Structure and reactivity of alkylsiloxane monolayers formed by reaction of alkyltrichlorosilanes on silicon substrates. Langmuir 5:1074–1087
12. Schwartz DK (2001) Mechanisms and kinetics of self-assembled monolayer formation. Annu Rev Phys Chem 52:107–137
13. Doudevski I, Hayes WA, Schwartz DK (1998) Submonolayer island nucleation and growth kinetics during self-assembled monolayer formation. Phys Rev Lett 81:4927–4930
14. Leitner T, Friedbacher G, Vallant T, Brunner H, Mayer U, Hoffmann H (2000) Investigations of the growth of self-assembled octadecylsiloxane monolayers with atomic force microscopy. Mikrochim Acta 133:331–336
15. Iimura K, Nakajima Y, Kato T (2000) A study on structures and formation mechanisms of self-assembled monolayers of n-alkyltrichlorosilanes using infrared spectroscopy and atomic force microscopy. Thin Solid Films 379:230–239

16. Wang Y, Lieberman M (2003) Growth of ultrasmooth octadecyltrichlorosilane self-assembled monolayers on SiO_2. Langmuir 19:1159–1167
17. Sugimura H, Ushiyama K, Hozumi A, Takai O (2000) Micropatterning of alkyl- and fluoroalkylsilane self-assembled monolayers using vacuum ultraviolet light. Langmuir 16:885–888
18. Hozumi A, Ushiyama K, Sugimura H, Takai O (1999) Fluoroalkylsilane monolayers formed by chemical vapor surface modification on hydroxylated oxide surfaces. Langmuir 15:7600–7604
19. Sugimura H, Nakagiri N (1997) Organosilane monolayer resists for scanning probe lithography. J Photopolym Sci Technol 10:661–666
20. Sugimura H, Nakagiri N (1996) Scanning probe anodization: nanolithography using thin films of anodically oxidizable materials as resists. J Vac Sci Technol A 14:1223–1223
21. Hozumi A, Sugimura H, Yokogawa Y, Kameyama T, Takai O (2001) ζ-potentials of planar silicon plates covered with alkyl- and fluoroalkylsilane self-assembled monolayers. Colloid Surf A 182:257–261
22. Hayashi K, Saito N, Sugimura H, Takai O, Nakagiri N (2002) Regulation of the surface potential of silicon substrates in micrometer scale with organosilane self-assembled monolayers. Langmuir 18:7469–7472
23. Sugimura H, Hanji T, Hayashi K, Takai O (2002) Surface potential nanopatterning combining alkyl and fluoroalkylsilane self-assembled monolayers fabricated via scanning probe lithography. Adv Mater 14:524–526
24. Siqueira PDF, Wenz G, Schunk P, Schimmel T (1999) An improved method for the assembly of amino-terminated monolayers on SiO_2 and the vapor deposition of gold layers. Langmuir 15:4520–4523
25. DePalma V, Tillman N (1989) Friction and wear of self-assembled trichlorosilane monolayer films on silicon. Langmuir 5:868–872
26. Barrelet CJ, Robinson DB, Cheng J, Hunt TP, Quate CF, Chidsey CED (2001) Surface characterization and electrochemical properties of alkyl, fluorinated alkyl, and alkoxy monolayers on silicon. Langmuir 17:3460–3465
27. Cicero RL, Linford MR, Chidsey CED (2000) Photoreactivity of unsaturated compounds with hydrogen-terminated silicon(111). Langmuir 16:5688–5695
28. Niu MN, Ding XF, Tong QY (1996) Effect of two types of surface sites on the characteristics of Si_3N_4-gate pH-ISFETs. Sens Actuators B 37:13–17
29. Birge RR (ed) (1994) Molecular electronics and bioelectronics. Am Chem Soc, Washington, DC
30. Bard AJ (1994) Integrated chemical systems. Wiley, New York
31. Spochiger-Kuller UE (1998) Chemical sensors and biosensors for medical and biological applications. Wiley, Washington
32. Schena M, Shalon D, Davis RW, Brown PO (1995) Quantitative monitoring of gene expression patterns with a complementary DNA microarray. Science 270:467–470
33. MacBeath G, Schreiber SL (2000) Printing proteins as microarrays for high-throughput function determination. Science 289:1760–1763
34. Wilson DS, Nock S (2003) Recent developments in protein microarray technology. Angew Chem Int Ed 42:494–500
35. Frutos AG, Smith LM, Corn RM (1998) Enzymatic ligation reactions of DNA “words” on surfaces for DNA computing. J Am Chem Soc 120:10277–10282
36. Lamture JB, Beattie KL, Burke BE, Eggers MD, Ehrlich DJ, Fowler R, Hollis MA, Kosicki BB, Reich RK, Smith SR, Varma RS, Hogen ME (1994) Direct detection of nucleic acid hybridization on the surface of a charge coupled device. Nucleic Acids Res 22:2121–2125
37. Souteyrand E, Cloarec JP, Martin JR, Wilson C, Lawrence I, Mikkelsen S, Lawrence MF (1997) Direct detection of the hybridization of synthetic homo-oligomer DNA sequences by field effect. J Phys Chem B 101:2980–2985
38. Niwa D, Omichi K, Motohashi N, Homma T, Osaka T (2004) Formation of micro and nanoscale patterns of monolayer templates for position selective immobilization of oligonucleotide using ultraviolet and electron beam lithography. Chem Lett 33:176–177

39. Osaka T, Matsunaga T, Nakanishi T, Arakaki A, Niwa D, Iida H (2006) Synthesis of magnetic nanoparticles and their application to bioassays. Anal Bioanal Chem 384:593–600
40. Osaka T, Komaba S, Seyama M, Tanabe K (1996) High-sensitivity urea-sensor based on the composite film of electroinactive polypyrrole with polyion complex. Sens Actuators B 36:463–469

[illegible]

Chapter 11
Electrochemical and Magnetic Technologies for Bio Applications

Tadashi Matsunaga and Tsuyoshi Tanaka

11.1 Electrochemical and Magnetic Probes for Biochip Applications

The electrochemical and magnetic biosensors have an advantage because of the easy miniaturization of electric device components as compared with photometric instruments. These technologies have been applied to develop portable, compact and inexpensive biochip devices. A commercially successful example is the glucose sensor using enzyme transducers, which was originally reported by Clark and Lyons [1] to measure glucose by detecting the decrease in oxygen by pO_2 electrode when glucose is converted to gluconic acid and hydrogen peroxide. Electrochemical biosensors can be separated into three typical assay systems using amperometric, potentiometric or conductometric transducers. Furthermore, various magnetosensors using magnetic particles have been developed over a decade in place of photometric biosensors. In this chapter, recent advances in electrochemical and magnetic bio sensors toward development of portable, compact and inexpensive biochip devices have been focused.

Electrochemical biosensors have been divided into two basic types: enzyme-based sensor and electrochemical probe-based sensor. Alkaline phosphatase (ALP) and horse radish peroxidase (HRP) have been often employed for enzyme-based biosensors using p-nitrophenyl phosphate (PNP), α-naphtyl phosphate, 3–3′,5,5′-tetramethylbenzidine (TMB) and 2,2′-azino-bis-(3-ethylbenzthiazoline-6-sulfonic acid) (ABTS) as substrates of electrochemically active species, and ferrocene (Fc) and methylene blue as the electrochemical mediators. In general, enzymatic amplification of electrochemical signals enables highly sensitive detection of analytes. On the other hand, a direct detection of analytes by using electrochemical probes allows a more rapid time-response onto the detector surface and needs no enzymatic reaction. Based on the reason, a direct detection of analytes by using electrochemical probes has been

T. Matsunaga (✉) and T. Tanaka
Department of Biotechnology, Tokyo University of Agriculture and Technology,
2-24-16 Naka-Cho, Koganei-city, Tokyo 184-8588, Japan

T. Osaka et al. (eds.), *Electrochemical Nanotechnologies*,
Nanostructure Science and Technology, DOI 10.1007/978-1-4419-1424-8_11,

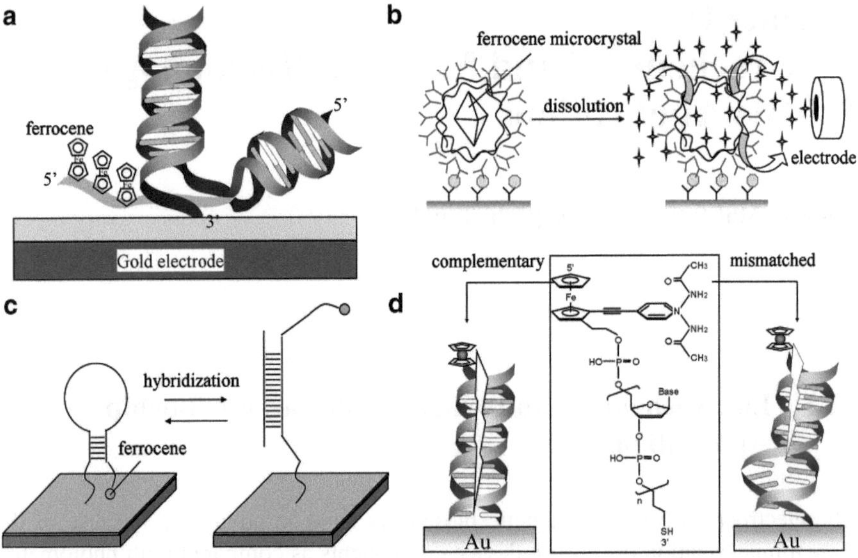

Fig. 11.1 Direct detection of electrochemical signals using ferrocene-labeled biomolecules. Adapted from (**a**) Umek et al. 2001; (**b**) Mak et al. 2005; (**c**) Fan et al. 2003; (**d**) Inouye et al. 2005

preferred to simplify the process towards miniaturized biochip devices. A CMOS-based electric DNA chip, eSensor™ with 16 gold electrodes manufactured by printed circuit board technology has been developed [2]. Thiolated oligonucleotide probes are immobilized directly onto the gold electrode surface. The detection of DNA hybridization is performed by alternating current voltametry using ferrocene-incorporated oligonucleotides as detection probes (Fig. 11.1a) [3]. However, the study of electrochemical probe-based biosensors has been limited by the sensitivity as compared with enzyme-based biosensors. To overcome the limitation, various approaches have been attempted.

Various probes for the electrochemical detection have been reported, such as ferrocene [4, 5], gold nano particles [6], intercalater [7] and semiconductor nanoparticles [8]. Among them, ferrocene derivatives have often been used as electrochemical probes for immunoassay [9–11] as well as the DNA hybridization assay [4, 5, 12–14]. Labeling of ferrocene derivatives to enzymes such as glucose oxidase has been intensively studied and used as mediators in biosensors [15–17]. Also, electroactive labeling of IgG with ferrocenemonocarboxylic acid (Fc-COOH) by chemical crosslinkers, sulfo-N-hydroxysulfosuccinimide (NHS) and 1-ethyl-3-[3-dimethylaminopropyl]-carbodiimide hydrochloride (EDC), has been commonly used [9, 10] (Fig. 11.2a). Only two to three ferrocene moiety have been stably introduced to IgG. Recently, labeling of ferrocenecarboaldehyde (Fc-CHO) to immunoglobulin G (IgG) via formation of Schiff-base and its reduction was investigated for construction of an electrochemical probe for miniaturized amperometric flow immunoassay [18] (Fig. 11.2b). Approximately, eight molecules of Fc–CHO

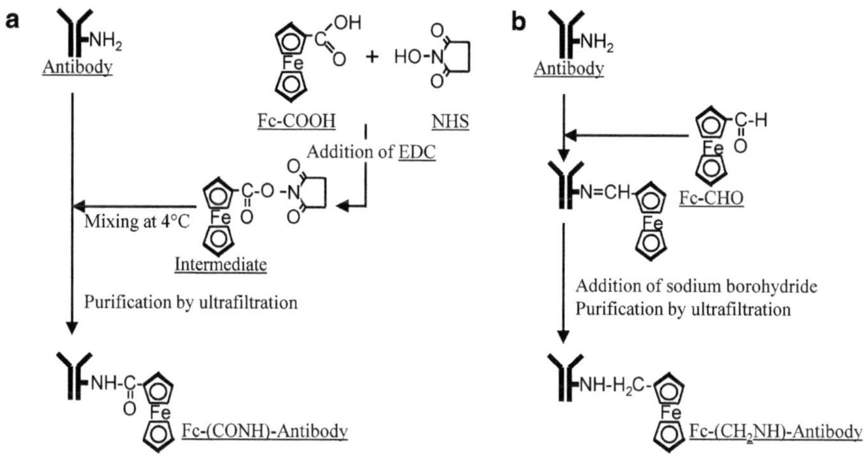

Fig. 11.2 Reaction schemes for labeling of IgG with ferrocenemonocarboxylic acid (Fc-COOH) (**a**) and ferrocenecarboaldehyde (Fc-CHO) (**b**)

were labeled to IgG and the reversible redox property of ferrocene was observed. Labeling efficiency improved by over three times as compared to the conventional method using ferrocenemonocarboxylic acid (Fc–COOH). IgG labeled with Fc–CHO that retained eight ferrocene moiety showed sufficient binding affinity to its antigen and the current response obtained in the flow electrochemical detection system increased by 14-fold as compared with IgG labeled with Fc–COOH. The minimum detectable concentration of IgG labeled with Fc–CHO was 60 pM. IgG labeled with Fc–CHO demonstrate biochemical and electrochemical properties, which are useful for electrochemical immunosensors (See also section 11.2). The utilization of encapsulated ferrocene microcrystal was proposed as another approach to increase the electroactive species in probe-based immunosensors (Fig. 11.1b) [19]. Ferrocene microcrystals encapsulated within a capsule, which provided a stable interface for antibody conjugation, were used as an electrochemical probe to perform a sandwich immunoassay. After the immunoreaction with ferrocene microcrystal-antibody complexes, the dissolution of the ferrocene microcrystals, and subsequent amperometric detection of the released ferrocene molecules were performed. The ferrocene microcrystal-based biosensor provided a high-signal molecule to antibody ratio of 10^4–10^5 (detection limit: 20 pM).

The direct detection of DNA via conformational changes of ferrocene-conjugated DNA stem-loop (or hair-pin) structure onto electrode surface induced by the hybridization was also proposed (Fig. 11.1c) [4, 20]. The conformational change induces the displacement of ferrocene molecules from the electrode surface, resulting in a drop in peak redox current measured by cyclic voltametry. The detection limits were 115 fM for 24-base oligonucleotide [20] and 10 pM for 17-base oligonucleotide [4], respectively. Furthermore, the difference between chemical structures of ferrocene-conjugated DNA hybridized with a complementary or a single-base mismatch was detected by hole transport (Fig. 11.1d). The presence of a single-base mismatch in the DNA duplexes caused a dramatic decrease in the electrochemical response [21].

Alternative approach for direct detection is magnetosensor. Magnetic particles are useful magnetosensor probes for quantitative detection of molecular interactions, including those between antigen-antibody, DNA–DNA and ligand-receptor. Measurements are traditionally performed using superconducting quantum interference device (SQUID) magnetometer [22–25], giant magneto-resistive (GMR) sensors [26–30], magnetic susceptometry [31–34] or magnetic force microscopy [35, 36]. The use of magnetic particles offers a great advantage for assays, because it allows the analytical signal to be measured in terms of intensity of magnetization. Furthermore, the use of magnetic particles enables separation of a trace amount of target in solution, which simplifies the process of sample preparation. These types of magnetosensors are highly sensitive and theoretically can detect single magnetic particles [27]. An ideal magnetic probe for use as a magnetosensor requires uniformity in size and magnetization. Ferromagnetic particles have been used rather than ferrimagnetic particles, since ferrimagnetic particles aggregate with each other in aqueous conditions. The ferromagnetic particles generally used consist of polymer containing dispersed nano-sized magnetic particles, such as maghemite (γ-Fe_2O_3) or magnetite (Fe_3O_4), and have highly uniform diameters. However, they are not sufficiently uniformly magnetic to be used as highly sensitive magnetosensor because of the variation of density of nano-sized magnetic particles [26]. For highly sensitive assay using magnetosensors, it is necessary to use homogenously sized magnetic particles. Furthermore, micro size magnetic particles or aggregates of nano size magnetic particles have been mainly used as magnetosensors. The detection of a single magnetic probe using small magneto-resistive spin valve sensor has been reported. The size of magnetic particle used was, however, micro size (2 μm) and the detection range was narrow, 6–20 particles were needed for magnetic signal detection [30]. These particles are too large for use as probes in place of fluorescent dyes or luminescent reagents. Therefore, the use of nano-sized magnetic particle as a magnetic probe is a preferable approach to develop a novel magnetosensor toward the miniaturization.

Magnetite (Fe_3O_4) particles synthesized by magnetotactic bacteria [37, 38], are more uniform in size and shape as compared with artificial magnetite particles. The bacterial magnetite particles (BacMPs) are small in size (50 to 100 nm) and consist of single crystal of magnetite having a single magnetic domain. Therefore, BacMPs are regarded as having uniform particle size and magnetization. Each BacMP is covered with a lipid bilayer membrane mainly of phosphatidylethanolamine. BacMPs are ferrimagnetic; however, they have excellent dispersion in aqueous solutions imparted by the lipid membrane [39]. The BacMPs have been studied with much interest with reference to many engineering applications, such as immunoassay, DNA detection, ligand screening, and cell separation techniques as magnetic carriers [40–45].

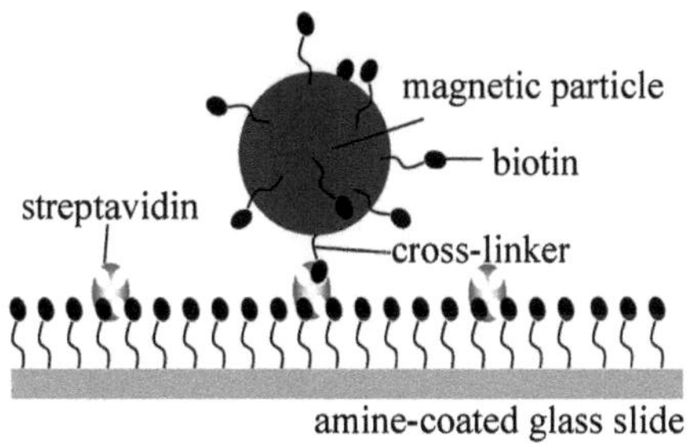

Fig. 11.3 Schematic illustration of biotin-SA interaction on biotinylated glass slide. Biotinylated magnetic particles were applied to the glass slide after treatment with various concentrations of SA

The utilization of BacMPs as a magnetic probe in biosensor has been examined by MFM imaging of single particles [35]. An investigation to determine the presence of a specific biomolecular interaction between biotin and streptavidin (SA) was done using biotinylated BacMPs and biotin-conjugated glass slide (Fig. 11.3). Magnetic detection of SA on biotin-labeled glass slides using biotin-labeled BacMPs was performed by direct BacMP counts using MFM. Biotin-BacMPs were applied to biotin immobilized on the glass slide after treatment with various concentrations of SA. The number of biotin-conjugated BacMPs (biotin-BacMPs) bound to SA immobilized on the glass slides increased with SA concentrations up to 100 pg/ml. The minimum detection limit for SA was 1 pg/ml. For a comparison, a fluorescent detection of Cy3-labeled SA binding onto a biotinylated glass slide was performed by a photomultiplier using a fluorescent scanner. The minimum detection limit of Cy3-SA was 100 pg/ml of SA, which corresponds to approximately 2,000 molecules of SA in the same area if all SA molecules were immobilized on their surface. Fluorescent signals at less than 10 pg/ml SA were not able to discriminate from background noise. This detection limit is almost the same with a previous report (150 pg/ml of IgG) by using a laser-scanning system with a photomultiplier [46]. The results by BacMPs-based assay are 100 times more sensitive for the detection of SA compared with fluorescent detection, thereby suggesting that its use has potential advantages for extremely sensitive biomolecule detection.

11.2 Electrochemical Flow Immunoassay Using Ion Exchange Chromatography

Miniaturized immunosensors, which combine the analytical power of microfluidic devices with the high specificity of antibody-antigen interactions, have been intensively developed [9–11, 47–51]. These platforms have proven to be highly suitable vehicles for conducting various immunoassay protocols. Our research groups have described a new approach to the performance of miniaturized electrochemical flow immunoassay system (on-chip typed flow immunoassay system) by using ferrocene-conjugated

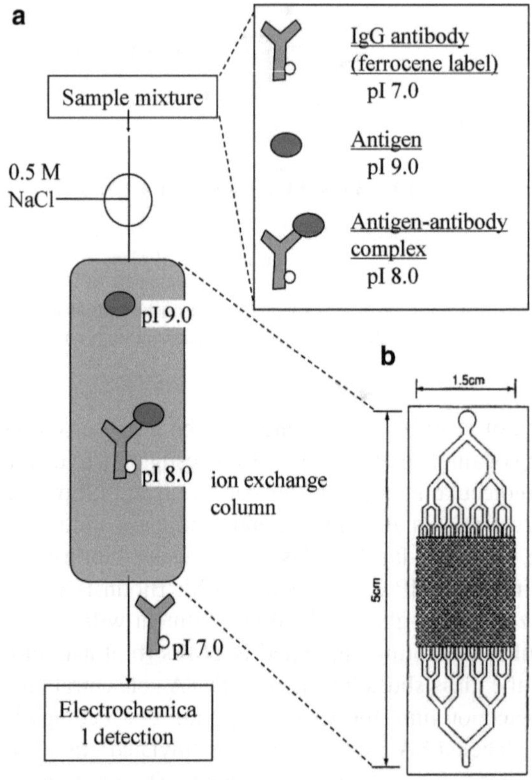

Fig. 11.4 (**a**) Principle of flow immunoassay using ion exchange column. (**b**) Multi-channeled matrix column

antibody as an electrochemical probe and ion exchange chromatography [9, 52] (Fig. 11.4). Antibody-antigen complexes are separated from unreacted antibodies (or antigens) on the basis of their differences in isoelectric point (pI) using an ion-exchange resin, and electrochemically detected using a three-electrode flow-cell system. This method does not require prior immobilization of the antibody onto a solid phase, and has several advantages of a much shorter assay time and a minimal sample volume. Recently, a multi-channeled matrix column coated with cation-exchange resin on PMMA (polymethyl methacrylate) plate, which is the same size as of a credit card, has been constructed as a novel reaction platform towards pregnancy and allergy tests (Fig. 11.4). The flow immunoassay system enables the generation of highly reproducible results using only minute quantities of whole blood samples within 2 min [10]. The integration of chromatography and electrochemical detection with the microchip technology enables us to expand the potentials of other clinical applications, such as diagnosis of diabetes.

Hemoglobin A_{1c} (HbA_{1c}), which is glycated on the N-terminal valine of the β-chain, is well known as the main diabetes marker protein for monitoring long-term glycemic control clinically. Minor hemoglobin (Hb) components including HbA_{1c}

have been originally separated from hemolysates of healthy adult by cation-exchange chromatography [53–55]. Because minor Hb components show lower isoelectric point (pI ; <6.9) than that of non-glycated Hb (HbA_0) due to the glycation [56], they move down a cation-exchange column faster than HbA_0 (pI ; 7.0). HbA_{1c} level in a healthy adult is ranged from approximately 4% to 5.8 % peak area of total Hb on the chromatogram. Based on these understandings, the Japanese Diabetes Society (JDS) has developed a high-resolution cation-exchange chromatography for the detection of HbA_{1c} (KO500 method) [57]. However, the method is time consuming, resulting in slow transfer of examination results from the medical laboratory to the patients. Until now, several POCT devices for simple and rapid HbA_{1c} detection have been proposed to overcome the above problems. In the application for the development of POCT devices, immunoassays are suitable due to the easiness in price-reduction, miniaturization and simplification. However, immunoassay requires an additional process for the evaluation of HbA_{1c} levels because HbA_{1c} levels are expressed as a percentage of total Hb. Total Hb amounts are commonly measured by colorimetry. Because of the difference in detection range between immunoassay and colorimetry, each sample is prepared by individual dilution series. These manual operations should be minimized as far as possible in POCT device for HbA_{1c} to increase the assay precision [58].

In the on-chip typed flow immunoassay system (Fig. 11.5a), manual operation is only mixing the hemolysate sample and ferrocene-conjugated antibody. Although a flow immunoassay system based on boronate affinity for detection of glycated hemoglobin has been proposed [59, 60], ion exchange chromatography is found more suitable to separate and measure each hemoglobin (HbA_0 and HbA_{1c}). This system enables to measure total Hb and HbA_{1c} simultaneously using the same sample. The currents in the eluted fractions were detected by the on-chip typed

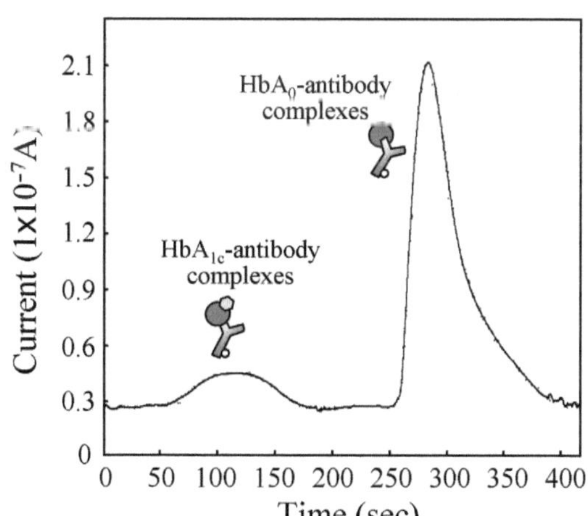

Fig. 11.5 Electrochemical detection of immunocomplexes using on-chip typed cation-exchange chromatography and ferrocene-conjugated antibody. Flow rate: 200 μl/min; Column volume: 30 μl, Buffer change point: 350, 650 s

cation exchange chromatography. Three peaks observed by a stepwise elution, were explained as minor Hbs, HbA_{1c}, HbA_0 and HbA_2. HbA_{1c} levels in hemolysates were calculated from the peak area. Good correlation of HbA_{1c} level between KO500 method and on-chip typed method was obtained (Fig. 11.5b). This method can be applied to POCT devices for clinic or bedside testing of diabetic patients.

11.3 Integrated Genetic Analysis System

Miniaturized biochip device has been recently attracting much attention due to their increasing applications to clinical diagnosis based on genomic analysis, consisting of DNA extraction, PCR, electrophoresis and DNA sequencing (Fig. 11.6). To design a total analysis system on a chip, DNA extraction step plays an important role in subsequent polymerase chain reaction (PCR). Furthermore, microchips to accelerate DNA hybridization and to attain rapid, accurate and high-throughput DNA detection are required. Especially, the detector compatibility with the micro-total analysis system is extremely important to solve the technical and cost problems. In this section, recent developments in DNA extraction and miniaturized photosensor are summarized.

11.3.1 Microchip for DNA Extraction Using Aminosilane-Modified Solid Supports

Magnetic particle (or bead)-based DNA extraction has been commonly used for molecular biology due to quick processing time, reduced chemical need and easy separation using a magnet. Several magnetic particles for DNA extraction have been already commercialized. Magnetic silica particles have been widely used for the DNA extraction based on the method invented by Boom [61]. A solid phase reversible immobilization (SPRI) based on DNA binding to the surface of carboxyl coated solid-phase under conditions of high polyethylene glycol and salt concentration [62, 63] has been developed for a novel DNA purification method specially for highly robust and cost-effective assay toward the complete human genome sequence [64].

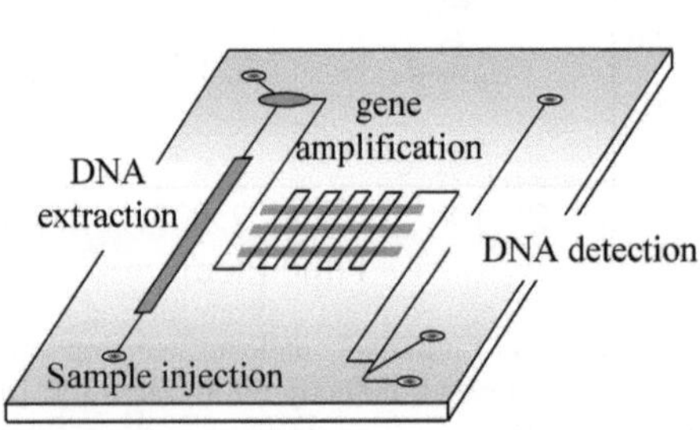

Fig. 11.6 Schematic illustration of integrated genetic analysis system

A separation process using magnetic particles has been commonly accepted as a microdevice or microchamber-friendly procedure. The magnetic separation has been adopted in a number of flow-based analytical operations, such as the purification of PCR products using SA-labeled magnetic particles [65], bound/free separations in immunoassay using antibody-labeled magnetic particles [66] and in pathogen detection using DNA-labeled magnetic particles [67, 68], in microchambers. Despite the utility of magnetic particles, microfabricated structures, such as packed beads, resins and pillars were preferred to magnetic particles [69–71] for flow-based DNA extraction using the Boom method and SPRI, since the magnetic particles settle down by spontaneous sedimentation in aqueous conditions, resulting in the need for suspending magnetic particles in microchamber. Especially, the mixing of magnetic particles in microchamber is difficult for the above-mentioned DNA extraction because highly viscous solutions were used in DNA adsorption process. Until now, the mixing of magnetic particles in a microchamber has not been optimized although several studies on efficient mixing in flat microchamber have been reported in DNA microarray analysis to enhance the hybridization efficiency by using air-driven bladder [72], cavitation microstreaming [73] and chaotic mixer [74]. Furthermore, magnetic particles suitable for flow-based DNA extraction have not been proposed.

Our research groups have demonstrated a novel DNA extraction using aminosilane-modified solid supports [75–77]. The principle is based on electrostatic interaction between amino groups on solid supports and nucleic acids, and subsequent DNA release under high salt or higher pH conditions. In this format, DNA adsorption under high viscous conditions was not required because cell lysates were directly used as DNA samples. At first, a simple microchip device for DNA extraction was constructed based on electrostatic interactions between surface amine groups and DNA. Microchannel (Fig. 11.7a) was fabricated on silicon wafer by photolithography and coated with 3-[2-(2-aminoethylamino)-ethylamino]-propyltrimethoxysilane (AEEA) to introduce amine groups on the surface. The amount of DNA captured in the microchip increased depending on surface amine density. Furthermore, DNA extraction using amine-coated microchip from whole blood was examined. Only DNA was effectively eluted by changing alkalinity of buffer from pH 7.5 to 10.6. The amount of DNA extracted from whole blood was approximately 10 ng and its recovery ratio was 27–40%. Performance of PCR for the eluted fraction indicates that DNA extracted from whole blood was well purified using amine-coated microchip. Recently, a cascading hyperbranched polyamidoamine dendrimer was successfully synthesized on the surface of bacterial magnetic particles (Fig. 11.7B), which were purified from magnetic bacterium, *Magnetospirillum magneticum* strain AMB-1 (See also Section 11.1), to enhance the efficiency of DNA extraction [78, 79]. The amine-dendrimer modified magnetic particles (amine-magnetic particles) show good dispersity in aqueous solutions and easy separation using a magnet after DNA capturing. The high dispersibility will be suitable for an effective mixing of magnetic particles and DNA extraction in microchamber. The potential utility of amine-magnetic particles in the DNA extraction from *Escherichia coli* cells in a polydimethylsiloxane (PDMS)-based microchamber has been investigated. DNA capturing by amine-magnetic particles

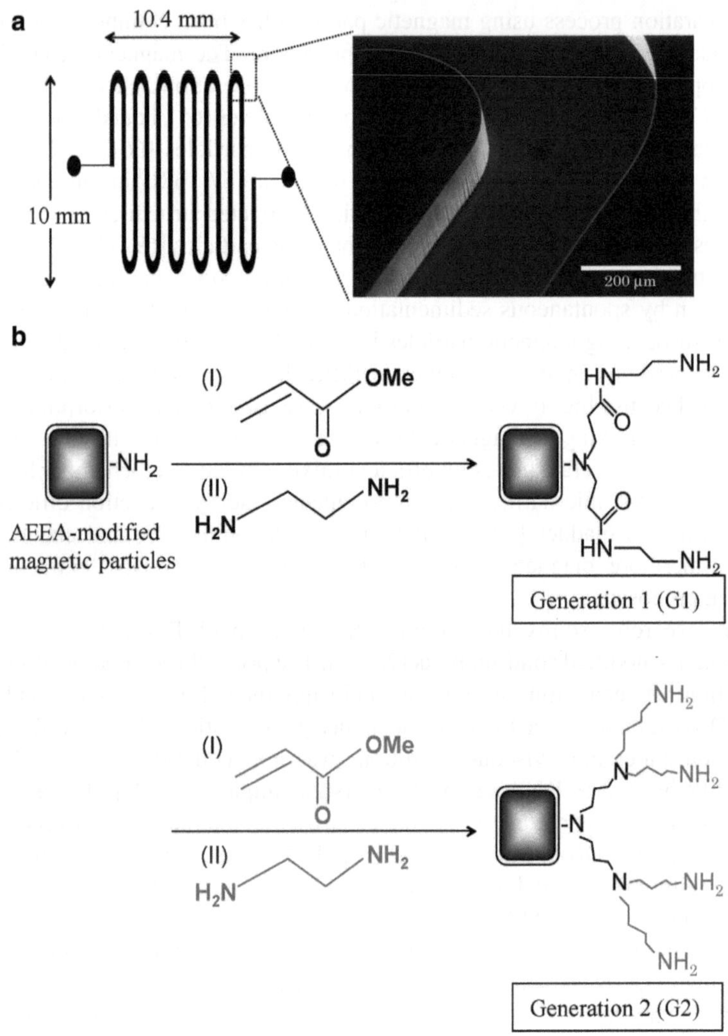

Fig. 11.7 (**a**) Layout of the aminosilane-modified microchip for DNA extraction. Channels: 300 μm wide × 100 μm deep; Total volume: 3.7 μl (adapted from Nakagawa et al. 2005). (**b**) Polyamidoamine dendrimer synthesis on the surface of aminosilane-modified magnetic particles. Dendrimer generation was initiated with AEEA coated magnetic particles in methyl acrylate (I). Then, ethylenediamine (II) was reacted with the modified particles (Generation 1). Stepwise growth was repeated until the desired number of generations was achieved. (adapted from Yoza et al. 2003)

from the cell lysate was more than 90% of released DNA in the PDMS microchamber. These results indicated that efficient cell lysis and DNA capturing was successfully achieved by amine-magnetic particles in the microchamber. Successful PCR amplification was performed using *E. coli* genomic DNA released from amine-magnetic particles. The peak based on PCR amplification was observed by capillary electrophoresis when more than 10^2 cells of *E. coli* were used.

11.3.2 Integrated Circuits for DNA Chip Technology

DNA chips (or microarrays) have been used widely in gene expression studies and genotyping. Photomultiplier tubes (PMTs), which are adopted for use in most commercialized detectors due to their high gain potential (in the order of 10^6) [80], have been one of the preferred detection systems for DNA chips. However, a laser-scanning system and confocal microscopy are required for two-dimensional (2D) measurement of fluorescent spots. Alternatively, charge-coupled device (CCD) arrays have been used for 2D measurements. Recent advances in optical sensing technology allow us to construct compact DNA chip devices. Various photodetectors, such as the PIN photodiode [81, 82], the microavalanche photodiode (μAPD) [83], and the miniature complementary metal oxide semiconductor (CMOS) sensor [84–87] have been proposed in place of PMTs and CCD [88] as they are portable, compact, and inexpensive photodetecting devices, which can be adapted easily to biochip systems. This integrated circuit (IC) technology is a promising technique for DNA chip systems in addition to biochip systems. Electric (or electrochemical) detection of DNA is also a promising technique; however, research on electric DNA sensors has focused more on label-free DNA hybridization [89–93] because electric detection shows good performance at qualitative assays, such as point-of-care testing, but not in quantitative assays as compared with fluorescence detection. In this section, miniaturized photodetectors towards DNA chip technology are summarized (Table 11.1).

In general, DNA probes have not been immobilized directly onto the above photo-detecting devices. Rather, probes have been fixed onto glass slides, optical fibers, membranes, and microchambers as separated reaction sites, because these devices are not disposable and do not have sufficient thermostability under relatively high temperature conditions required for the DNA hybridization and DNA denaturation process. Correspondingly, external optical systems are required to collect the emitted fluorescence efficiently from the separated reaction sites. Thin film transistor (TFT) photosensor was an ideal choice as a disposable and thermostable (tolerant to more than 100°C) photodetecting device [94]. A novel DNA chip system was demonstrated using a TFT photosensor that was fabricated by semiconductor IC technology. The oligonucleotide-arrayed TFT photosensor was applied to single

Table 11.1 Applications of miniaturized photosensor to biosensor

Photosensor	Form at	Application	References
CCD	Luminescence	DNA hybridization	Lamture et al (1994) [88]
PIN photodiode	Fluorescence	Size fractionation of DNA	Kamei et al (2003, 2005) [81, 82]
micro-APD	Fluorescence	–	Chabinyc et al (2001) [83]
CMOS	Fluorescence	–	Vo-Dinh et al (1999) [85]
	Luminescence	Glucose sensor	Ho et al. (2007) [87]
	Luminescence	DNA chip	Mallard et al (2005) [86]
IFT	Fluorescence	DNA chip	Tanaka et al (2006) [94]

nucleotide polymorphisms (SNPs) detection. A SNP in the aldehyde dehydrogenase 2 (ALDH2) gene on chromosome 12, which has significant implications for the evaluation of susceptibility of human organs to damage induced by alcohol [95, 96] was used as a target. DNA hybridization with biotinylated DNA and subsequent binding of fluorescently labeled SA was detected on the TFT photosensor surface. The TFT photosensor consisted of a 200×240 pixel array (50 μm×50 μm each) with a 50-μm pitch. To develop a photosensing system with optimized performance, the spectral responses of the TFT photosensor were measured. The non-coated TFT photosensor detected light over a broad wavelength range with peak sensitivity at 450 nm. A TFT photosensor with a cut-off wavelength shorter than approximately 300 nm was prepared by coating with TiO_2 film. Correspondingly, the wavelength of the excitation source was set at 11 nm below the cut-off wavelength to eliminate excitation light and to detect only the emission radiation on the TFT photosensor (Fig. 11.8a). The UV light source with a band pass filter of 289 nm and cut-off filters for more than 300 nm and 350 nm was used as excitation source. Two fluorophores, AlexaFlour 350 (ex: 346 nm, em: 442 nm) and Qdot 565

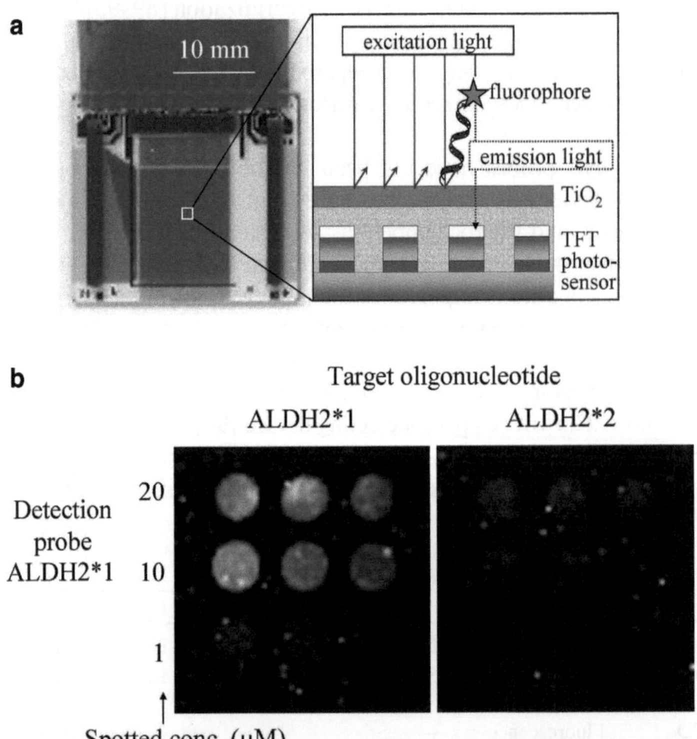

Fig. 11.8 (**a**) Topside view and schematic cross-sectional view of the TFT photosensor. (**b**) TFT photosensor images of oligonucleotide array. Oligonucleotides immobilized on TFT photo sensor were reacted with biotin-labeled ALDH2*1 (21 mer), and subsequently reacted with Alexa Fluor 350-labeled SA. Spot diameter: 500–600 μm

(ex: <565 nm, em: 565 nm) were selected for use because these fluorophores have shorter excitation wavelengths and showed little overlap of the emission signal with the excitation signal. Alexa Fluor 350 is the only organic fluorescent dye, which is commercially available, excited in the ultraviolet wavelength, and shows little overlapping of the emission signal with the excitation signal. DNA hybridization with biotinylated DNA and subsequent binding of fluorescently labeled SA was detected on the TFT photosensor surface. Approximately, 100 of the TFT elemental devices were used to observe one spot. The signal was converted to an electric current, amplified, and digitized for 2D imaging. Fig. 11.8b shows a TFT photosensor image of the oligonucleotide-array after target DNA was exposed to various concentrations of immobilized oligonucleotide. Initially, biotinylated target DNA for ALDH2*1 was hybridized, and then AlexaFluor-SA was reacted with the biotin-labeled DNA on the surface of the photosensor. Fluorescent spots were observed only on the ALDH2*1 detection oligonucleotide-arrayed photosensor and *not* on the ALDH2*2 detection oligonucleotide-arrayed photosensor (Fig. 11.8b). The use of the TFT photosensor will allow the development of a disposable photodetecting device for DNA chip systems.

11.4 Conclusions

In this chapter, the role of electrochemical and magnetic biosensors towards development of portable, compact and inexpensive biochip devices has been demonstrated. Direct measurement of electrochemical signals, such as ferrocene molecule, is a preferred approach to simplify the process towards miniaturized biochip devices. Furthermore, the use of magnetic probes promises to increase the sensitivity. Future research may be in the direction of developing new probes, e.g. nanomaterials for differing bio-applications.

References

1. Clark LCJ, Lyons C (1962) Electrode systems for continuous monitoring in cardiovascular surgery. Ann NY Acad Sci 102:29–45
2. Vernon SD, Farkas DH, Unger ER et al (2003) Bioelectronic DNA detection of human papillomaviruses using eSensor: a model system for detection of multiple pathogens. BMC Infect Dis 3:12
3. Yu CJ, Wan Y, Yowanto H et al (2001) Electronic detection of single-base mismatches in DNA with ferrocene-modified probes. J Am Chem Soc 123:11155–11161
4. Fan C, Plaxco KW, Heeger AJ (2003) Electrochemical interrogation of conformational changes as a reagentless method for the sequence-specific detection of DNA. Proc Natl Acad Sci USA 100:9134–9137
5. Umek RM, Lin SW, Vielmetter J et al (2001) Electronic detection of nucleic acids: a versatile platform for molecular diagnostics. J Mol Diagn 3:74–84
6. Burmeister J, Bazilyanska V, Grothe K et al (2004) Single nucleotide polymorphism analysis by chip-based hybridization and direct current electrical detection of gold-labeled DNA. Anal Bioanal Chem 379:391–398

7. Kara P, Ozkan D, Kerman K et al (2002) DNA sensing on glassy carbon electrodes by using hemin as the electrochemical hybridization label. Anal Bioanal Chem 373:710–716
8. Wang J, Liu G, Merkoci A (2003) Electrochemical coding technology for simultaneous detection of multiple DNA targets. J Am Chem Soc 125:3214–3215
9. Lim TK, Imai S, Matsunaga T (2002) Miniaturized amperometric flow immunoassay system using a glass fiber membrane modified with anion. Biotechnol Bioeng 77:758–763
10. Lim TK, Ohta H, Matsunaga T (2003) Microfabricated on-chip-type electrochemical flow immunoassay system for the detection of histamine released in whole blood samples. Anal Chem 75:3316–3321
11. Wang J, Ibanez A, Chatrathi MP (2002) Microchip-based amperometric immunoassays using redox tracers. Electrophoresis 23:3744–3749
12. Takenaka S, Yamashita K, Takagi M, Uto Y, Kondo H (2000) DNA sensing on a DNA probe-modified electrode using ferrocenylnaphthalene diimide as the electrochemically active ligand. Anal Chem 72:1334–1341
13. Takenaka S, Ohtuka K, Miyahara H, Nojima T, Takagi M (2002) An anthracene derivative carrying ferrocenyl moieties at its 9 and 10 positions as a new electrochemically active threading intercalator. Nucleic Acids Res Suppl; 291–2
14. Wang J, Li J, Baca AJ et al (2003) Amplified voltammetric detection of DNA hybridization via oxidation of ferrocene caps on gold nanoparticle/streptavidin conjugates. Anal Chem 75:3941–3945
15. Degani Y, Heller A (1988) Direct electrical communication between chemically modified enzymes and metal electrodes. 2. Methods for bonding electron-transfer relays to glucose oxidase and D-amino-acid oxidase. J Am Chem Soc 110:2615–2620
16. Gleria KD, Hill HA, Mcneil CJ, Green MJ (1986) Homogeneous ferrocene-mediated amperometric immunoassay. Anal Chem 58:1203–1205
17. Suzawa T, Ikariyama Y, Aizawa M (1994) Multilabeling of ferrocenes to a glucose oxidase-digoxin conjugate for the development of a homogeneous electroenzymatic immunoassay. Anal Chem 66:3889–3894
18. Okochi M, Ohta H, Tanaka T, Matsunaga T (2005) Electrochemical probe for on-chip type flow immunoassay: immunoglobulin G labeled with ferrocenecarboaldehyde. Biotechnol Bioeng 90:14–19
19. Mak WC, Cheung KY, Trau D et al (2005) Electrochemical bioassay utilizing encapsulated electrochemical active microcrystal biolabels. Anal Chem 77:2835–2841
20. Jenkins DM, Chami B, Kreuzer M et al (2006) Hybridization probe for femtomolar quantification of selected nucleic acid sequences on a disposable electrode. Anal Chem 78:2314–2318
21. Inouye M, Ikeda R, Takase M, Tsuri T, Chiba J (2005) Single-nucleotide polymorphism detection with "wire-like" DNA probes that display quasi "on-off" digital action. Proc Natl Acad Sci U S A 102:11606–11610
22. Enpuku K, Minotani T, Gima T et al (1999) Detection of magnetic nanoparticles with superconducting quantum interference device (SQUID) magnetometer and application to immunoassays. Jpn J Appl Phys 38:L1102–L1105
23. Enpuku K, Minotani T, Hotta M, Nakahodo A (2001) Application of High Tc SQUID magnetometer to biological immunoassays. IEEE Trans on Appl Supercond 11:661–664
24. Chemla YR, Grossman HL, Poon Y et al (2000) Ultrasensitive magnetic biosensor for homogeneous immunoassay. Proc Natl Acad Sci U S A 97:14268–14272
25. Grossman HL, Myers WR, Vreeland VJ et al (2004) Detection of bacteria in suspension by using a superconducting quantum interference device. Proc Natl Acad Sci U S A 101:129–134
26. Baselt DR, Lee GU, Natesan M, et al (1998) A biosensor based on magnetoresistance technology. Biosens Bioelectron 13:731–739
27. Edelstein RL, Tamanaha CR, Sheehan PE et al (2000) The BARC biosensor applied to the detection of biological warfare agents. Biosens Bioelectron 14:805–813
28. Schotter J, Kamp PB, Becker A et al (2004) Comparison of a prototype magnetoresistive biosensor to standard fluorescent DNA detection. Biosens Bioelectron 19:1149–1156
29. Megens M, Prins M (2005) Magnetic biochips: a newoption for sensitive diagnostics. J Magn Magn Mater 293:702–708

30. Graham DL, Ferreira HA, Freitas PP, Cabral JM (2003) High sensitivity detection of molecular recognition using magnetically labelled biomolecules and magnetoresistive sensors. Biosens Bioelectron 18:483–488
31. Kriz CB, Rådevik K, Kriz D (1996) Magnetic permeability measurements in bioanalysis and biosensors. Anal Chem 68:1966–1970
32. Kriz K, Gehrke J, Kriz D (1998) Advancements toward magneto immunoassays. Biosens Bioelectron 13:817–823
33. Kriz K, Ibraimi F, Lu M, Hansson LO, Kriz D (2005) Detection of C-reactive protein utilizing magnetic permeability detection based immunoassays. Anal Chem 77:5920–5924
34. Lu M, Ibraimi F, Kriz D, Kriz K (2006) A combination of magnetic permeability detection with nanometer-scaled superparamagnetic tracer and its application for one-step detection of human urinary albumin in undiluted urine. Biosens Bioelectron 21:2248–2254
35. Amemiya Y, Tanaka T, Yoza B, Matsunaga T (2005) Novel detection system for biomolecules using nano-sized bacterial magnetic particles and magnetic force microscopy. J Biotechnol 120:308–314
36. Arakaki A, Hideshima S, Nakagawa T et al (2004) Detection of biomolecular interaction between biotin and streptavidin on a self-assembled monolayer using magnetic nanoparticles. Biotechnol Bioeng 88:543–546
37. Matsunaga T, Sakaguchi T, Tadokoro F (1991) Magnetite formation by a magnetic bacterium capable of growing aerobically. Appl Microbiol Biotechnol 35:651–655
38. Sakaguchi T, Burgess JG, Matsunaga T (1993) Magnetite formation by a sulphate-reducing bacterium. Nature (London) 365:47–49
39. Nakamura N, Matsunaga T (1993) Highly sensitive detection of allergen using bacterial magnetic particles. Anal Chim Acta 281:585–589
40. Kuhara M, Takeyama H, Tanaka T, Matsunaga T (2004) Magnetic cell separation using antibody binding with protein a expressed on bacterial magnetic particles. Anal Chem 76:6207–6213
41. Tanaka T, Matsunaga T (2000) Fully automated chemiluminescence immunoassay of insulin using antibody-protein A-bacterial magnetic particle complexes. Anal Chem 72:3518–3522
42. Matsunaga T, Maeda Y, Yoshino T et al (2007) Fully automated immunoassay for detection of prostate-specific antigen using nano-magnetic beads and micro-polystyrene bead composites, 'Beads on Beads'. Anal Chim Acta 597:331–339
43. Matsunaga T, Takahashi M, Yoshino T, Kuhara M, Takeyama H (2006) Magnetic separation of CD14+ cells using antibody binding with protein A expressed on bacterial magnetic particles for generating dendritic cells. Biochem Biophys Res Commun 350:1019–1025
44. Yoshino T, Takahashi M, Takeyama H et al (2004) Assembly of G protein-coupled receptors onto nanosized bacterial magnetic particles using Mms16 as an anchor molecule. Appl Environ Microbiol 70:2880–2885
45. Yoshino T, Tanaka T, Takeyama H, Matsunaga T (2003) Single nucleotide polymorphism genotyping of aldehyde dehydrogenase 2 gene using a single bacterial magnetic particle. Biosens Bioelectron 18:661–666
46. Wacker R, Schroder H, Niemeyer CM (2004) Performance of antibody microarrays fabricated by either DNA-directed immobilization, direct spotting, or streptavidin-biotin attachment: a comparative study. Anal Biochem 330:281–287
47. Saleh OA, Sohn LL (2003) Direct detection of antibody-antigen binding using an on-chip artificial pore. Proc Natl Acad Sci U S A 100:820–824
48. Sato K, Yamanaka M, Takahashi H et al (2002) Microchip-based immunoassay system with branching multichannels for simultaneous determination of interferon-gamma. Electrophoresis 23:734–739
49. Soo Ko J, Yoon HC, Yang H et al (2003) A polymer-based microfluidic device for immunosensing biochips. Lab Chip 3:106–13
50. Wu J, Tang J, Dai Z et al (2006) A disposable electrochemical immunosensor for flow injection immunoassay of carcinoembryonic antigen. Biosens Bioelectron 22:102–108
51. Zeravik J, Ruzgas T, Franek M (2003) A highly sensitive flow-through amperometric immunosensor based on the Peroxidase chip and enzyme-channeling principle. Biosens Bioelectron 18:1321–1327

52. Lim TK, Matsunaga T (2001) Construction of electrochemical flow immunoassay system using capillary columns and ferrocene conjugated immunoglobulin G for detection of human chorionic gonadotrophin. Biosens Bioelectron 16:1063–1069
53. Allen DW, Schroeder WA, Balog J (1958) Observation on the chromatographic heterogeneity of normal adult and fetal human hemoglobin: a study of the effects of crystallization and chromatography on the heterogeneity and isoleucine content. J Am Chem Soc 80:1628–1634
54. Clegg MD, Schroeder WA (1959) A chromatographic study of the minor components of normal adult haemoglobin including a comparison of haemoglobin from normal and phenylketamine individuals. J Am Chem Soc 81:6065–6069
55. Schneck AG, Schroeder WA (1961) The relation between the minor components of normal adult haemoglobin sa isolated by chromatography and starch block electrophoresis. J Am Chem Soc 83:1472–1478
56. Bunn HF, Haney DN, Gabbay KH, Gallop PM (1975) Further identification of the nature and linkage of the carbohydrate in hemoglobin A1c. Biochem Biophys Res Commun 67:103–109
57. Hoelzel W, Weykamp C, Jeppsson JO et al (2004) IFCC reference system for measurement of hemoglobin A1c in human blood and the national standardization schemes in the United States, Japan, and Sweden: a method-comparison study. Clin Chem 50:166–174
58. St John A, Davis TM, Goodall I, Townsend MA, Price CP (2006) Nurse-based evaluation of point-of-care assays for glycated haemoglobin. Clin Chim Acta 365:257–63
59. Tanaka T, Matsunaga T (2001) Detection of HbA(1c) by boronate affinity immunoassay using bacterial magnetic particles. Biosens Bioelectron 16:1089–1094
60. Tanaka T, Tsukube S, Izawa K et al (2007) Electrochemical detection of HbA1c, a marker [correction of maker] for diabetes, using a flow immunoassay system. Biosens Bioelectron 22:2051–2056
61. Boom R, Sol CJ, Salimans MM et al (1990) Rapid and simple method for purification of nucleic acids. J Clin Microbiol 28:495–503
62. Hawkins TL, O'Connor-Morin T, Roy A, Santillan C (1994) DNA purification and isolation using a solid-phase. Nucleic Acids Res 22:4543–4544
63. Lis JT (1980) Fractionation of DNA fragments by polyethylene glycol induced precipitation. Methods Enzymol 65:347–353
64. Hawkins TL, Mckernan KJ, Jacotot LB et al (1997) A magnetic attraction to high-throughput genomics. Science 276:1887–1889
65. Liu RH, Yang J, Lenigk R, Bonanno J, Grodzinski P (2004) Self-contained, fully integrated biochip for sample preparation, polymerase chain reaction amplification, and DNA microarray detection. Anal Chem 76:1824–1831
66. Hayes MA, Polson TN, Phayre AN, Garcia AA (2001) Flow-based microimmunoassay. Anal Chem 73:5896–5902
67. Zhao W, Yao S, Hsing IM (2006) A microsystem compatible strategy for viable Escherichia coli detection. Biosens Bioelectron 21:1163–1170
68. Zaytseva NV, Goral VN, Montagna RA, Baeumner AJ (2005) Development of a microfluidic biosensor module for pathogen detection. Lab Chip 5:805–811
69. Christel LA, Petersen K, Mcmillan W, Northrup MA (1999) Rapid, automated nucleic acid probe assays using silicon microstructures for nucleic acid concentration. J Biomech Eng 121:22–27
70. Cady NC, Stelick S, Batt CA (2003) Nucleic acid purification using microfabricated silicon structures. Biosens Bioelectron 19:59–66
71. Xu Y, Vaidya B, Patel AB et al (2003) Solid-phase reversible immobilization in microfluidic chips for the purification of dye-labeled DNA sequencing fragments. Anal Chem 75:2975–2984
72. Adey NB, Lei M, Howard MT et al (2002) Gains in sensitivity with a device that mixes microarray hybridization solution in a 25-microm-thick chamber. Anal Chem 74:6413–6417
73. Liu RH, Lenigk R, Druyor-Sanchez RL, Yang J, Grodzinski P (2003) Hybridization enhancement using cavitation microstreaming. Anal Chem 75:1911–1917

74. Mcquain MK, Seale K, Peek J et al (2004) Chaotic mixer improves microarray hybridization. Anal Biochem 325:215–226
75. Yoza B, Matsumoto M, Matsunaga T (2002) DNA extraction using modified bacterial magnetic particles in the presence of amino silane compound. J Biotechnol 94:217–224
76. Nakagawa T, Hashimoto R, Maruyama K et al (2006) Capture and release of DNA using aminosilane-modified bacterial magnetic particles for automated detection system of single nucleotide polymorphisms. Biotechnol Bioeng 94:862–868
77. Nakagawa T, Tanaka T, Niwa D et al (2005) Fabrication of amino silane-coated microchip for DNA extraction from whole blood. J Biotechnol 116:105–111
78. Yoza B, Arakaki A, Maruyama K, Takeyama H, Matsunaga T (2003) Fully automated DNA extraction from blood using magnetic particles modified with hyperbranched polyamidoamine dendrimer. J Biosci Bioeng 95:21–26
79. Yoza B, Arakaki A, Matsunaga T (2003) DNA extraction using bacterial magnetic particles modified with hyperbranched polyamidoamine dendrimer. J Biotechnol 101:219–228
80. Voss KJ (2000) Physics of low light level detectors. Methods Enzymol 305:53–61
81. Kamei T, Paegel BM, Scherer JR et al (2003) Integrated hydrogenated amorphous Si photodiode detector for microfluidic bioanalytical devices. Anal Chem 75:5300–5305
82. Kamei T, Toriello NM, Lagally ET et al (2005) Microfluidic Genetic Analysis with an Integrated a-Si:H Detector. Biomed Microdevices 7:147–152
83. Chabinyc ML, Chiu DT, Mcdonald JC et al (2001) An integrated fluorescence detection system in poly(dimethylsiloxane) for microfluidic applications. Anal Chem 73:4491–4498
84. Song JM, Culha M, Kasili PM, Griffin GD, Vo-Dinh T (2005) A compact CMOS biochip immunosensor towards the detection of a single bacteria. Biosens Bioelectron 20:2203–2209
85. Vo-Dinh T, Alarie JP, Isola N et al (1999) DNA biochip using a phototransistor integrated circuit. Anal Chem 71:358–363
86. Mallard F, Marchand G, Ginot F, Campagnolo R (2005) Opto-electronic DNA chip: high performance chip reading with an all electric interface. Biosens Bioelectron 20:1813–1820
87. Ho WJ, Chen JS, Ker MD et al (2007) Fabrication of a miniature CMOS-based optical biosensor. Biosens Bioelectron 22:3008–3013
88. Lamture JB, Beattie KL, Burke BE et al (1994) Direct detection of nucleic acid hybridization on the surface of a charge coupled device. Nucleic Acids Res 22:2121–2125
89. Fritz J, Cooper EB, Gaudet S, Sorger PK, Manalis SR (2002) Electronic detection of DNA by its intrinsic molecular charge. Proc Natl Acad Sci U S A 99:14142–14146
90. Kim DS, Jeong YT, Park HJ et al (2004) An FET-type charge sensor for highly sensitive detection of DNA sequence. Biosens Bioelectron 20:69–74
91. Guiducci C, Stagni C, Zuccheri G et al (2004) DNA detection by integrable electronics. Biosens Bioelectron 19:781–787
92. Wong ELE, Gooding JJ (2006) Charge transfer through DNA: a selective electrochemical DNA biosensor. Anal Chem 78:2183–2244
93. Zhang Q, Subramanian V (2007) DNA hybridization detection with organic thin film transistors: toward fast and disposable DNA microarray chips. Biosens Bioelectron 22:3182–3187
94. Tanaka T, Hatakeyama K, Sawaguchi M et al (2006) Oligonucleotide-arrayed TFT photosensor applicable for DNA chip technology. Biotechnol Bioeng 95:22–28
95. Chen WJ, Loh EW, Hsu YP, Cheng AT (1997) Alcohol dehydrogenase and aldehyde dehydrogenase genotypes and alcoholism among Taiwanese aborigines. Biol Psychiatry 41:703–709
96. Maruyama K, Takeyama H, NEMOTO E et al (2004) Single nucleotide polymorphism detection in aldehyde dehydrogenase 2 (ALDH2) gene using bacterial magnetic particles based on dissociation curve analysis. Biotechnol Bioeng 87:687–694

Chapter 12
Nano-Bio Electrochemical Interfacing–Linking Cell Biology and Micro-Electronics

Y. Shacham-Diamand, R. Popovtzer, and Y. Rishpon

Abstract Integration of biological substance within electronic devices is an innovative and challenging area combining recent progress in molecular biology and micro technology. First, we introduce the concept of integrating living cells with Micro Electro Mechanical Systems (MEMS). Following a brief overview on "whole cell based biosensors" we describe the design, fabrication, and process of a biocompatible electrochemical "Lab-on-a-Chip" system. Demonstrating the application of electrochemical interfacing based whole cell bio chips, we present two different configurations: a. integration of prokaryotic cells (bacteria) for water toxicity detection, and b. integration of eukaryotic cells (human colon cancer cells) for rapid evaluation of the effectiveness of drug treatments. Both applications, with either microbes or mammalian cells integrated onto MEMS based biochips with liquid volume in the range of 100 nL–1 μL, function well and yield a detectable signal much higher than noise level after few minutes.

12.1 Introduction

In the last couple of decades, both Micro-electro-mechanical systems (MEMS) and microbiology achieved remarkable progress. MEMS provide very small systems, made of very small electrical and mechanical components. These systems range in size from sub micrometer level to millimeter level, and can include any component number, from a few to millions, in a particular system. MEMS extend the fabrication

Y. Shacham-Diamand (✉)
Faculty of Engineering, Tel Aviv University, Tel Aviv, 69978, Israel
e-mail: YosiSh@tauex.tau.ac.il

R. Popovtzer
School of Engineering, Bar-Ilan University, Ramat Gan 52900, Israel.
e-mail: rachelap@eng.biu.ac.il

Y. Rishpon
Faculty of Life Sciences, Tel Aviv University, Tel Aviv, 69978, Israel

T. Osaka et al. (eds.), *Electrochemical Nanotechnologies*,
Nanostructure Science and Technology, DOI 10.1007/978-1-4419-1424-8_12,

techniques developed for the integrated circuit industry to many fields like micro-mechanics, micro-fluidics, optoelectronics, and magnetism.

The intensive researches and developments in microbiology, biotechnology, and genetic engineering over the last decade have demonstrated that life science is a central scientific discipline. The introduction of genetic tools such as transposes, coupled with the ability to clone genes and determine DNA sequences, and the subsequent explosion of techniques based upon these methods made it possible to dissect the molecular genetics of any organism. These developments allowed us to answer questions that were previously deemed impossible and that have a major impact on every discipline of biology, with practical applications in medicine, agriculture, bioremediation, and biotechnology.

The integration between the two different disciplines, micro-biology and micro-electronics, poses an exciting and challenging goal for the present decade. The developments related to this area led to the invention of novelty devices and smart biosensors.

A general definition for the term "biosensor" is "the coupling of a biological material with a microelectronic system or a device to enable rapid, accurate, low-level detection of various substances in body fluids, water, and air" [1]. Cell-based biosensors are devices that contain living biological cells that monitor physiological changes induced by exposure to environmental perturbations such as toxicants, pathogens, or other agents [2]. Recent progress in cell culture and micro-fabrication technologies has suggested the development of cell-based sensors for the functional characterization and detection of drugs, pathogens, and toxicants. Unlike other biosensors such as nucleic acid or antibody-based sensors, cell-based biosensors are not specific for certain compounds but are capable of responding to a wide range of biologically active compounds and offer the potential to gather greater information content than bio-molecular-based sensors. Cell-based biosensors measure physiologic, metabolic, or network processes and responses that integrate many biological components [3, 4]. The cell-based biosensors can be used for high-throughput drug discovery and clinical diagnostics, and for the detection of toxic agents and certain odorants. For evaluating toxicology and efficiency of potential pharmaceuticals, the cell-based sensor can substitute for the use of live animals, which is expensive, cumbersome, and is probably going to be restricted in the future.

12.2 Electrochemical "Lab on a Chip" for Biological Applications

Lab-on-a-chip technology characterize in network of channels and wells mainly on silicon substrate that enables faster, better ,and cheaper multi sample handling and detection, on a single integrated system. Basic electrochemical studies and biological researches can be beneficial utilizing electrochemical "lab-on-a-chip" systems, which are easy to use, require low sample and reagent consumption, and enable high reproducibility due to standardization and automation. In the following section we describe the development of a miniaturized electrochemical "Lab-on-a-Chip" system for multi functional use and for biological based applications.

12.2.1 Device Fabrication

The chip was produced on silicon wafers and includes arrays of eight independent electrochemical cells, which are temperature-controlled. Each electrochemical cell can hold 100 nL of solution and consists of three embedded electrodes: (1) Gold working electrode, (2) Gold counter electrode, and (3) Ag/AgCl reference electrode. The electrodes are made by gold sputtering, microlithography, and by selectively depositing Ag and anodizing it in a chloride containing solution for the reference electrode. The wall of the chambers is constructed from photopolymerized polyimide (SU-8). The silicon chip is wire bonded to a plastic chip which interfaces the electronic circuit (Fig. 12.1).

The device was manufactured as two parts: the first part is a disposable silicon chip – with the electrochemical cells arrays. The silicon chip was wire bonded to a special printed circuit board (PCB) platform, which was directly connected to the data processing units. The second part of the device is reusable, which includes a multiplexer, potentiostat, temperature control and a pocket PC for sensing and data analysis (for more details see [5]). This design enables performance of multi experiments simultaneously and each electrochemical cell can be measured independently. The total weight of the entire system is ~900 g, making it ideal for medical applications.

This new design of nano chambers array on chips allows a broad band of measurements. We can simultaneously test eight different toxicant types with the general stress responsive promoter by introducing to each chamber a different toxicant, or, in order to specify unknown aqueous sample, we can test the sample with eight different stress responsive promoters. Thus we obtain an indication of the toxicant type. In addition, the array configuration enables the addition of positive and negative control chambers for each experiment.

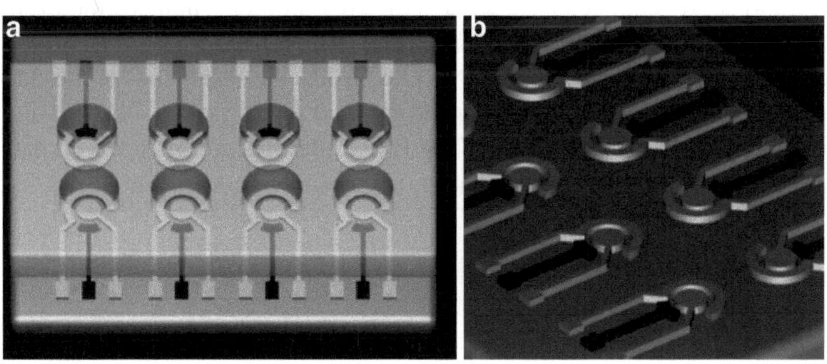

Fig. 12.1 Images of the electrochemical chip. (**a**) Silicon chip contains an array of eight miniaturized electrochemical cells with external pads. (**b**) The electrodes without the top layer (SU-8). Each electrochemical-cell consists of three circular-shaped electrodes: gold working and counter electrodes and Ag/AgCl reference electrode

12.3 Water Toxicity Detection

Here we present a novel method for water toxicity detection by integrating living cells with microelectronic device. Bacteria, which have been genetically engineered to respond to environmental stress, act as a sensor element and trigger a sequence of processes, which lead to generation of electrical current. The specific design and process of the electrochemical "Lab on a chip", which integrates the recombinant bacteria provide highly accurate, sensitive, and rapid detection of acute toxicity in water.

The most important reason for utilizing living cells as biosensors lies in their capability to provide functional information, i.e. information about the effect of a stimulus on living organisms. The aim of the present biochip is to offer the unique "functionality" sensing capability. It answers the question "Is the water toxic?" and it does not intend to perform chemical analysis or identify the nature of the toxicant. These whole-cell sensing systems can be visualized as an environmental switch, which is turned on in the presence of toxins or stressful conditions. In many cases, functional rather than analytical information is ultimately desired. In other applications, it complements and adds a special feature to a system by emulating the behavior of living entities.

The vast development in genetic engineering of living cells enables the use of recombinant cells as cell-based sensing systems [6, 7]. The cascades of mechanisms by which *E.coli* bacterial reactions to toxic chemicals or to stressful condition are electrochemically converted into electronic signals have been previously reported [8–10].

12.3.1 Recombinant Bacteria

Genetically engineered bacteria were used as whole cell sensors for acute toxicity in water. The recombinant bacteria react to the presence of toxin by activating specific promoter (regulatory DNA sequence). This promoter induces the production of the reporter enzyme β-galactosidase. This enzyme reacts with the PAPG substrates (molecules that where initial placed inside the chambers) to produce two different products: electrochemical active product p-aminophenol (PAP), and inactive product β-d-galactopyranoside. The PAP molecules are oxidized on the working electrode at 220 mV. This oxidation is converted to a current signal using an amperometric technique.

12.3.2 Experimental

Recombinant *E.coli* bacteria bearing plasmid with one of the following promoters: *dnak*, *grpE*, or *fabA* were used. These promoters were fused to the reporter enzyme β-galactosidase [5, 7]. Ethanol [1%] or phenol [1.6 ppm] was introduced to the bacterial samples in the presence of the substrate PAPG. Immediately after (~1 s), the suspensions were placed in the electrochemical cells. The response of the bacteria to the toxic chemicals was measured on-line by applying a potential of 220 mV. The substrate, PAPG, was added to a final concentration of 0.8 mg/ml (100 nL total volume). The product of the enzymatic reaction (PAP) was monitored by its oxidation current. Additional measurements in the absence of the bacteria were performed to exclude the possibility of electroactive species in the LB medium, in the substrate, or in the substrate and the LB medium mixture, which can contribute to the current response. Bacteria concentration at 3×107 cells/ml was used for all experiments.

12.3.3 Water Toxicity Measurements

Real time detection of the response of recombinant *E. coli* bacteria, with one of the promoters *dnak*, *grpE*, or *fabA*, to ethanol and phenol are shown in Figs. 12.2 and 12.3, respectively.

The different *E. coli* reporters are *fabA, dnaK* and *grpE*. Measurement was performed immediately after the ethanol addition (~1 min) at 220 mV working potential vs Ag/AgCl reference electrode. The LB curve represents the bacterial response to the LB medium with the substrate PAPG without ethanol.

The results show that concentration of 1% ethanol could be detected within less than 10 min, and concentration as low as 1.6 ppm phenol could be detected within less than 6 min. Different intensity response of the various bacterial sensors, *dnaK*,

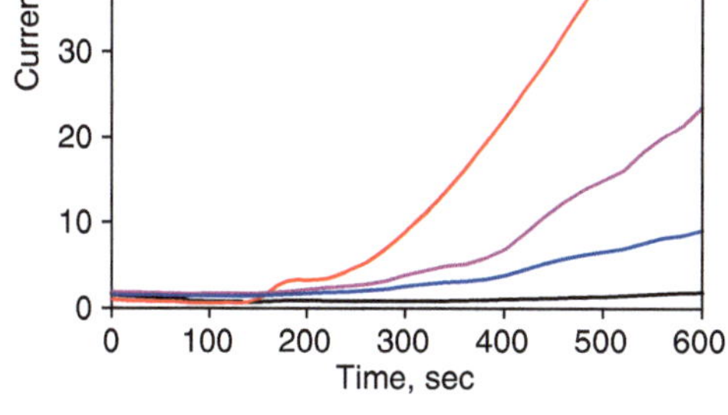

Fig. 12.2 Amperometric response curves for on line monitoring of different *E. coli* reporters in response to the addition of 1% ethanol, using the nano-bio-chip

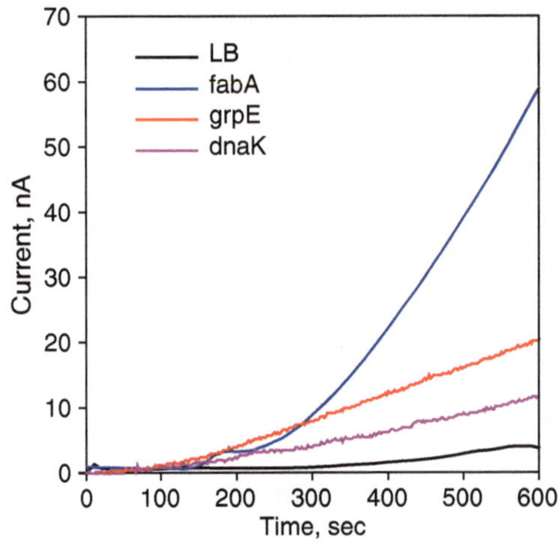

Fig. 12.3 Amperometric response curves for on line monitoring of different *E. coli* reporters in response to the addition of 1.6 ppm phenol, using the nano-bio-chip

grpE, and *fabA*, to ethanol and phenol is due to the specific activation of each promoter to the type of the toxicant. The promoters *dnaK* and *grpE* are sensitive to protein damage (SOS system); thus, they were induced in response to ethanol which is known as protein damage agent [6] (Fig. 12.2). *grpE* showed high induction activity in response to ethanol, *dnaK* showed reduced enzyme activity, and *fabA* was only slightly induced above the background level. *fabA* promoter is sensitive to membrane damage, and thus, reacts to phenol exposure, which is a known membrane damage chemical. As expected, *grpE* and *dnaK* promoters were less activated by phenol (Fig. 12.3).

The different *E. coli* reporters are *fabA, dnaK* and *grpE*. Measurement performed immediately after the ethanol addition (~1 min) at 220 mV working potential vs Ag/AgCl reference electrode. The LB curve represents the bacterial response to the LB medium with the substrate PAPG without phenol.

In comparison to equivalent optical detection methods using whole cell biosensors for water toxicity detection, these results proved to be more sensitive and produce faster response time. Concentrations as low as 1% of ethanol and 1.6 ppm of phenol could be detected in less than 10 min of exposure to the toxic chemical, whilst a recent study [11] which utilized bioluminescent *E.coli* sensor cells, detected 0.4 M (2.35%) ethanol after 220 min. An additional study [1] based on fluorescent reporter system (GFP), enabled detection of 6% ethanol and 295 ppm phenol after more than one hour. Cha et al [12] used optical detection methods of fluorescent GFP proteins, detected 1 g of phenol per liter (1,000 ppm) and 2% ethanol after 6 hours. Other studies [13] could not be directly compared due to different material used; however their time scale for chemicals identification is hours.

These results emphasize the advantages of merging electrochemical detection methods with adjusted design and process of MEMS, which result in fast response time and low detection limit. Enhanced sensitivity and high signal to noise ratio is achieved by optimizing the ratio between working electrode area and cell volume. The larger the ratio, the higher the signal.

In order to prevent false alarms, all arrays include positive and negative controls chambers. In the positive control chamber, other than adding the tested sample with the unknown chemicals to the bacterial solution, pure water was added. In case a current signal was generated, it is a false alarm. A negative control chamber includes w.t. (MG1655) *E.coli* bacteria that constitutively expresses β-galactosidase; thus, current should be generated in all cases. When no current is generated, measurement is incorrect due to bacterial death from highly toxic chemicals added or from other unknown reason. However, chemicals can produce only constant DC current signal, while the enzymatic reaction act as an intrinsic amplifier, and generates increasing current signal.

Biochemical process, which intends to produce a measurable signal, has immense benefit while utilizing enzymatic activity. Since enzymes form continuously, and each enzyme reacts with many substrate molecules successively, this enzymatic mechanism serves as an intrinsic amplifier; consequently the signal is produced faster, more sensitively, and increasing with time [14]. Combining enzymatic system with electrochemical detection methods enables measurements in turbid solutions and under anaerobic conditions [9].

In addition to the aforementioned capabilities, this "lab on a chip" system could be easily adapted to different applications, including specific identification of chemicals by using binding techniques, i.e., each electrochemical cell in the array can incorporate different biosensors. Thus, large amount of analytes can be detected simultaneously and independently. Similarly, in experiments aiming to analyze physiological reactions, bacteria harboring different types of promoters can be introduced to the chambers, and thus, this lab on a chip can detect a variety of toxicant types simultaneously.

12.4 High-Throughput Detection of Colon Cancer Cells Response to Drug Treatments

A new method is presented for rapid, sensitive, and high-throughput detection of colon cancer cells (HT-29) response to differentiation therapy, using our novel electrochemical Lab on a Chip system. Differentiation inducing agents such as butyric acid (BA) and its derivatives were introduced to miniature colon cancer samples (few to hundreds cancer cells in each sample) within the chip chambers. The efficacy of each of the differentiation inducing agents to elevate alkaline phosphatase in colon cancer cells was evaluated through electrochemical detection of the cellular enzymatic activity level, while reappearance of normal enzymatic activity denoted effective differentiation therapy drug treatment. The results demonstrate the array based device ability to evaluate simultaneously multiplex drugs effect on miniature tumor sample (~15 cells) rapidly (5 min) and sensitively, with quantitative correlation between the cancer cell number and the induced current signal. Utilizing nano-volume analytical device is of special interest in clinically relevant samples since it requires less tissue for diagnostics, and enables high-throughput analysis and comparison of various drugs effect on one small tumor sample, while keeping uniform biological and environmental conditions. In addition, this new method can help tailor cancer drugs and treatments to individual patients towards "personalized medicine".

High-throughput detection of nano volume tissue and cell samples offers great potential for increasing the amount and quality of biomedical data and has a wide range of applications. In cancer research, this technology can greatly improve diagnostics and therapy, including basic research, detection of cancer markers, and testing patient's response to different treatment modalities and thereby tailoring the treatment to an individual patient. Currently it is known that tumor treatment response cannot be predicted only from its type and anatomical location, but from its own overall individual parameters [15, 16]. Examination of the particular tumor responses to several drug types simultaneously and in exactly the same conditions, will be invaluable in providing optimal treatment to each patient.

In this study we present a unique high-throughput electrochemical system for the detection of colon cancer cells response to different differentiation inducer agents. Current cancer therapeutic strategies focus predominantly on achieving the removal or death of cancer cells within the patient, through three basic approaches: surgery, chemotherapy [17], and gamma irradiation [18]. These methods are aggressive, highly toxic, and often nonspecific. [19]. "Differentiation therapy" is an alternative less toxic approach for cancer treatment, which employs agents that modify cancer cell differentiation [20]. Upon appropriate treatment, cancer cells restrain their own growth and resume their normal growth rate [21, 22]. In addition, the differentiation therapy is utilized as a complementary treatment to conventional surgery treatment, as the latter can eliminate most, but not completely cancer cells within the patient, leading to remission of the disease.

BA is a potent differentiation agent in a wide variety of cancer cells in-vivo and in-vitro [23, 24]. Butyrate has been shown to specifically affect genes regulation by transcriptional and post transcriptional modifications. It induces dose dependent differentiation and inhibits proliferation of various malignant cells types including erythroleukemia, embryonal carcinoma and colon carcinoma [25, 26]. BA induced the expression of specific differentiation-associated genes when used at concentration between 0.5–10 mM. Differentiated cells are characterized by appearance of regulatory enzymes such as alkaline phosphatase [27].

Butyrate has potential for use in differentiation therapy but is limited by the requirement of millimolar concentrations and short metabolic half-life for efficacy. In order to overcome these problems BA derivatives (prodrugs) have been synthesized and screened. Among them, pivaloyloxymethyl butyrate (named AN-9) has demonstrated impressive anticancer activity in preclinical and clinical studies [23, 24, 28]. AN-9, metabolized intracellularly to acids and aldehyde, affects and penetrates cancer cells about 100-fold faster than BA [25].

In this work we chose HT-29 human colon cancer cells (ATCC) as a model system for high throughput screening of human cancer cells response to differentiation therapy.

HT-29 cells are sensitive to differentiation therapy agents such as BA and its derivatives (AN-7 and AN-9) by enhancement of the alkaline phosphatase activity [25, 29]. According to the enzymatic activity level of the treated cancer cells, the efficiency of the particular drug treatment was evaluated. In general, normal enzymatic activity denotes that the cells differentiate properly as a consequence of the particular drug treatment, whilst lack of enzymatic activity denotes ineffectual drug treatment for the particular cancer tumor and for the particular patient.

12.4.1 Example to Electrochemical Sensing of Cancerous Cells Response to Drug Treatment

Alkaline phosphatase activity measurements: The activity of the enzyme alkaline phosphatase is determined by using the substrate p-aminophenylphosphate (PAPP). The product of the enzymatic reaction, PAP, is oxidized on the working electrode at 220 mV. This oxidation current is monitored.

An array of eight-channeled 100 nL electrochemical chambers was loaded with HT-29 human colon cancer cells that were treated with BA or its derivatives. The HT-29 cells were grown in DMEM medium in the presence of fetal bovine serum for 3 days prior to drugs treatment. The measurements were performed in PBS with the intact cells and without additional treatment of the cancer cells such as lysis.

The treated cells were placed into the electrochemical chambers and the substrate PAPP was added to a 1 mg/ml final concentration at a total volume of 100 nL. Alkaline phosphatase activity was measured by monitored the PAP oxidation current. The electrochemical chips are disposable and were replaced every experiment.

BA at increasing concentrations of 0, 0.078, 0.156, 0.3125, 0.625, 1.25, 2.5, 5, and 10 mM was added to the HT-29 cell cultures and incubated for 72 h for optimization. Optimal butyrate concentration, LC_{50} and viability were calculated accordingly. The optimal BA concentration was found to be 2.5 mM.

12.5 Results and Discussion

The effect of BA, AN-9, and AN-7 on the HT-29 colon cancer cells was examined after incubation for 96 h, by measuring the induced alkaline phosphatase activity. Each electrochemical chamber on the array was loaded with cells exposed to different agent type. BA, AN-7, and AN-9 were tested. The results are shown in Fig. 12.4.

Normal enzymatic activity denotes that the cells differentiate properly as a consequence of the particular drug treatment. As shown in Fig. 12.4, BA and AN-7 induced enzymatic activity of alkaline phosphatase, whilst AN-9 did not induced any enzyme activity. AN-9 showed no induction compared to AN-7 at 50 μM concentration exposure, which may be related to its reduced potency regarding

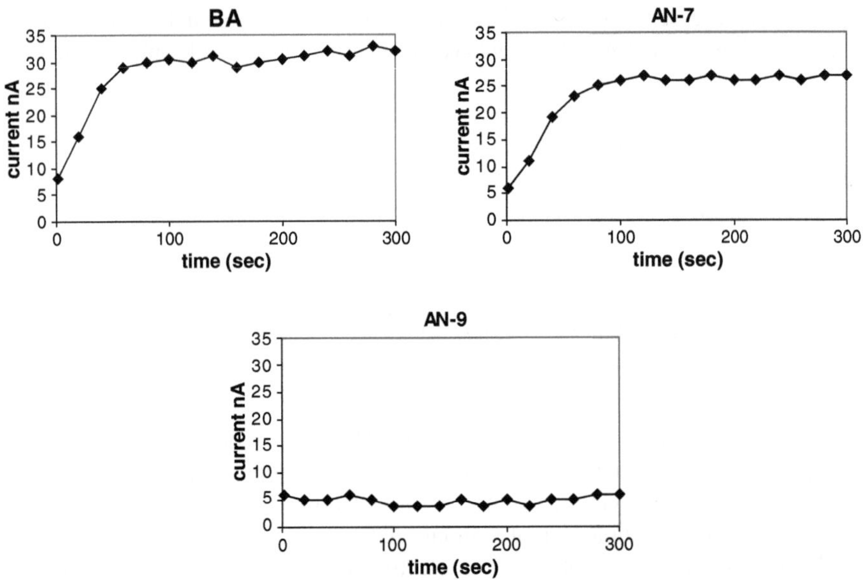

Fig. 12.4 HT-29 colon cancer cells response to BA, AN-7, and AN-9. Amperometric response curves for monitoring of alkaline phosphatase activity using the electrochemical array chip. The HT-29 colon cancer cells were exposed to the differentiation agents: Butyric acid (2.5 mM), AN-7 and AN-9 (50 μM). The HT-29 cells with the substrate PAPP were placed into the 100 nL volume electrochemical chambers on the chip. Current was measured using the amperometric technique at 220 mV

HT-29 cancer cells [30, 31]. In addition, positive and negative controls were performed (data not presented); It is important to note that AN-7 exerted similar differentiation effect on the BA although its concentration was lower by 1.5 orders of magnitude - 50 μM Vs. 2.5 mM for AN-7 and BA, respectively.

To exclude the possibility of false positive and false negative signals, all arrays included controls chambers (data not shown); for false positive control we first amperometrically measured each of the components alone: the untreated HT-29 cells, the treated HT-29 cells, and the drugs. Next we measured the combinations of untreated HT-29 cells with and without PAPP and the different drugs with PAPP. False negative control necessitates current generation in all experiments; we loaded the chamber with purified alkaline phosphatase and PAPP.

12.5.1 Quantification of Cancer Cells by the Biochip

We further examined the correlation between the current density and the number of cells that were presented in the chip chamber. The HT-29 cells were counted under the microscope and the induced current was measured as shown in Fig. 12.5.

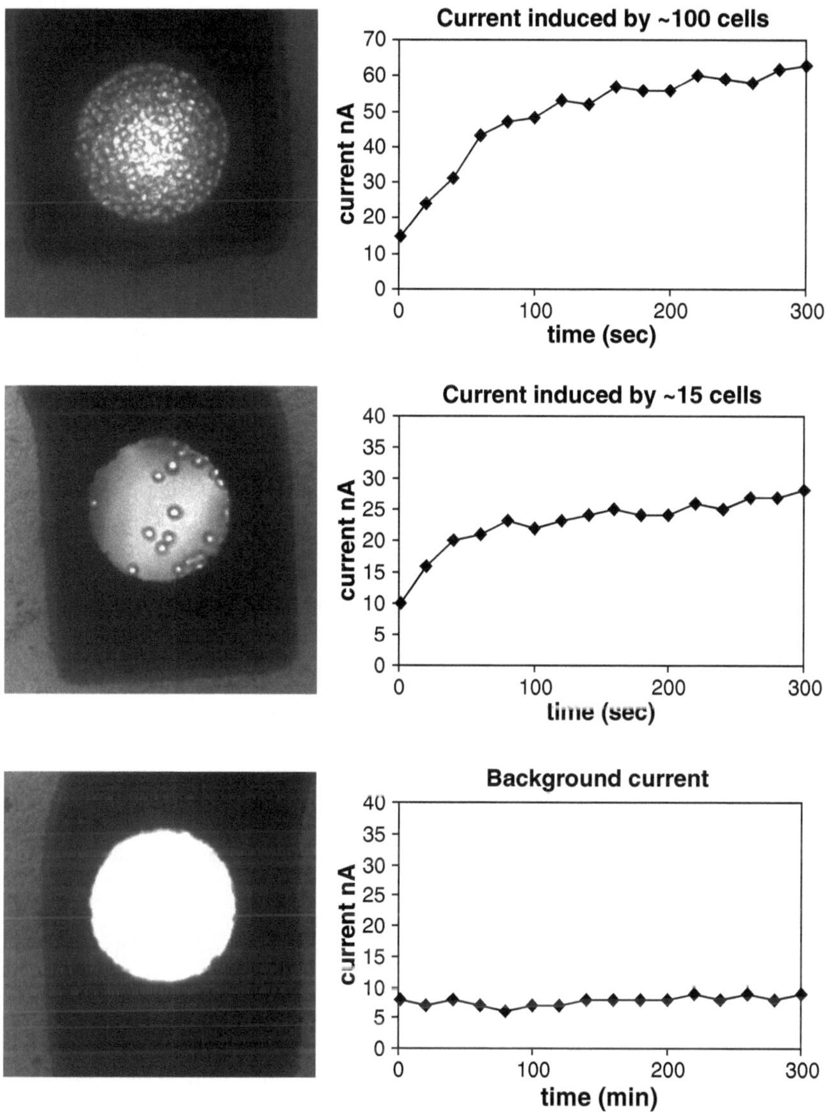

Fig. 12.5 Amperometric response curves for monitoring of alkaline phosphatase activity using the electrochemical array chip. We show the deposited cells with PAPP (*left*) and the output as a function of time on the right correspondingly

That figure shows Amperometric response curves for the monitoring of alkaline phosphatase activity using the electrochemical array chip. HT-29 colon cancer cells were exposed to Butyric acid (2.5 mM). The HT-29 cells with the substrate PAPP were placed into the 100 nL volume electrochemical chambers on the chip (left column). Current was measured using the amperometric technique at 220 mV.

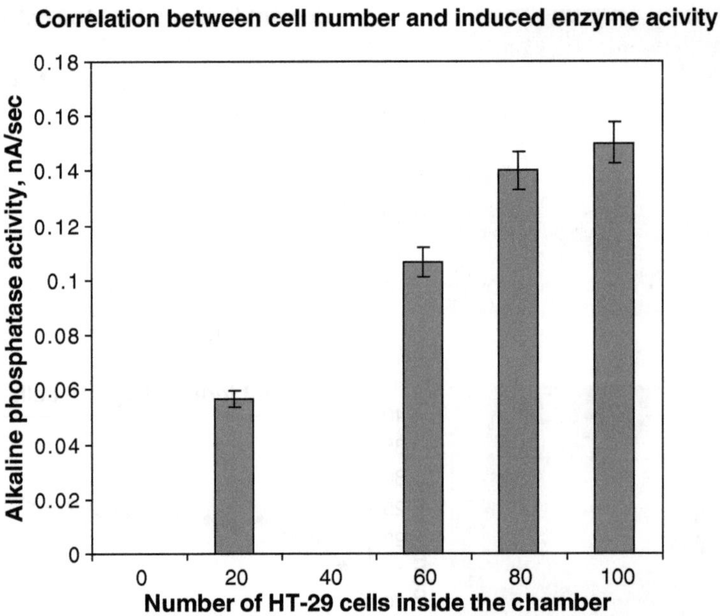

Fig. 12.6 Correlation between HT-29 colon cancer cell number and the induced alkaline phosphatase enzymatic activity. Activity is presented by Δcurrent/Δtime. Each result represents the mean of three measurements. Current was measured using the amperometric technique at 220 mV

Upper middle and lower curves represent the current response of about 100, 15, and 0 cells counted inside the chamber, respectively.

Multiple measurements demonstrated high correlation between cell number counted inside the chamber and alkaline phosphatase activity. The results are shown in Fig. 12.6.

In this study, we have demonstrated a new electrochemical method for highly sensitive, accurate, and rapid multiplex detection of colon cancer cells response to differentiation therapy. The development of an electronic biochip array system, adjusted to biological experiments can help tailor cancer treatment to individual patients. Human colon cancer cells, HT-29, were treated with the differentiation therapy drug agents BA, AN-7, and AN-9, and the cells response was simultaneously measured on line and compared. This microarray technology provides the ability to test on line the affect of multi-drug agents, and to tailor effective therapy to the individual.

We found significant quantitative correlation between the induced current signals and the number of cancer cells counted inside the nano-volume electrochemical chambers. The ability to quantitate the enzymatic reaction of few living cells is due this new geometry, in which electrochemical cell dimensions were reduced to nano-scale. The construction of an array of nano chambers on one silicon chip leads to high throughput in addition to the capability of performing multi experiment simultaneously and independently.

12.6 Summary and Conclusion

In this study, a miniaturized and portable electrochemical analytical device was fabricated, studied, and characterized. The benefit of this new geometry, in which electrochemical cell dimensions were reduced to nano-scale, and the ratio between the working electrode area and cell volume was optimized, is demonstrated by the results presented here, which showed high sensitivity and extremely fast response time. The construction of an array of nano chambers on one silicon chip leads to high throughput in addition to the capability of performing multi experiment simultaneously and independently. The total weight of the entire system is ~900 g, making it ideal for field environmental monitoring and for medical applications.

A novel integration between microelectronic devices and living organisms for electrochemical detection of toxicity in water has been demonstrated. The electrochemical "Lab on a chip" provides rapid and sensitive real-time electrochemical detection of acute toxicity in water. A clear signal is produced within less than 10 min of exposure to various concentrations of toxicants, or stress conditions, with a direct correlation between the toxicant concentration and the induced current. During measurement period the bacteria remained active and were capable of performing cellular gene expression and enzymatic activity, which demonstrate the chip biocompatibility.

The use of MEMS technology in the health care arena leads to the developments of indispensable sophisticated intelligent devices. The miniaturization of these analytical devices is critical since it will enable the analysis of large number of drugs per sample and may be used directly with small biopsy's samples or small volumes of body fluids. This in turn leads to fast response time, sensitive and cost-effective analysis.

Future extension of the present research project includes in vivo and clinical applications. Interfacing biological materials with microelectronics may make it possible for chips to be inserted as reporter elements within humans. It may be utilized as nano-laboratories, on-chip data acquisition and partial processing as well as communicating with bio-systems. These technologies will change the scope of our abilities to monitor biochemical reactions in multicultural organisms, and may lead to interesting pharmaceutical and clinical developments.

The power of the microelectronic technology, and its usefulness to extreme and precise miniaturization enable the construction of highly sensitive, fast, and robust devices, which can provide exciting opportunities in the biosensors field as well as in the basic research on microorganisms.

References

1. Belkin S (2003) Microbial whole-cell sensing systems of environmental pollutants. Curr Opin Microbiol 6(3):206–212
2. Stenger DA et al (2001) Detection of physiologically active compounds using cell-based biosensors. Trends Biotechnol 19(8):304–309
3. Rudolph AS, Reasor J (2001) Cell and tissue based technologies for environmental detection and medical diagnostics. Biosens Bioelectron 16(7–8):429–431
4. Pancrazio JJ et al (1999) Development and application of cell-based biosensors. Ann Biomed Eng 27(6):697–711
5. Popovtzer R, Neufeld T, Ron EZ, Rishpon J, Shacham-Diamand Y (2006) Electrochemical detection of biological reactions using a novel nano-bio-chip array. Sensor Actuator B-Chem 119(2):664–672
6. Daunert S et al (2000) Genetically engineered whale-cell sensing systems: coupling biological recognition with reporter genes. Chem Rev 100(7):2705–2738
7. Kohler S et al (2000) Detection of 4-chlorobenzoate using immobilized recombinant Escherichia coli reporter strains. Sensor Actuator B-Chem 70(1–3):139–144
8. Biran I et al (1999) On-line monitoring of gene expression. Microbiology-Sgm 145:2129–2133
9. Paitan Y et al (2003) On-line and in situ biosensors for monitoring environmental pollution. Biotechnol Adv 22(1–2):27–33
10. Paitan Y et al (2004) Monitoring aromatic hydrocarbons by whole cell electrochemical biosensors. Anal Biochem 335(2):175–183
11. Premkumar JR et al (2001) Antibody-based immobilization of bioluminescent bacterial sensor cells. Talanta 55(5):1029–1038
12. Cha HJ et al (1999) Green fluorescent protein as a noninvasive stress probe in resting Escherichia coli cells. Appl Environ Microbiol 65(2):409–414
13. Nivens DE et al (2004) Bioluminescent bioreporter integrated circuits: potentially small, rugged and inexpensive whole-cell biosensors for remote environmental monitoring. J Appl Microbiol 96(1):33–46
14. Sagi E et al (2003) Fluorescence and bioluminescence reporter functions in genetically modified bacterial sensor strains. Sensor Actuator B-Chem 90(1–3):2–8
15. Grigsby PW, Corn BW (1992) Localized urethral tumors in women–Indications for conservative versus exenterative therapies. J Urol 147(6):1516–1520
16. McLeod H (2007) Graying anatomy? Toward molecular tumor characterization. Pharmacogenomics 8(1):15–16
17. Guillemard V, Saragovi HU (2004) Novel approaches for targeted cancer therapy. Current Cancer Drug Targets 4(4):313–326
18. Konski A, Feigenberg S, Chow E (2005) Palliative radiation therapy. Semin Oncol 32(2):156–164
19. Cao T, Heng BC (2005) Differentiation therapy of cancer. Potential advantages over conventional therapeutic approaches targeting death of cancer/tumor cells. Med Hypotheses 65(6):1202–1203
20. Miller SJ (2004) Cellular and physiological effects of short-chain fatty acids. Mini-Reviews in Medicinal Chemistry 4(8):839–845
21. Sell S, Pierce GB (1994) Maturation arrest of stem-cell differentiation is a common pathway for the cellular-origin of teratocarcinomas and epithelial cancers. Lab Invest 70(1):6–22
22. Bell DR, Van Zant G (2004) Stem cells, aging, and cancer: inevitabilities and outcomes. Oncogene 23(43):7290–7296
23. Patnaik A et al (2002) A phase I study of pivaloyloxymethyl butyrate, a prodrug of the differentiating agent butyric acid, in patients with advanced solid malignancies. Clin Canc Res 8(7):2142–2148
24. Rephaeli A, Zhuk R, Nudelman A (2000) Prodrugs of butyric acid from bench to bedside: Synthetic design, mechanisms of action, and clinical applications. Drug Dev Res 50(3–4):379–391

25. Boren J et al (2003) The stable isotope-based dynamic metabolic profile of butyrate-induced HT29 cell differentiation. J Biol Chem 278(31):28395–28402
26. Newmark H, Lipkin M (1994) Omega-3-fatty-acids and rectal epithelial-cell proliferation. Gastroenterology 107(6):1892–1894
27. Comalada M et al (2007) The effects of short-chain fatty acids on colon epithelial proliferation and survival depend on the cellular phenotype (vol 132, pg 487, 2006). J Cancer Res Clin Oncol 133(3):211–211
28. Reid T et al (2004) Phase II trial of the histone deacetylase inhibitor pivaloyloxymethyl butyrate (Pivanex, AN-9) in advanced non-small cell lung cancer. Lung Cancer 45(3):381–386
29. Pawitan Y et al (2005) Gene expression profiling spares early breast cancer patients from adjuvant therapy: derived and validated in two population-based cohorts. Breast Cancer Res 7(6):R953–R964
30. Entin-Meer M et al (2005) Butyric acid prodrugs are histone deacetylase inhibitors that show antineoplastic activity and radiosensitizing capacity in the treatment of malignant gliomas. Mol Canc Therapeut 4(12):1952–1961
31. Engel D et al (2005) Novel prodrugs of tegafur with antiangiogenic properties and improved anticancer activity. Clin Canc Res 11(24):9052S–9052S

Part IV
Nanotechnology for MEMS/NEMS and Advanced Packaging

Chapter 13
Electrochemically Fabricated Microelectromechanical Systems/ Nanoelectromechanical Systems (MEMS/NEMS)

Carlos M. Hangarter, Thomas George, and Nosang V. Myung

13.1 Introduction

Microelectromechanical Systems (MEMS) [1] and its more recent extension Nanoelectromechanical Systems (NEMS) [2] are successful offshoots of the semiconductor revolution, which was the hallmark of the latter half of the previous century [3]. Both MEMS and NEMS have established themselves as successful fields of endeavor in their own right. In fact, it is fair to say that all nonintegrated circuit (IC) technologies are included under the MEMS/NEMS umbrella. Although the vast majority of MEMS/NEMS technologies deal with the design and fabrication of novel sensors and actuators [3], ancillary technologies such as interconnects and packaging form an important part of MEMS and NEMS [4]. In many cases, MEMS/NEMS also find applications, not as stand-alone devices but as the key enabling subcomponents of otherwise conventionally fabricated systems [2]. This chapter will provide a survey of MEMS/NEMS fabrication and device technologies with particular emphasis on electrochemistry- based fabrication techniques and novel devices/instruments technologies that can be developed using electrochemistry.

13.2 Electrochemical Fabrication Techniques in MEMS/NEMS

Electrochemical fabrication techniques offer tremendous versatility as cost-effective methods to generate MEMS/NEMS materials and structures. The batch processing nature of these methods carried out at near room temperature in ambient pressure,

C.M. Hangarter and N.V. Myung (✉)
Department of Chemical and Environmental Engineering,
University of California-Riverside, Riverside, CA, 92521, USA
e-mail: myung@engr.ucr.edu

T. George
ViaLogy Corporation, Altadena, CA, 91001, USA

T. Osaka et al. (eds.), *Electrochemical Nanotechnologies*,
Nanostructure Science and Technology, DOI 10.1007/978-1-4419-1424-8_13,

also allows for easy integration into Integrated Circuit (IC) and MEMS manufacturing technology. However, to date, the diverse capabilities of electrochemical processes have been largely under-utilized in the creation of novel MEMS/NEMS devices. Fabrication is still dominated by vacuum processes and silicon technology, not only because silicon has been so well characterized in terms of processing and as a material, being continually fueled by the semiconductor industry, but also because silicon is inherently predictable. By comparison, electrochemically fabricated materials have lagged, with more recent thrusts to correlate deposition conditions to crystalline structure and material properties. Structural properties of electroformed materials, such as film stress, adhesion, and elastic modulus, are further complicated as they depend not only on deposition parameters but also on previous and subsequent deposits [5]. This gap between synthesis and properties is also due to the complexity of electrochemical processes. While an electrolytic solution with an applied electric current as a reducing or oxidizing medium is simple in theory, electrochemical systems can become complicated with a potentially active carrier phase, current distribution effects, functional additives, and mass transfer phenomena. With the exception of electroless deposition, these systems are controlled by both mass transport and current distribution phenomena [6], which, though outside the scope of this chapter, are critical for implementing electrochemical microfabrication of MEMS and NEMS. In the MEMS/NEMS context, electrochemistry has nevertheless made significant progress, functioning as an important tool for deposition, removal, and surface modification of microstructures described herein.

13.2.1 Electrodeposition

Simply stated, electrodeposition, also known as electroplating, is the use of electric current to deposit metal ions from a liquid solution onto a surface. More specifically, aqueous, organic, and ionic solutions can be used as the electrolyte for electrodeposition with the latter two being restricted primarily by cost and safety considerations. The substrate must also be conductive; a limitation that often necessitates physical deposition as an alternate fabrication technique, especially for dielectric materials. Nevertheless, electrodeposition has proven particularly robust, as a "workhorse" electrochemical technique for MEMS fabrication. Electrodeposition is particularly cost effective for producing thick films (tens of microns to millimeters), which is not practical using vacuum deposition processes. Electrodeposited films also conform extremely well to the surface topography, filling recesses and covering complex geometries that are not covered easily with line-of-sight physically deposited coatings. In addition, the crystallographic structure and orientation can be controlled for desired material properties through adjustments to the solution composition, deposition conditions, and the type and amount of additives and complex agents [7]. Electrodeposition is also a powerful tool for alloying, being able to deposit a wide range of solid solutions and nonstoichiometric alloys that are not attainable with other methods. Recent advances in electrodeposition for MEMS have focused on deposition optimization for copper interconnects, magnetic media storage devices, and magnetically-actuated MEMS with particular interest in tuning physical properties including magnetic properties, elastic properties, and microstructure.

Since 1997 the damascene copper electroplating process has been utilized in silicon chip manufacturing for "super-conformal" plating of holes and trenches. This market has revitalized copper plating research, focusing on the influence of additives, and elucidating their roles as suppressors, levelers, and accelerators [8, 9]. In the magnetic storage industry, soft magnetic materials have gained considerable attention for their use as recording heads. These materials, also commonly used for magnetic-MEMS, require excellent magnetic properties, (high magnetic saturation, low coercivity, high permeability, and zero magnetostriction), good adhesion, low-stress, low electrical resistance, good thermal stability, and high corrosion resistance [10]. These specifications are more demanding for electroplated materials, which are vulnerable to corrosion due to anodic surface activation by organic additives [10]. To meet these requirements, investigators have turned to ternary and quaternary alloys of CoNi and CoFe for their enhanced magnetic properties with uncompromised mechanical properties [11]. Although hard magnetic materials have not been typically used in magnetic-MEMS, they have several advantages for magnetic actuation, including low power requirements, high remanence, and high coercivity, which have motivated investigators to develop new electrodeposition processes for these materials. Many of these permanent or hard magnetic materials are Co-based with the inclusion of a nonmagnetic material, typically P, to increase coercivity [10, 12].

13.2.2 Electroless Deposition

Electroless deposition is an autocatalytic reaction utilizing a chemical reducing agent (e.g. sodium hypophosphite, amineboranes, hydrazine, or sodium boron hydride) to reduce metal ions onto surfaces. This method can be used to plate insulating materials and fabrics as well as metals; the surface need only be catalytically active to initiate the self propagating process [13]. Nonconducting surfaces can be rendered catalytic by a series of chemical treatments [14]. Film composition and efficiency of these baths can be controlled by electrolyte recipes and operating temperature, typically ranging between 70 and 90°C [14]. The main benefit of electroless deposition is a uniformly thick layer over any profile due to the absence of an electric field [15]. The downside to electroless deposition is the introduction of several modes of stress, which if not mitigated, can lead to fatigue and cracking. The most common type of stress associated with electroless deposition is due to differences in thermal expansion coefficients between the deposited material and the substrate. The resulting stress can be either compressive or tensile depending on which coefficient is greater [16]. These systems are also prone to voids and significant hydrogen evolution, which can result in tensile stress and hydrogen embrittlement [17].

Currently, electroless materials for MEMS and Ultra Large Scale Integration (ULSI) include Cu, Co, Ni, Ag, and their alloys [18]. The uniform films, nanometer-size crystals, and ability to coat nonconducting materials have allowed the use of electroless deposits for interconnects, barrier layers, packaging, and potential nano-electronics [19]. In this regard, electroless copper plating has been primarily investigated for IC applications and was utilized in semiconductor manufacturing as the predecessor to the damascene process. Electroless nickel is the other extensively studied material, and is selected primarily for its corrosion resistance and mechanical properties.

13.2.3 Electrochemical Etching

Electrochemical etching is a facile technique for the topographical dissolution of solids with nanometer precision through control of current density or voltage, fluid flow, etching time, and electrolyte composition. Electrochemical etching can also be used to create porous oxides by anodic etching processes or even porous metals by de-alloying, a process by which the less thermodynamically stable or more kinetically active component is preferentially etched away from an alloy [20]. Scanning tunneling microscopy (STM) also relies on electrochemical etching processes in order to synthesize the atomically sharp apex of the tips used in STM instruments, with the ease of fabrication and excellent reproducibility of this method being the key factors leading to its rapid commercialization [21]. For traditional MEMS devices, electrochemical etching serves as an alternative or complementary technique to the popular chemo-mechanical planarization (CMP) as the means to brighten the surface and diminish surface irregularities such as bumps, dishes, and rough morphology [22]. Electrochemical planarization methods operate at the limiting current density condition, where the rate of dissolution is limited by the mass transfer rate of a reactant or product. This planarization process etches surface protrusions at an accelerated rate because of their increased proximity to the bulk [22]. Electrochemical polishing or brightening of a surface is also achieved under rate limited conditions, where dissolution is independent of structure [6]. A more site-specific and commonly used form of electrochemical etching is electrochemical micromachining for through-mask passivation or anodic dissolution of a material. This selective etching method is used to fabricate precision inkjet nozzles with specified nozzle angle and undercut as shown in Fig. 13.1 [23]. Mask-free electrochemical micromachining has been demonstrated by the use of ultra short voltage pulses, for appreciable polarization of the double layer only at small electrode separations (microns), resulting in highly localized dissolution of the counter electrode [24].

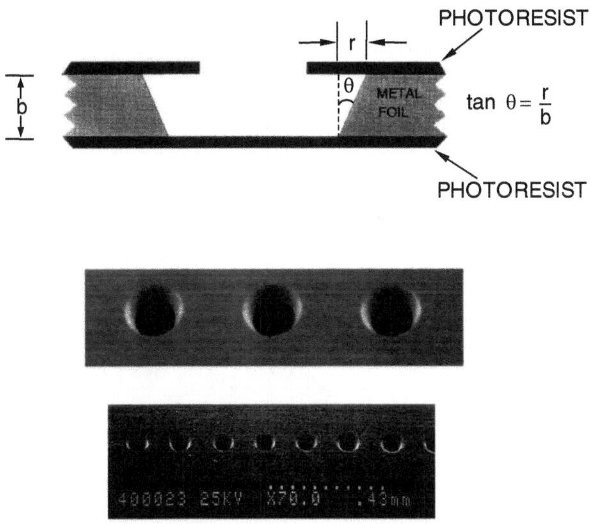

Fig. 13.1 Ink-jet nozzle plate fabricated by through-mask electrochemical micromachining, Schematic of the concept (*top*), and a photograph showing an array of nozzles (*bottom*) [23]

13.3 Top Down Fabrication Approaches

"Top Down" fabrication approaches in the context of MEMS and NEMS technology usually refers to planar, lithographic techniques that are a legacy from Si-based integrated circuit manufacturing [25]. This is where a pre-designed pattern is transferred onto the wafer or other substrate of choice using primarily photolithographic techniques. Optical lithography has made major strides in terms of technological advancement with the recent announcement by IBM's Almaden Research Center of achieving 30 nm features using 193-nm lithography (http://nanotechweb.org/articles/news/5/2/10/1). With regard to electrochemistry-based processing, we have selected the following top–down fabrication topics in MEMS and NEMS.

13.3.1 LIGA

LIGA [26] is a German acronym, short for Lithographie Galvanoformung Abformung. As shown in Fig. 13.2, LIGA is used to create high aspect ratio structures with width to height ratios as high as 1:1000. Recent work at NASA's Jet Propulsion Laboratory (B. Eyre Private Communication 2004) has resulted in the development of axially-symmetric structures with the use of rotating exposure stages. LIGA structures straddle the gap between MEMS and conventionally-machined parts and fall into the so-called "Meso" size range. As shown in Fig. 13.2, LIGA employs collimated synchrotron X-rays that require much more stringent control than optical lithography.

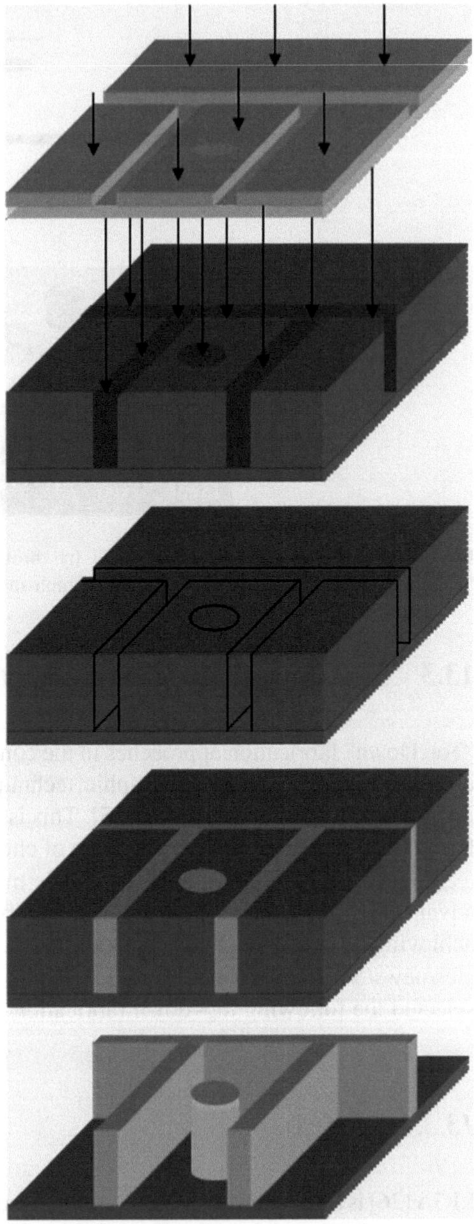

Fig. 13.2 The LIGA process consists of first exposing a thick PMMA resist collimated, synchrotron x-rays. The resist is subsequently developed and a MEMS structure is electroplated within the developed "mold"

An optically generated x-ray mask is itself one of the key features of LIGA, producing the necessary high contrast with a thick (>10 μm) gold absorber on materials membrane consist of low atomic number elements [27]. The absorber mask is patterned using polymethyl methacrylate (PMMA) resists exposed by electron-beam lithography (e-beam) or from an intermediate mask that relies on an e-beam exposure using a chromium thin film as an adhesion layer. The absorber is then electroplated using a gold plating solution, followed by removal of the PMMA. The critical properties for the electroplated gold absorber are good adhesion and low stress. Intermediate masks are subsequently exposed to x-rays to pattern thicker resists that will ultimately serve as molds for thicker master masks.

LIGA also uses PMMA as a resist to pattern MEMS devices with thicknesses as high as 8 mm. However, in practice, the dimensional tolerances achievable by very thick PMMA resists is degraded significantly by the swelling and thermal expansion of the PMMA caused by absorption of the electrolyte, thereby limiting the maximum height and minimum sizes of LIGA-fabricated features [28, 29]. Although the strain resulting from thermal expansion and absorption of water are minimal for PMMA - 0.2% and 0.4% respectively - dimensions vary significantly for high-aspect ratio LIGA molds. Vertical structures have exhibited lateral offsets as high as 14 μm per mm, with curved features being further distorted [29]. Some of the proposed solutions for reducing linear strain in PMMA are targeted at minimizing thermal expansion effects by using electroplating baths at ambient temperatures or by bonding the resist at elevated temperatures in order to compensate for the thermal expansion and partially for electrolyte absorption. Cross-linked PMMA and alternative resists have also been suggested as alternatives to remedy the absorption problem.

Following exposure, the PMMA is developed much like in conventional photolithography. The conducting back plate for the PMMA is then used as the electrode to create primarily Ni-based electroplated structures within the PMMA mold. These Ni baths have traditionally used sulfur-bearing additives, such as saccharin, as grain refiners that enhance strength and hardness of the plated material. Unfortunately, the mechanical properties of these deposits are rapidly diminished at elevated temperatures, ~600°C, as sulfur groups migrate to grain boundaries causing intergranular fracture.

To alleviate this problem many researchers have turned to Ni–Mn alloys with Mn content of ~1 wt%. While Ni–Mn deposits have been reported with high yield strengths, of which up to 85% can be retained after high temperature annealing, the addition of Mn also increases the residual stress in the deposit. This residual stress problem has been mitigated by pulse plating Ni/Ni–Mn films to create thick structures which have also exhibited increased ductility [30–33]. LIGA fabricated structures have been used for numerous applications including for scroll pump for miniature mass spectrometers (Fig. 13.3) [156].

13.3.2 EFAB™ (Electrochemical Fabrication)

One of the major challenges facing the commercialization of MEMS devices is the cost associated with fabrication of devices with complex geometries requiring tens or potentially hundreds of layers. In this regard, LIGA has fallen short of demonstrating such intricate lamellar fabrication, hindered by the cost associated with numerous layers. Vacuum processes are also unpractical for the fabrication of such multi-layer structures [34]. A new micromachining technology entitled EFAB™

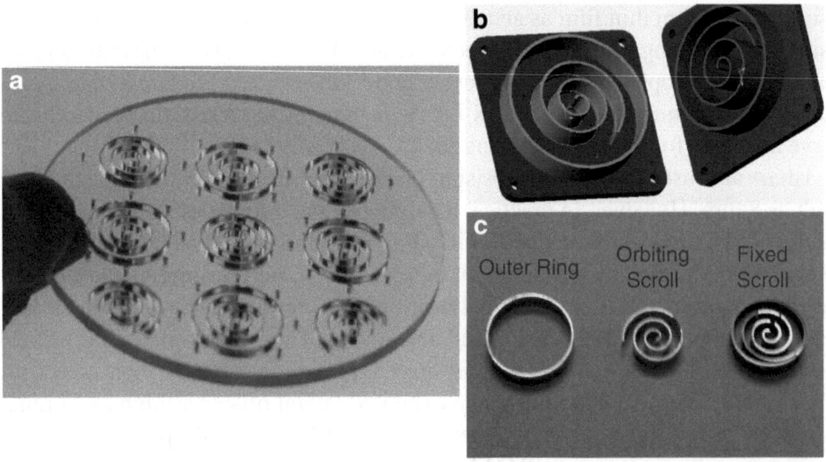

Fig. 13.3 A prototype scroll pump for a miniature GCMS system fabricated using the LIGA process. (**a**) 3 mm PMMA mold, (**b**) schematic of scroll pump, (**c**) electroformed scroll pump

was introduced in 1999, to address the issue of constructing high aspect ratio, three-dimensional, multi-layer microstructures from electroplated metals using a rapid, automated, batch fabrication technique. The core of EFAB™ technology includes an "Instant Mask" process, followed by planarization of each deposited layer. The Instant Mask process is an in-situ masking technique consisting of a pre-patterned insulator that conforms to the surface of the substrate, to allow electroplating of sacrificial material, usually copper, through channels in the mask. This step is followed by blanket-deposition of the structural material, nickel, and planarization, which reduces the need for morphological uniformity, provides precise control over vertical dimensions, and permits the deposition of countless subsequent layers [34, 35]. The device is completed by removal of the sacrificial copper layers through selective etching to release the nickel structure (Fig. 13.4). Although EFAB™ can purportedly create devices from any electrically conductive deposits, material compatibility appears to be an issue as only nickel-based devices have been demonstrated to date.

Since both the sacrificial and structural layers are metallic, EFAB™ can create overhangs, extruded shapes, and disconnected features, but lacks the ability to integrate dielectrics [34]. This is detrimental for the production of fluidic and bio-MEMS which benefit from the low-cost, biocompatible interfaces of plastic microstructures [36]. EFAB™ can currently achieve lateral minimum feature sizes of 20 microns - significantly larger than minimum feature sizes possible using other microfabrication techniques - but has demonstrated vertical feature sizes as low as 5 microns, an advantage of its layer by layer synthesis. However, the key advantage of EFAB™ is that it does not require a cleanroom with stringent environmental and particulate control requirements. Its fabrication process strengths include mild

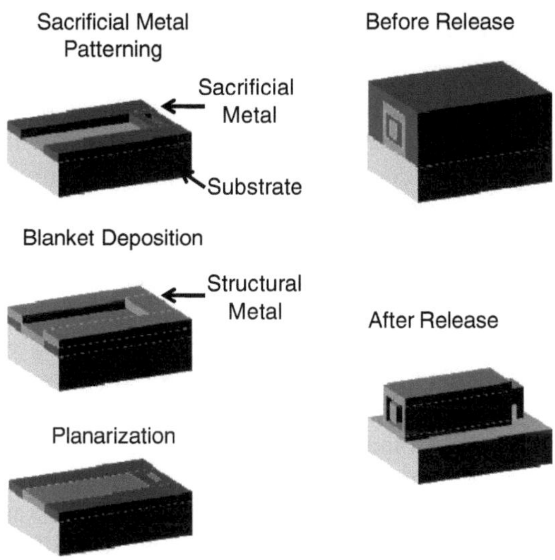

Fig. 13.4 The EFAB process flow used to fabricate a hybrid coupler (Permission from Microfabrica Inc.)

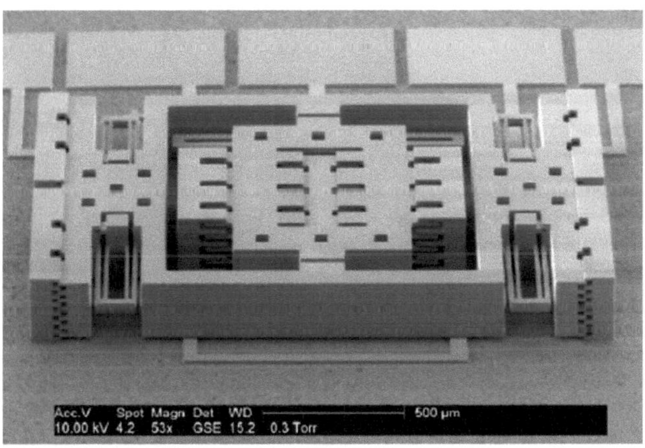

Fig. 13.5 SEM picture of the fabricated nickel gyroscope. The gyroscope occupies an area of 2.1×1.3 mm^2 [37]

electroplating conditions, 60°C ambient, and in-situ masking, all of which are amenable with the low temperature requirements for IC processing. Despite its relative "youth" as a fabrication technique, EFAB™ has already demonstrated its versatility in creating a number of otherwise challenging structures such as rectangular coaxial conductors, helical inductors, accelerometers, toroidal inductors, variable capacitors, and a microgyroscope (Fig. 13.5), [37, 38].

13.3.3 Electron Beam Lithography of NEMS Structures

Electron Beam Lithography (EBL) has gained considerable recognition as a very capable "top–down" nanofabrication technique when coupled with electrochemical processes, allowing precise control of nanostructures that are not possible using complex self-assembly techniques. In-situ fabrication techniques are possible by combining lift-off processes with EBL in order to prepare the substrate for directed nanowire growth. In fact, using directed growth techniques, electrochemically deposited nanowires can be made to laterally bridge the gap between two micro-electrodes by confining the nanowire to grow within an EBL fabricated channel in polymethyl methacrylate (PMMA) [39] (Fig. 13.6). These nanowires can be grown laterally, simultaneously from each micro-electrode, either potentiostatically or galvanostatically. Completion of the bridge is indicated by a jump in the corresponding current or potential when contact is made between the two ends of the nanowire. Synthesized nanowires are individually addressable and can span lengths of micrometers with diameters limited only by EBL techniques [40]. Various sensing materials such as palladium, polyanaline, and polypyrrole have been investigated for chemical and biological sensor applications demonstrating the potential for high density arrays of nanowires [40–42]. Precise four-point conductivity measurements of nanowires have also been conducted using built in micro electrodes [43].

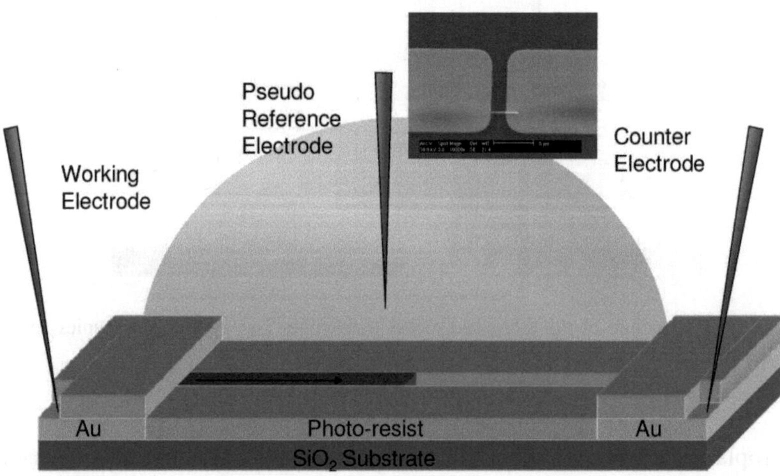

Fig. 13.6 In-situ nanowire synthesis: The working electrode and counter electrode are contacted by probe tips. A probe tip is also used as a pseudo reference electrode, submerged in the solution containing the analyte. The nanowire (*red*) is growing in the direction of the arrow, from the working electrode to the counter electrode, confined to a predetermined path by the photoresist. The inset shows an SEM micrograph of a polypyrrole nanowire fabricated by in-situ synthesis

13.4 Bottom Up Fabrication Approaches

The "bottom up" approach to NEMS and MEMS is still very much in developmental stages with few commercially available devices. With current MEMS/NEMS products relying significantly on top–down processes, the bottom up approach is being integrated into certain application niches where they provide cost reduction and performance enhancement over conventional techniques. While current research has actively pursued the investigation of nanowires and carbon nanotubes (CNTs), spatial manipulation and addressability remain crucial hurdles that must be overcome in order to realize their full potential. The construction of novel MEMS/NEMS devices from nanostructures remains the subject of many research activities, aimed at arranging these nanowires in some ordered fashion, with either an applied energy field or through surface functionalization, and thereby enabling the use of nanowires and CNTs for sensing functions. In this section we present current electrochemical techniques for bottom up nanostructure synthesis and methodologies for assembling these nanowire structures.

13.4.1 Template-Based Nanostructure Synthesis

The bottom up approach to MEMS and NEMS has been significantly impacted by electrochemistry-based template synthesis techniques. A popular approach is the use of nanoporous templates as working electrodes by sputtering or evaporating a thin conductive layer on one side of the template [44–46]. The template acts as a scaffold to confine the electrodeposited material within the nanopores thereby producing nanostructures ranging from nanodots to high aspect ratio nanowires. Following deposition, the seed layer is removed and the template is etched to release the nanowires in suspension (Fig. 13.7). The power of template synthesis approach lies in both its simplicity and highly parallel manufacturability, with template pore densities in the range of 10^9–10^{11} cm^{-1} and also in providing a platform for array applications or more complex structures, ultimately through template removal and assembly [47].

A diverse spectrum of materials is amenable to the production of electrodeposited nanowires. This includes magnetic materials, semiconductors, metals, alloys, and conducting polymers [44–46]. Complex, multicomponent nanostructures such as superlattice nanowires can also be deposited from a multielectrolyte solution by modulating the voltage or current. The requirement for these electrolytes is that they should have a substantial difference in their deposition potentials starting with the nobler elements at lower concentrations. Segmented nanowires can also be produced by switching the electrolyte baths during the deposition process [47]. The switched-bath method has been used to successfully produce metal/semiconductor, conducting polymer/semiconductor, and conducting polymer/metal heterostructures [48–50]. Surface modified templates have even been used to investigate core/

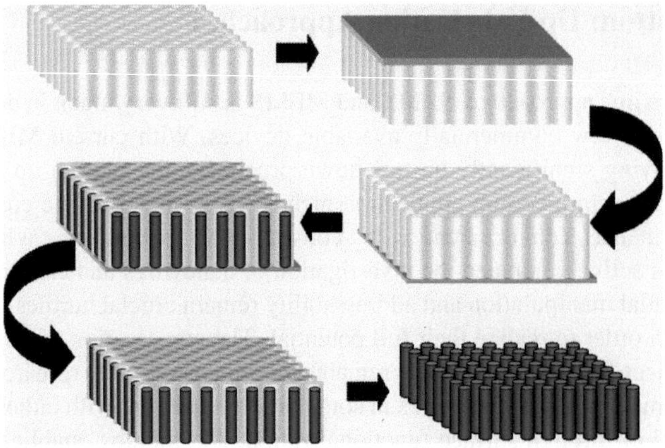

Fig. 13.7 Template Synthesis: A conductive seed layer is sputtered on the backside of the template. The nanowires are electrodeposited in the template. The seed layer is removed by physical or chemical means. The template is dissolved to yield a colloid of nanowires in solution

shell structures as well as molecular imprinting on the surface of conducting polymer nanowires [51, 52].

All these nanostructures, though simple by design, are crucial for performing complex functions. A polypyrrole nanowire, which is essentially an n-type semiconductor, could be a resistor in a single component form. However, when coupled with a Cd segment the nanowire acts as a p-n junction, and can be used as a diode-rectifier. For good ohmic contacts, Au segments can be electrochemically added to both ends of the same wire. Such a device was recently demonstrated and characterized by Mirkin's group, as shown in Fig. 13.8a and b [53]. Similarly Ni/Cu superlattice nanowires can also be produced for one-dimensional Giant Magneto Resistive (GMR) properties by adjusting layer thickness and composition. The core shell design has been used to create coaxial nanowires, consisting of an Au/CdS/Au nanowire sheathed within SiO_2, and implemented in a field effect transistor (Fig. 13.8c and d) [54]. In addition to mimicking traditional electronic components with nanowires, electrodeposited materials can also be incorporated for spatial manipulation via applied energy fields or surface functionalization. However, as these nanostructures increase in complexity, they also become more susceptible to mechanical breakage and poor interfacial adhesion. These mechanical limitations often result in these structures being demonstrated first in the submicron range, precluding quantum confinement effects. Similarly, many materials fabricated by template synthesis technique are characterized in array form, whereby the template provides mechanical support to the nanowires of interest. Although many of the materials and structures studied are precursors to further miniaturized, complex systems, investigations of these one-dimensional entities have elucidated new fundamental phenomena and stimulated creative, nanowire-based device design and fabrication. Therefore, template synthesis and post synthesis manipulation are important techniques which merit further discussion in the section below.

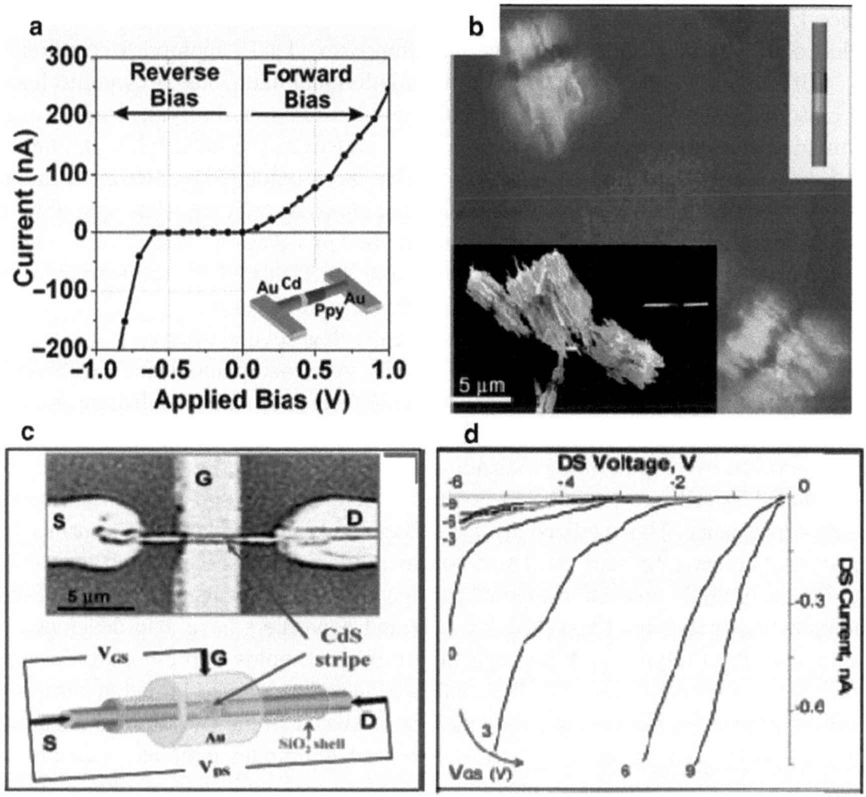

Fig. 13.8 (**a**) *I-V* characteristics for a single Au–Ppy–Cd–Au rod at room temperature. (**b**) Optical microscope image of Au–Ppy–Cd–Au rods. The lower left inset shows the corresponding field emission scanning electron microscopy (FESEM) image [53]. (**c**) Optical micrograph and schematic drawing of the test structure and Au/CdS/Au@(SiO2)10 nanowire aligned for measurement of electrical properties. Letters S, D, and G indicate source, drain, and gate electrodes, respectively. (**d**) Typical *I*DS *V*DS characteristics of in-wire TFTs for different values of gate voltage (*V*GS), 11 devices measured. Reprinted with permission from ref. [54]. Copyright (2004) American Chemical Society

13.4.1.1 Anodic Alumina Templates

Anodic alumina has been the most rigorously investigated material for template synthesis, with early work dating as far back as the 1950's [55]. Characterized by highly ordered, vertically parallel pores, it is one of the most densely packed porous structure at approximately 10^{11} pores/cm^2. Anodic alumina templates can be fabricated to produce pore sizes ranging from a few nanometer to several hundred nm in diameter [56]. The pores are formed by anodizing the surface of high purity (99.999%), and often preannealed aluminum. The pore nucleation is initiated locally by the electric field distribution, with pore growth being determined by the chemical equilibrium at the electrolyte/oxide interface and the oxide/metal interface [55]. The alumina pore sizes and interstitial spacings are controlled by the electrolyte type, concentration, voltage, and temperature. Under appropriate conditions, a "self assembly" process takes place, with individual nanopores growing to form a close-packed assembly of vertical honeycomb structures. Individual pores with this assembly are separated from each other and from the underlying aluminum substrate by an insulating barrier layer of aluminum oxide [57]. After anodization is complete the barrier layer can be penetrated either by an ion milling step or by a two step etching process [58]. The etching method involves removal of the residual, elemental Al with either $HgCl_2$ or $CuCl_2$ solution followed by the barrier layer removal or pore widening step using phosphoric acid [59, 60].

The order of the pores over large areas greater than several microns is limited by the heterogeneity in the pore initiation defects. A two-step anodization technique has been employed in order to mitigate these defects [60]. The Al substrate surface is first electropolished using a strong acid mixture such as perchloric acid and ethanol followed by long period anodization. A first, long duration anodization step is conducted in order to produce an ordered structure at the substrate interface via pore coalescence. The anodized layer is subsequently removed to yield a surface of patterned concave formations. These concave indentations serve as starting points to initiate a highly ordered hexagonal configuration of pores over large areas of the template surface (Fig. 13.9) [60]. Masuda and coworkers have also developed a similar method relying on lithographically produced molds to pre-texture the surface with an array of dimples. These indentations serve as pore initiation sites in a similar manner to the concave indentations produced in the two step anodization process [61]. The versatility of these anodized aluminum templates extends to three-dimensional structures, including cylinders and multifaceted objects [62]. Alumina templates can also be easily integrated into existing CMOS technologies, by utilizing a wet etching technique for creating complex shapes with vertical side walls [63] Lateral anodic oxidation of Al thin films has also been established on SiO_2 substrates. [64]. These pores were observed to be ordered and parallel to the substrate surface with diameters as low as 3 nm, thus paving the way for more complex configurations.

13.4.1.2 Track-Etched Templates

Membranes suitable for "track etching," such as mica and polycarbonate (Fig. 13.10a), use heavy charged particles from a nuclear radiation source to generate tracks of radiation damage within the material. These high-energy particle tracks within these membranes are subsequently chemically etched to produce uniform pores with diameters determined by etching time. Side-wall tapering for the pores is minimal because of the high etching rate selectivity along the tracks in comparison to the lateral etch rate [65]. Mica is a unique membrane material in that its pores are diamond shaped with the same size and orientation for all pores. This feature has been shown to be consistent across a wide range of mica samples, with the aspect ratio being governed primarily by the crystal structure of material [65, 66].

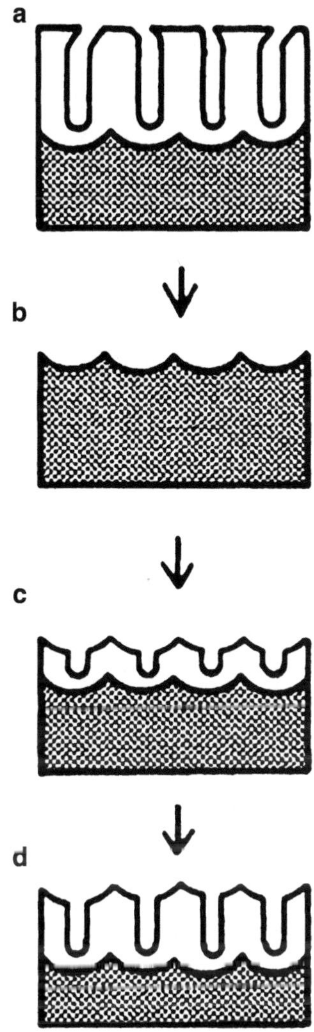

Fig. 13.9 Schematic diagram for two step anodization: (**a**) porous alumina after first anodizing, (**b**) removal of porous alumina layer, (**c**) initiation of hole formation in second anodizing, (**d**) porous alumina after second anodizing. Reprinted with permission from ref. [60]

The primary drawback to the use of these mica templates are the very low pore densities, 10^9 pores/cm, and randomness in the lateral ordering of the pores [47]. Polycarbonate templates, which also exhibit track-etching behavior, have the further complication of being susceptible to swelling during electrodeposition, consequently giving rise to large variances in the pore diameter, resulting ultimately in cigar-shaped or conical nanowires [67]. Track-etched pores generated on commercially available polycarbonate templates have been shown to have random orientations resulting in poor properties for nanowire arrays fabricated within these templates. However, it is possible to control pore quality and inter-pore separation distances

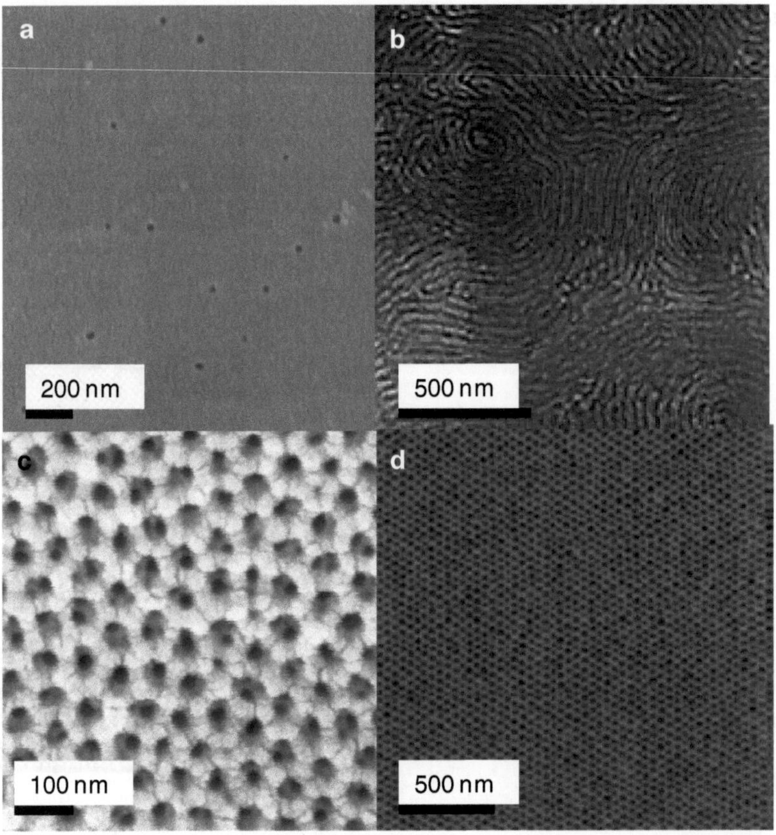

Fig. 13.10 Various materials used in template synthesis; (**a**) Polycarbonate (**b**) Silica [74] (**c**) Alumina (**d**) Diblock Copolymers. Figure 13.10b reprinted with permission from ref. [74]. Figure 13.10d reprinted with permission from ref. [155]

by optimizing the distance between the template and the particle source [65]. These templates have been fabricated with pores as small as 8 nm and commercially available templates with 15 nm pores [68].

13.4.1.3 Silica Templates

Considerable amount of research has gone into the development and characterization of surfactant templated silica, in order to produce various mesophases and nanowire geometries with well-controlled pore diameters within the quantum confinement regime (i.e. 2–20 nm). Lu and coworkers have characterized one of the more promising methods, a sol-gel dip-coating method for rapid (tens of seconds) template generation. The fabrication process involves evaporation-induced self assembly of liquid crystal domains dominated by inward growth from the solid-liquid and liquid-vapor interfaces [69–71]. The resulting mesophase is manipulated by controlling the surfactant type and concentration whereby a progression of structures from hexagonal through cubic to lamellar is observed with surfactant enrichment. These mesophases are capable of yielding either one-dimensional pores or three-dimensional interconnected pore structures (Fig. 13.10b). The surfactant/silicate thin films are subjected to calcination at 400°C to remove the surfactants producing the nanoporous silica [72]. By incorporating a conductive substrate for the sol-gel dip-coating process, nanowire arrays or a nanomesh can be electrodeposited within the one-dimensional and three-dimensional pore structures [73, 74]. Lu's group has also extended this fabrication method to hexagonal, cubic, and vesicalar structured spherical nanoparticles by aerosol-assisted self assembly [70].

13.4.1.4 Diblock Copolymer Templates

Immiscible polymers have attracted significant interest as the materials for self assembled templates for nanowire production (Fig. 13.10d). By covalently bonding together polymers with different interfacial energies, phase separations occur at the nanoscale level with highly oriented and periodic domains [75, 76]. Thin films of lamellar, cylindrical, and spherical microdomains have all been formed with tunable dimensions controlled by the polymer chain lengths or molecular weights [75, 77]. Common copolymers include polystyrene (PS) and polymethylmethacrylate (PMMA) denoted PS-*b*-PMMA, and polystyrene and polyethyleneoxide (PEO) denoted PS-*b*-PEO. Current PS-*b*-PMMA copolymer thin films with cylindrical domains are synthesized by spin coating followed by annealing for 12–24 h under an electric field of 30–40 V/μm at a temperature above the glass transition temperature of both polymers. The applied voltage induces alignment of the cylindrical domains with the orientation of the electric field [78]. Xu and co-workers have shown that small concentrations of Li ions contribute to the electric field alignment of the PMMA cylinders [79]. The ensuing matrix is subsequently exposed to ultraviolet light, which degrades the PMMA while simultaneously cross-linking the PS [80]. These self-assembled polymer templates are a promising platform for the creation of future MEMS and NEMS structures, by facilitating both lateral and vertical growth of nanowires with diameters down to 5 nm [81] (Table 13.1).

13.4.1.5 Alternative Nanostructure Synthesis

Biotemplates have been fabricated from two-dimensional crystalline surface proteins for the subsequent electrochemical deposition of nanowire structures [82]. These surface layer proteins encapsulate certain bacterial cells, controlling extracellular transport. They form especially robust thin films over cells, and are resistant to conditions that normally denature proteins, such as low pH and heat, thus making them ideal as template materials for electrodeposition. Schwartz and coworkers investigated a hexagonally packed intermediate surface layer protein from

Table 13.1 Comparison of different templates parameters for synthesis of nanowires and references for various materials fabricated with each template

	Alumina	Polycarbonate	Mica	Silica	Copolymers	Biotemplate
Pore diameter (nm)	4–500	15–1,000	8–1,000	2–40	5–50	2–3
Pore density	10^{11}	10^{9}	10^{9}	10^{11}	10^{11}	10^{11}
Pore packing	Hexagonal	Random	Random	Cubic, Hexagonal	Hexagonal	Hexagonal
Pore tilting	No	Yes	Yes	No	No	No
Thickness (μm)	10–100	6–20	10–70	2.5	1–5	0.005–0.006
Macroscopic range	Centimeters	Centimeters	Centimeters	Microns	Microns	Microns
IC Compatability	Yes	No	No	Yes	Yes	Yes
Lateral Growth	Yes	No	No	No	Yes	No
Swelling	No	Yes	No	No	No	No
Geometry	Cylindrical	Cylindrical	Diamond cross section	Cylindrical, Mesh	Cylindrical, spherical	Cylindrical, ring
Solubility	Strong Acids or Bases	Chloroform, Dichloromethane	Strong bases, HF	HF	Organic Solvents, Toluene	
Metals	Martin 1995 [45]	Staufer 1997 [67]	Possin 1969 [157]	Lu 2003 [74]	Russell 2000 [80]	Schwartz 2005 [82]
Magnetic materials	Martin 1995 [45]	Staufer 1997 [67]	Chien et al. 1999 [66]	Luo et al. 2005 [158]	Russell 2000 [80]	Schwartz 2005 [82]
Semiconductors	Klein et al. 1993 [159]	Sima et al. 2004 [160]	Mukherjee and Chakravorty 2004 [161]	Wang et al. 2006 [162]		Schwartz 2005 [82]
Conducting polymer	Martin 1995 [45]	Staufer 1997 [67]				
Metal oxides	Li et al. 2000 [163]					Schwartz 2005 [82]

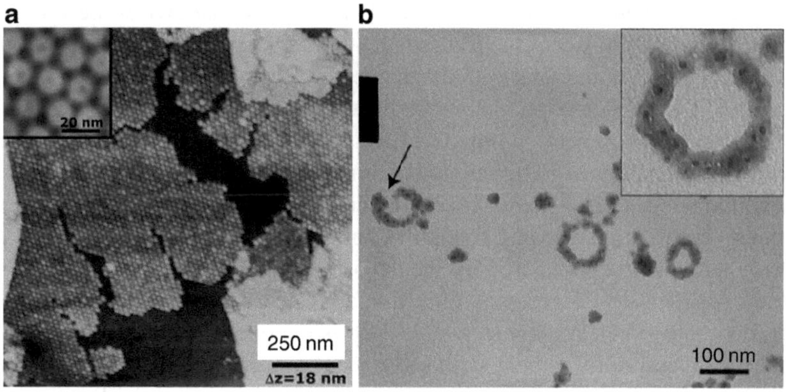

Fig. 13.11 (**a**) Tapping mode AFM image of HPI layer proteins on a polished steel disk reveals typical coverage when proteins are contacted for several minutes before washing. Inset: higher resolution image on freshly cleaved mica revealed patterned arrangement of channels. (**b**) TEM images of Cu2-TraIi1753:CN225 nanoparticles self organized on circular X174 RFII DNA. Nanoparticle loops are never observed in the absence of circular DNA additions. The arrow identifies an incompletely decorated loop. Figure 13.11a reprinted with permission from ref. [82]. Copyright (2005) American Chemical Society. Figure 13.11b reprinted with permission from ref. [83]. Copyright (2005) American Chemical Society

Deinococcu radiodurans, with 2–3 nm pores and widths of 5–6 nm (Fig. 13.11). The surface layer proteins were bound to the cells by hydrophobic interactions and were extracted using a detergent, sodium dodecyl sulfate (SDS). The resulting templates have been randomly dispersed on conductive substrates for through-mask deposition of Cu_2O, Ni, Pt, Pd, and Co [82].

Schwartz and co-workers have also investigated hybrid inorganic-protein moieties for the assembly of unique three-dimensional nanostructures [83–85]. The DNA binding protein TraI was engineered with a polypeptide binding sequence for Cu_2O, and optimized using cell adhesion assays on a quartz crystal microbalance. The binding sequence was also designed with an absence of cysteine groups to eliminate the possibility of thiol-mediated metal center binding, resulting in loop formation by oxidation of the flanking cysteine thiols. The binding affinity (K_d) of the engineered TraI for Cu_2O was determined to be 1.2×10^{-8} M. The DNA binding sequence was also preserved, as demonstrated by the self assembly of protein bound Cu_2O particles binding to cyclic DNA [83]. This technique shows promise for the assembly of multi-dimensional nanostructures based on DNA-positive templating.

Selective electrochemical deposition (SED) at the edge planes of highly oriented pyrolytic graphite (HOPG), or electrochemical step edge decoration (ESED), has been utilized to fabricate nanowires from metals, metal oxides, semiconductors, and bimetallics by Penner and co-workers [86–90]. Although this method does not fall under the class of template synthesis, still, the step edges of HOPG could be considered as a positive template. Nucleation is preferentially initiated on the step edges, where the sp^2 bonds of the basal plane are terminated and a line of higher surface energy/chemical reactivity occurs [91, 92]. The nanowires are formed by the coalescence of adjacent nucleation sites on the step edge, with the minimum nanowire diameter coinciding with the diameter of the nuclei at the point of coalescence [93]. These step-edge nanowires have hemi-cylindrical cross sections and can reach lengths of several millimeters [94]. Metal oxide nanowires can be directly deposited on the step edge by employing small overpotentials. The selectivity of preferential nucleation at step-edge sites is sensitive to the overpotential value, with greater overpotentials leading to significant deposition on the terraces as well. A "tri-potential" process is used for the deposition of noble metals such as Ni, Cu, Ag, and Au. The step edge defects are first activated with a positive potential for several seconds; nucleation at step-edge sites is then initiated with a high overpotential for several milliseconds, with subsequent reduction in the overpotential for the coalescence of adjacent nuclei to grow the nanowires (Fig. 13.12) [88]. Arrays of metal nanowires can be extracted from the HOPG substrate using a polystyrene or cyanocrylate adhesive. Metal oxide nanowires however, must first be reduced to metal (annealing under hydrogen) to have the necessary mechanical integrity required to survive the transfer from the HOPG substrate [95].

Carbon nanotubes are responsible for spawning the age of nanotechnology, and creating a paradigm shift in our views on materials and their applications. In the field of MEMS and NEMS, CNTs have found applications as electromechanical switches or as pliable nano-actuators, among others. The high elastic modulus of

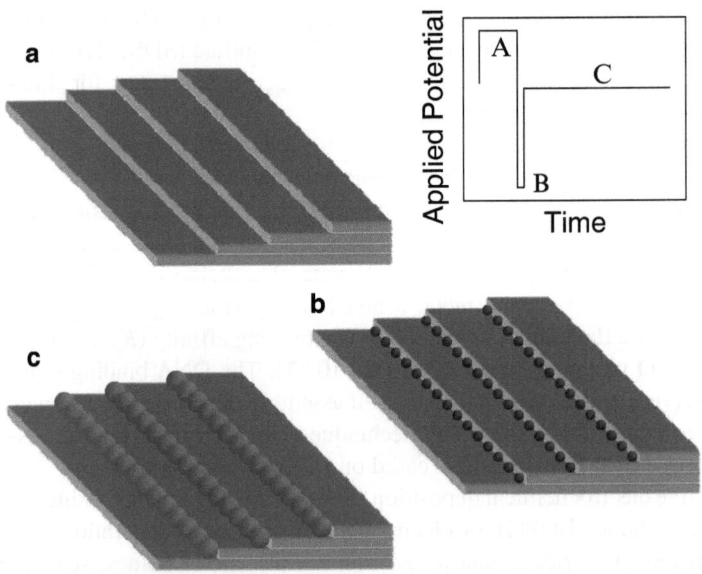

Fig. 13.12 ESED of noble metals is a tri-potential sequence (inset). (**a**) First the HOPG step edges are oxidized to form carboxyl groups, disrupting edge sp^2 bond, (**b**) deposition is initiated with nucleation sites at the edge plane from a low potential pulse, (**c**) followed by a low overpotential step for subsequent growth of the nucleation sites to the coalescing point

CNT has already been exploited in NEMS as the basis for a rotary element for a magnetically actuated nano-plate [96] and additional NEMS devices relying on both the novel mechanical and electrical properties of CNTs will continue to be developed in the years to come. Composite materials containing CNTs, especially inorganic material/CNT blends, have been demonstrated to have enhanced mechanical and tribological properties, unattainable with current metallurgical techniques [97]. These electroless and electrodeposited composites will provide a new class of materials available for MEMS and NEMS devices.

In addition to codeposition, metal/CNT composites can also be formed by electrochemically coating CNTs. Recent efforts to characterize various electrochemical coatings on CNTs have discovered unique charge transport behaviors in hybrid inorganic/CNT structures, which can be exploited for sensing applications [98]. However, to date, only a handful of materials have been electrochemically coated on CNTs (e.g. Ni, [97] Au, Ag, and Pt [99]). These studies have elucidated the mechanism of nucleation for electrodeposited materials on networks of single wall carbon nanotubes (SWNT) to create nanoparticle and nanowire coatings [99]. Both the type of electrodeposited materials and the potential gradient along SWNT networks were shown to influence the number of nucleation sites created and the subsequent coalescence and growth of the deposit. In all cases, electrodeposits on CNT networks were shown to form nanowire coatings with higher density nucleation and hence smaller nuclei closer to the working electrode, moving to larger particles, on the order of microns, that became sparser with increased distance from the electrode [99]. In addition, the ends of the nanotubes as well as defects within the nanotubes have been shown to be areas of high electrochemical and chemical reactivity, in a similar fashion as the edge planes in ESED, with enhanced electrochemical deposition occurring at these sites [91, 92]. These mechanisms of nucleation and growth are the subject of in-depth studies of CNT surface chemistry affecting electrodeposition. Thus, electrochemistry has played a key role as an important fabrication technique for CNT-based coatings, composites, and also as a means for providing conductive contacts for electrically addressing CNTs.

13.4.2 Template-Free Nanowires

Although a large majority of electrochemically fabricated nanowires use templates, there has also been considerable work reported on the generation of freestanding nanowire networks by either chemical or electrochemical means [100–103]. While chemical methods rely on polymerization at an aqueous/organic interface [101] or nonselective coverage of a substrate, electrochemical deposition involves site-directed growth between electrodes, and has been used successfully for creating conducting polymer nanowire electrode junctions (CPNEJs) [104]. A continuous, three-step galvanostatic method is utilized to fabricate the electrodeposited nanowire junctions. The first step requires the use of a high current density - 0.08 mA/cm^2 - for 30 min for the deposition of polyaniline, in order to initiate nucleation of polymer nanoparticles on the working electrode. During the second and third steps, nanowire growth propagates at lower current densities, 0.04 mA/cm^2 for 3 h and 0.02 mA/cm^2 for 3 h, respectively. The resulting nano-mesh can span a 2 μm gap with ligament diameters in the range of 40–80 nm, which correspond to the diameters of the nuclei produced during the first step. Polyaniline, polypyrrole, and polyethylenedioxythiphene (PEDOT) have all been used to make CPNEJs with high reproducibility and scalability for applications involving the detection of gaseous analytes [104].

13.4.3 Novel Assembly Techniques

13.4.3.1 Langmuir Blodgett Techniques

To date, a proven and robust technique for assembling large arrays of nanowires is the Langmuir-Blodgett (LB) technique. These nanowire-based thin films can be fabricated with controlled pitches and are also amenable to layering for the creation of more complex structures [105–108]. The LB process begins by surface functionalization of the nanowires with long alkyl chains. Once these functionalized nanowires are introduced into the sub-phase, or aqueous component, they selectively partition to the water-air interface because of their hydrophobically rendered surfaces. At low pressure, the nanowires are aligned isotropically on the surface as side-by-side aggregates called "nano-rafts" formed as a result of directional capillary forces and van der Waals forces [109]. With increased surface pressure created by compressing the trough container, the nanowires begin to align parallel to the trough wall, creating a two-dimensional nematic phase. Further compression pushes the monolayer into a smectic arrangement with uniaxial alignment and high precision spacing (Fig. 13.13). Above a certain critical pressure, the maximum areal density of the monolayer is surpassed, transitioning the nanowire film into multilayers or a three-dimensional nematic phase containing singularities [109]. The close-packed LB monolayers in the smectic regime are preferred for device

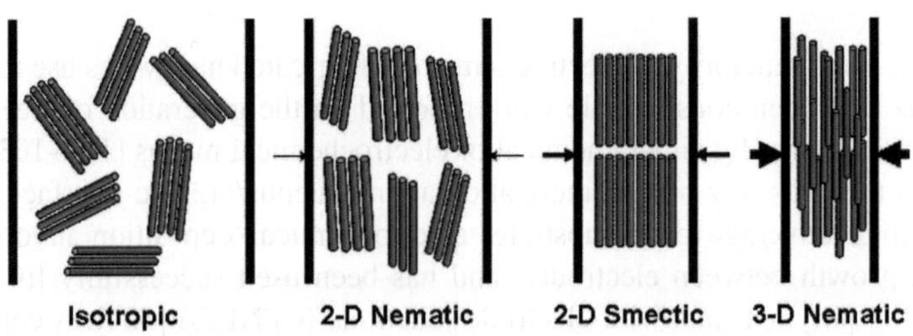

Fig. 13.13 Schematic of Langmuir–Blodgett monolayers. As the trough wall are pushed in, indicated by the arrows, the surface pressure is increased and the nanowires transition through the sequence of phases; isotropic, 2-D nematic, 2-D smectic, 3-D nematic

fabrication, as they have well-defined, uniform pitches with axial length scales of up to a centimeter [105–108].

13.4.3.2 Magnetic Alignment

The high aspect ratios of magnetic nanowires impose a shape anisotropy that only allows magnetization along the axial direction with only two possible orientations for the polarity [65]. The resulting magnetically polarized nanowire can be manipulated with external magnetic fields, ferromagnetic substrates, and neighboring magnetic nanowires. Magnetic interactions were first investigated as an approach to manipulate nanowires by Searson and co-workers [110]. Using ferromagnetic nanowires they demonstrated real time end-to-end alignment in solution, with the velocity of alignment being directly proportional to attractive forces between nanowires and inversely proportional to viscous drag in solution [110]. Searson later demonstrated the entrapment of a multi-segmented Pt/Ni/Pt nanowire between two Ni pads with a 135 Oe remnant magnetization from previous exposure to a 10G magnetic field. Although the remnant magnetization was sufficient to entrap nanowires, the success rate was greatly enhanced with an applied magnetic field, as the applied field facilitated alignment of the nanowires between the electrodes and reduced aggregation [111]. Crone's group successfully demonstrated the entrapment of an inversed nanowire structure to Searson's magnetically entrapped nanowire, consisting of a bronze midsection with ferromagnetic caps and aligned between Ni stripes [112]. Subsequently, Myung and coworkers have investigated the effect of ferromagnetic pad geometry and magnetic field strength for controlling the directionality of nanowire assembly (Fig. 13.14). Nanowire chaining and agglomeration was minimized by dilution (Fig. 13.15). A solder thin film was electrodeposited on Ni electrodes for the purpose of making robust electrical contacts to the Ni/Bi/Ni nanowires [113]. Myung and coworkers were the first to demonstrate magnetic assembly of true nanowires, with diameters of 30 nm, on gold and nickel electrodes, using an annealing step to achieve ohmic contact for magnetic and electrical measurements [114]. Magnetic alignment has also been demonstrated for structurally modified CNTs, where the modified CNT has a magnetic catalyst nanoparticle on one terminal end and a thermally evaporated Ni cap on the opposite end [115].

13.4.3.3 Electric Field Alignment

Electric field assisted assembly is applicable to a wide range of materials including semiconductors, metals, and CNT's, as opposed to being restricted to only ferromagnetic materials for magnetic alignment [116–118]. Electric field assisted assembly utilizes an alternating current to polarize nanostructures for assembly purposes or for other dynamic functions. Mallouk and coworkers were the first to demonstrate electrical alignment of Au nanowires, citing the voltage and the frequency as the main parameters controlling the alignment time.

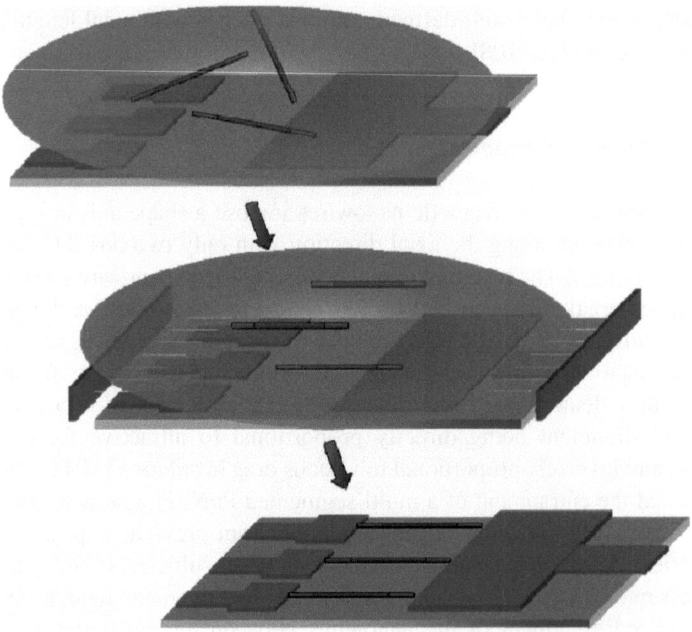

Fig. 13.14 Schematic for magnetic alignment. (**a**) A colloid initially contains randomly oriented nanowires. (**b**) The nanowires align parallel to an applied magnetic field (**c**) and position between ferromagnetic substrates

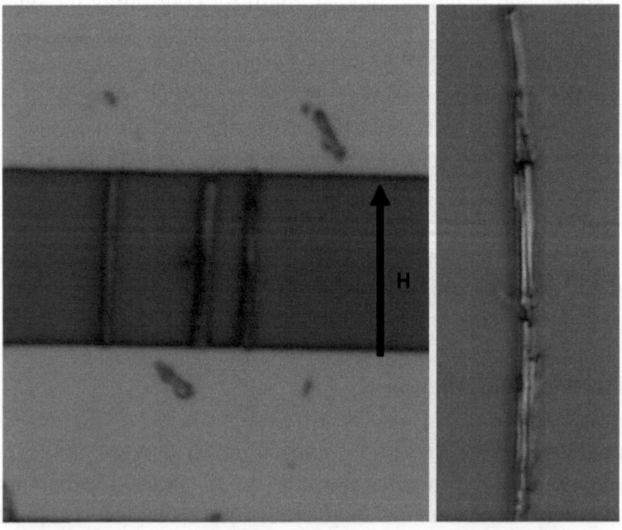

Fig. 13.15 Optical images of (**a**) Ni/Au/Ni nanowires aligned parallel to an applied magnetic field, inplane and adjacent to the nickel electrodes. (**b**) Ni/Au/Ni nanowires form chains in the absence of ferromagnetic electrodes to direct placement. The arrow indicates the direction of the applied magnetic field

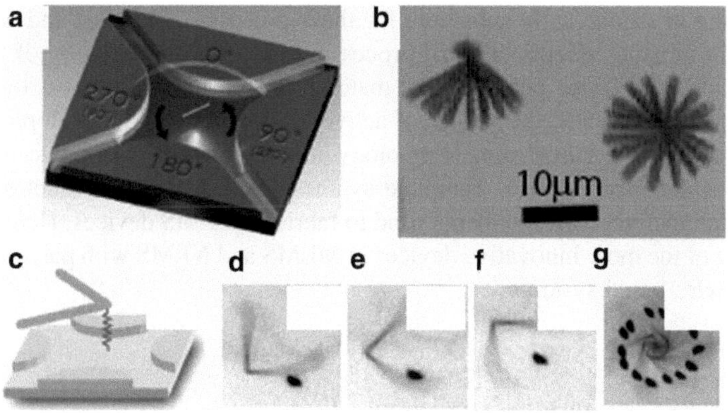

Fig. 13.16 (**a**) Schematic of a nanowire suspended in DI water set to rotation by quadruple electrodes, at which four phase-shifted ac voltages are simultaneously applied but with a sequential phase shift of 90°. (**b**) Overlapped images at 1/30 s interval of free (*right*) and on end fixed (*left*) rotating Au nanowire at 2.5 V, 80 kHz. (**c**) Schematic of a bent nanowire attached to the surface. Snap shots (**d**)–(**f**) of rotating bent nanowires taken every 1/30 sec and (**e**) overlapped images taken within 1.8 s under 10 V at 20 kHz illustrating a bent nanowire as a micromotor driving a dust particle. Reprinted with permission from ref. [121]. Copyright (2005) by the American Physical Society

The movement of nanowires within an electric field is dependent on its polarization within the surrounding dielectric medium. The electric field induces a dielectrophoretic force on the nanowire in the direction of the electric field [117]. Studies on dielectrophoretically controlled nickel silicide nanowires also revealed mechanisms for "chaining" and branching interactions among nanowires. The induced dipoles orient the nanowires parallel to the electric field while simultaneously creating localized coulombic attractive forces between adjacent nanowires resulting in their forming chains. Pre-aligned nanowires positioned on top of electrodes experience an enhanced electric field, consequently attracting other nanowires in solution along the radial direction, thereby stimulating branching structures [119]. Controlled rotation of nanowires has also been carried out by Chien's and Bhiladvala's groups using a quadruple electrode setup with simultaneously applied voltages with sequential phase shifts of 90°, thereby creating a "nanomotor". These devices have been demonstrated for free standing nanowires as well as for fixed nanowires (Fig. 13.16a and b) [120, 121]. A bent Au nanowire motor tethered to a thiolated substrate is also shown in Fig. 13.16.

13.5 Applications

From a practical standpoint, electrochemical fabrication is currently used primarily in LIGA or EFAB, the predominant electrochemical techniques for fabricating MEMS devices. Electroplating can be used to mimic certain forms of silicon processing or can actually substitute for these processes. The cost and associated trade-offs in using electrochemical processing, are a strong function of the exact process, desired device performance, material properties, and process limitations. Electrochemistry is also the subject of active research aiming at developing greater understanding of electroplating technology and the relationships between material properties and structure [6]. Template synthesis is the dominant electrochemical fabrication method currently being used to fabricate NEMS devices. Here we present some of the more innovative devices in MEMS and NEMS with particular focus on nanowire-based systems.

13.5.1 MEMS Example: Force-Detected Nuclear Magnetic Resonance Spectrometer

NMR spectroscopy can be conducted using two, very different approaches: the conventional, Faraday-law detection technique and the force-detection technique described below. In both cases, the NMR signal is derived from RF excitation of the sample's nuclear magnetic moment. In the conventional technique, detection occurs by virtue of the induced current in a detector coil generated by the cyclic inversion of the nuclear magnetic moment. FDNMR on the other hand relies on measuring the dipole-dipole force interaction between the sample's nuclear spin magnetic moment and a small detector magnet (of equivalent size to the sample) located in the vicinity of the sample (Fig. 13.17). The FDNMR spectrometer is constructed using MEMS fabrication techniques [122]. The detector magnet is mounted on a microfabricated Si beam making up a mechanical resonator. The detector magnet sits within an annular magnet (Fig. 13.18), thus providing a uniform magnetic field over the entire sample volume. RF pulses applied to the sample cyclically invert the nuclear spins of the target isotopes, thereby modulating the dipole-dipole interaction between the detector magnet and the net nuclear magnetic moment of the sample, at the former's mechanical resonance frequency. The resulting motion of the mechanical resonator is detected using a fiber-optic interferometer. The electronics driving the RF coil are capable of producing the desired complex pulse sequences, allowing both single and double resonance NMR experiments. The displacement of the resonator driven by cyclic inversion of the sample's nuclear magnetic moment is recorded using fiber-optic interferometry. A "double" Fourier Transform process is subsequently employed to derive the NMR spectrum from the acquired data. A significant strength of the FDNMR technique over the conventional technique is that it is capable of high-resolution, multi-nuclear analysis [123]. It has been shown previously [122] that the force detection technique is uniquely suited for MEMS sizes and has superior sensitivity over the conventional, inductive detection technique for sample sizes in the range of 10–100 microns. Thus, the MEMS FDNMR spectrometer is the only choice for in-line detection of aqueous samples, as well as dissolved organic samples, in a miniaturized, multi-instrument suite type application.

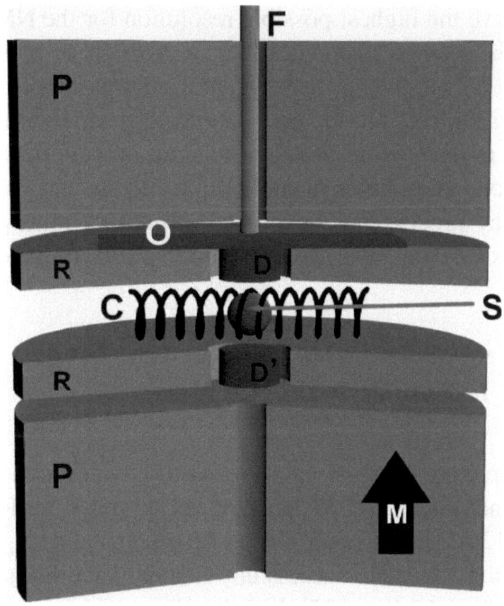

Fig. 13.17 Working principle of the MEMS-based FDNMR spectrometer. The sample (S) sits within an RF excitation coil (C), which modulates the NMR signal at the mechanical resonance frequency of a resonator made up of the detector magnet (D) and silicon beam (O). The detector magnet sits within an annular magnet (R) and the entire assembly is within the pole piece of an external field magnet (M). Detector magnet motion is measured using a fiber optic interferometer (F)

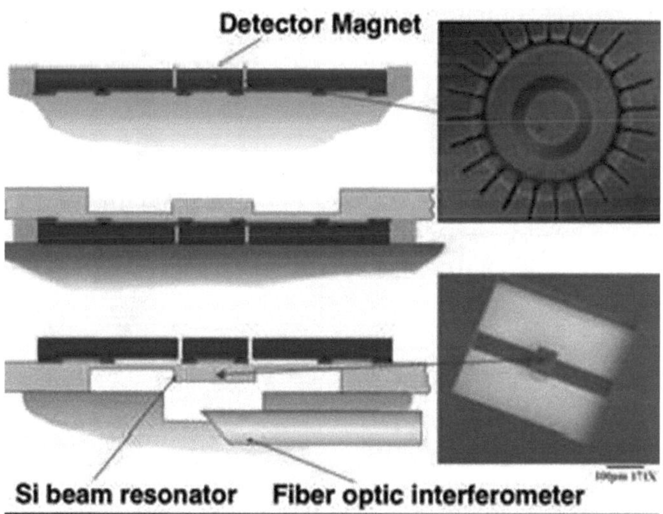

Fig. 13.18 Simplified processs schematic showing the major steps for fabrication and assembly of the MEMS FDNMR spectrometer. The steps involve electropating of a thin-film, multi-component, soft magnet alloy, fabrication of the Si beam resonator, and the final assembly of the spectrometer. Shown to the right are an optical micrograph of the electroplated 60 μm detector magnet, and an SEM image of the 400-μm-long Si beam resonator fabricated using Deep Reactive Ion Etching

In order to achieve the highest possible resolution for the NMR spectroscopy, it is important that the applied magnetic field be as high as possible. Our research determined that it is possible to electroplate a ternary, soft magnetic alloy of Fe–Ni–Co that is capable of producing a saturated, magnetic field of over 2 Tesla. The challenge lies in developing an electroplating process for a 10-μm-thick film with low stress, good morphology, and compositional uniformity. In addition, a precision lithography process was developed to produce thin-walled (1 μm), high aspect ratio (10:1) sacrificial layer "molds" for electroplating the soft magnet film.

13.5.2 NEMS Example: Nanogap Devices

Nanogaps have attracted attention as a means to contact individual molecules or nanocrystals for generating novel NEMS devices as well as for fundamental characterization studies [124]. Highly-controlled electrochemical nanogap fabrication with sub-angstrom precision has been successfully demonstrated using a gap-impedance-tuning mechanism (Fig. 13.19) [125]. Metal is deposited simultaneously on microfabricated electrodes separated by a several micron-sized gap. The deposition is conducted using a very low frequency feedback signal to monitor the electron tunneling current across the gap as it narrows. The current exhibits a stepwise increase when the gap is below 10 nm and subsequently an exponential increase in

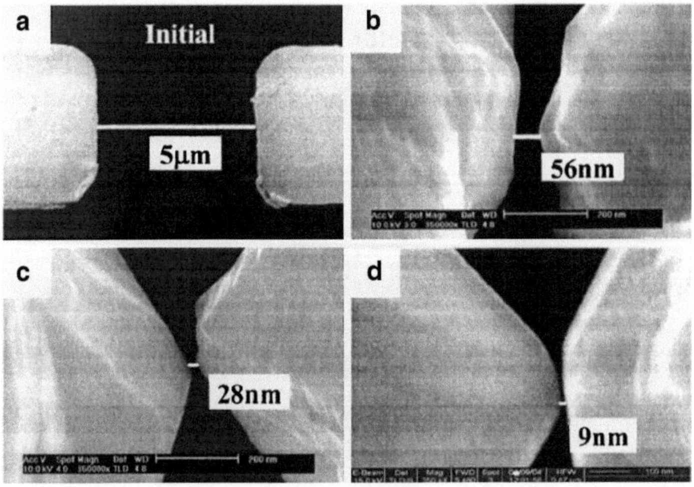

Fig. 13.19 SEM images of the gap electrodes: (**a**) the initial electrode pairs with the spacing of 5 mm fabricated by conventional photolithography; (**b**) the nanogap with a separation of 56 nm obtained at an ac sources frequency $f=260$ Hz and the series resistances $R_1=R_2=0.1$ kV; (**c**) the nanogap with a separation of 28 nm obtained at $f=260$ Hz and $R_1=R_2=1$ kV; (**d**) the nanogap with a separation of 9 nm at $f=820$ Hz and $R_1=R_2=1$ kV. Reprinted with permission from ref. [125]. Copyright 2005, American Institute of Physics

the tunneling current regime, corresponding to a gap size below 1 nm. The high level of precise control achieved by this fabrication technique allows for molecular manipulation of the nanogap [126, 127]. A key drawback of this technique is that although it is very efficient at producing angstrom size gaps, the precision in gap control falls sharply outside of the ~1 nm width required for the onset of electron tunneling. An alternative method uses one of the electrodes as a reference to detect the sharp potential drop when the electrical double layer is formed. This feedback enhancement has extended the generation of precisely controlled gaps to widths in the range of 10 nm [128]. Using a high frequency impedance feedback system similar to Tao et al, Liu et al have also demonstrated nanometer-level precision in controlling gaps in the range of 1–30 nm [129].

13.5.3 NEMS Example: Nanowire Barcodes

Keating and co-workers first investigated the use of segmented nanowires (nanowire "barcodes") for the detection of proteins and DNA using a sandwich immunoassay detection system [130]. The barcode nanowires consisted of alternating segments of Ag, Au, Ni, and Pt, with the Au segments functionalized with either specific nucleotide sequences for DNA detection or antibodies for protein detection. The lengths of the functionalized segments were also varied and alternated for creating the unique "barcodes" corresponding to specific proteins or oligonucleotides. Following exposure to the target analytes, the nanowire barcodes were introduced to the corresponding fluorophore-labeled markers, i.e. either complementary nucleotide sequences or antibodies.

The specificity of the detection scheme was demonstrated by correlating optical images of the nanowire barcodes acquired with and without fluorescence excitation. In this manner, multiple "probes" could be used simultaneously for the detection of a diverse set of analytes.

13.5.4 NEMS Example: Spintronics

Potential applications of ferromagnetic (e.g. nickel, cobalt, and permalloy) nanowires in high density magnetic recording and magnetoelectronics devices have attracted significant interest in understanding their magnetotransport properties [131–143]. Because of the extremely small dimensions and difficulties in addressing single nanowires, magnetotransport properties of nanowires have been mainly studied by measurements made on either lithographically fabricated ferromagnetic nanowire arrays [131, 134, 135, 139] or bundles of nanowires embedded within a suitable template [132, 136, 138, 140, 141]. Recently, Myung and coworkers [113, 114] have developed a facile technique for producing functional ferromagnetic devices based on single nanowires. This was achieved by a combination of processes.

Good electrical contacts for individual ferromagnetic nanowires were obtained by combining template-based electrodeposition techniques, with magnetic assembly, followed by post-annealing in a chemically reducing environment. The above fabrication and assembly technique allowed Myung et al to investigate the magnetotransport properties of single nanowires without interference from dipolar interactions between multiple nanowires. The magnetotransport properties of single electrodeposited ferromagnetic nanowires have been studied not only for nanowires with homogenous chemical composition but also for multi-segmented nanowires consisting of alternating lengths of ferromagnetic and nonmagnetic materials (e.g. NiFe/Cu, Ni/Cu, and Co/Cu). Measurements were aimed at characterizing the current-perpendicular-to-plane-giant-magnetoresistance (CPP-GMR) for these single nanowires [144–146]. Early results show that this approach requires additional work to overcome challenges that essentially limit reproducibility and high throughput. Problems notwithstanding, the approach shows great promise as a powerful platform to study the influence of different materials, dimensions, and structures (segmented, superlattice, and core/shell), on the magnetotransport properties of single nanowires that are crucial for developing a comprehensive knowledge base for spintronics applications.

13.5.5 NEMS Example: Nano Sensor

A diverse set of sensor devices including resistive elements and field effect transistors for chemical and biological detection have been fabricated using either top down or bottom up electrochemical synthesis. Equally diverse electrochemical fabrication methods have been used to generate these sensors, including ESED [95, 147], electrochemically coated CNT networks [97], template-free nanowires [103, 104], e-beam directed in-situ growth [39, 41, 42], and nanowire arrays [148]. In particular, high density nanowire arrays have gained attention because of their potential to detect multiple analytes within a compact and low mass device. Nanowires offer significant benefits over thin films in solid state sensor applications that exploit conductance changes in metals and metal oxides in the presence of gas analytes. Because of their significantly higher surface-to-volume ratio, nanowire-based devices have faster responses, lower power consumption, and higher sensitivities than their thin film counterparts. The large surface area to volume ratio enhances surface reactions and adsorption processes, driving the vastly improved properties of nanowire sensors. This is particularly true for biosensor applications in which antigens and proteins can be covalently attached to metal oxide nanowires using succinimidyl or carboxyl acid linkages [110, 149]. Zhou and coworkers have demonstrated biosensor systems binding prostate specific antigen antibodies (PSA-AB) to indium oxide nanowires via 3-phosphonopropionic linkages, and to single walled carbon nanotubes via 1-pyrenebutanoic acid succinimidyl ester groups (Fig. 13.20). Thiol groups, which are known to form self assembled monolayers (SAMs) on Au, can also facilitate the attachment of bio-molecules to Au nanowires with high binding affinities. Biocompatible conducting polymers can

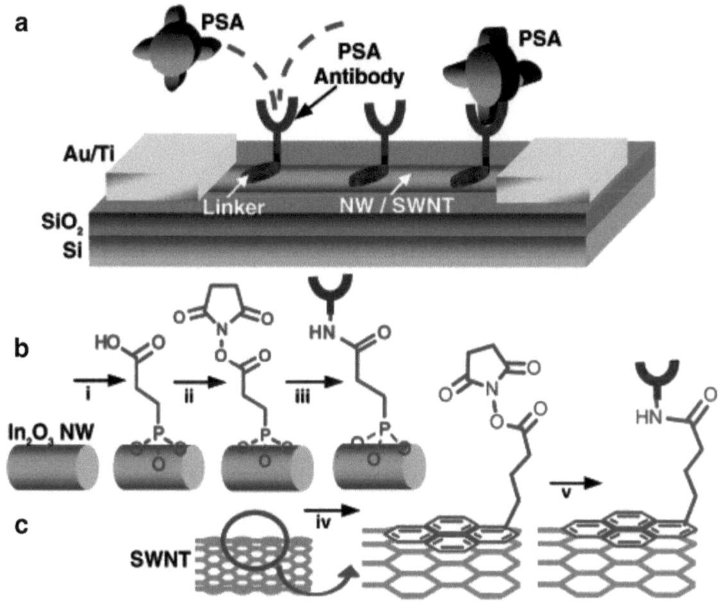

Fig. 13.20 (**a**) Schematic diagram of the nanosensor. PSA–ABs are anchored to the NW/SWNT surface and function as specific recognition groups for PSA binding. (**b**) Reaction sequence for the modification of In2O3 NW: *i*, deposition of 3-phosphonopropionic acid; *ii*, DCC and *N*-hydroxysuccinimide activation; *iii*, PSA-AB incubation. (**c**) Reaction sequence for the modification of SWNT: *i*V, deposition of 1-pyrenebutanoic acid succinimidyl ester; V, PSA-AB incubation. Reprinted with permission from ref. [149]. Copyright (2005) American Chemical Society

be functionalized through entrapment during electrodeposition [42], molecular imprinting [51, 52], and via covalent attachment to monomer precursors. These sensing schemes have the potential to satisfy the need for robust and easy-to-use diagnostic biomedical sensors by significantly reducing the time and complexity of bioassays. They provide a platform for label-free, real-time monitoring, resulting from the highly sensitive charge accumulation or depletion processes within the one-dimensional structures [150]. The sensitivity, selectivity, and rapid response for nanowire-based biosensors have been conclusively demonstrated by Zhou and coworkers for the detection of PSA (Fig. 13.21).

13.5.6 NEMS Example: Thermoelectric Devices

Thermoelectric materials have been used extensively for thermal sensing, energy conversion (heat to electricity), and for cooling (Peltier effect). In general, thermoelectric devices consist of n-type and p-type semiconductor constituents connected electrically in series and thermally in parallel [151]. While energy conversion efficiencies for thermoelectric devices are not yet competitive with conventional refrigerator or power generation systems, it is possible to achieve enhanced efficiencies

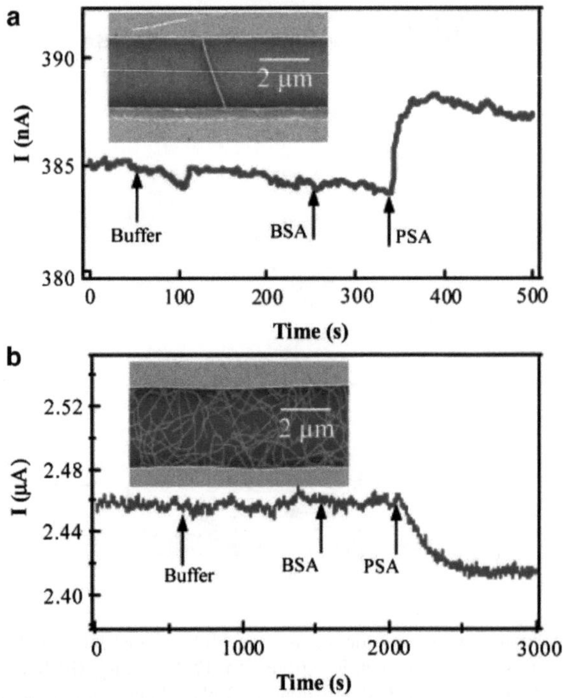

Fig. 13.21 Current recorded over time for an individual In2O3 NW device (**a**) and a SWNT mat device (**b**) when sequentially exposed to buffer, BSA, and PSA. Insets: SEM images of respective devices. Reprinted with permission from ref. [149]. Copyright (2005) American Chemical Society

through choices of novel materials and improved geometries. In particular, the nanowire geometry is particularly amenable to increase carrier mobility and decreased phonon transport arising from quantum confinement effects at the nanometer scale [152]. The key limitation standing in the way of achieving the promise of improved performance by thermoelectric nanowires is the ability to fabricate and couple together alternating n-type and p-type thermoelectric elements with high precision. Previous efforts at MEMS (micron) scales have successfully demonstrated thermoelectric microdevices containing over 100 electrochemically deposited Bi_2Te_3 (n-type) and Sb_2Te_3 (p-type) elements, each 60 μm in diameter, and fabricated using a LiGA technique [151]. This electrochemistry- based fabrication technique is highly scalable and inexpensive; however, the dimensions are still several orders of magnitude greater than the ideal dimensions (~10 nm) required to observe quantum confinement effects. While template based synthesis can certainly produce thermoelectric nanowires in this dimensional range, the key difficulty lies in ensuring that adjacent pores have alternating n-type and p-type semiconductor materials. In an effort to circumvent this problem, several researchers have used lithography to produce microzones of n-type and p-type nanowires and couple them to create the thermoelectric devices as shown in Fig. 13.22 [153, 154].

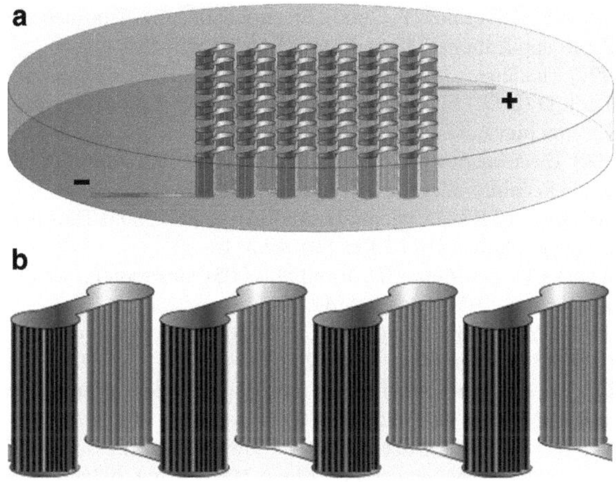

Fig. 13.22 (**a**) Proposed thermoelectric nanowire-based device within an alumina nanotemplate. (**b**) enlarged view of a few n-type and p-type nano-bundle elements connected electrically in series and arranged thermally in parallel (nanotemplate removed). Reprinted with permission from ref. [153]

13.6 Future Trends

Electrochemistry has gained recognition within the MEMS and NEMS community as an economical, high throughput means of producing a diverse range of materials and structures. As our understanding and requirements of materials have progressed, electrochemistry has grown to meet these demands, with tailored alloys, novel composites, and nanostructured features. However, considerable work remains to be done to adapt this centuries-old technology to the stringent demands of MEMS and NEMS. Specifically, a strong theoretical foundation resulting in robust modeling schemes for mass transport, current distribution, additive kinetics, and deposit nucleation needs to be developed in support of the largely empirical work in electrochemistry-based microfabrication that has been carried out thus far [6]. Beyond the actual fabrication process itself, considerable effort is required for developing an in-depth understanding of key electrodeposited materials properties including mechanical stress and interfacial adhesion, thermal, electrical, and chemical properties, all of which have a direct impact on the performance of MEMS and NEMS devices.

References

1. Mehragany M (1993) Microelectromechanical systems. IEEE Cuicuits Dev 8755–3996:14–22
2. Craighead HG (2000) Nanoelectromechanical systems. Science 290:1532–1535
3. Judy JW (2001) Microelectromechanical systems (MEMS): fabrication, design and application. Smart Mater Struct 10:1115–1134

4. Shacham-Diamand Y, Sverdlov Y (2000) Electrochemically deposited thin film alloys for ULSI and MEMS applications. Microelectron Eng 50:525–531
5. Peeters E (1997) Challenges in commercializing MEMS. IEEE Comput Sci Eng 4:44–48
6. Datta M, Landolt D (2000) Fundamental aspects and applications of electrochemical microfabrication. Electrochim Acta 45:2535–2558
7. Osaka T (2000) Electrodeposition of highly functional thin films for magnetic recording devices of the next century. Electrochim Acta 45:3311–3321
8. Vereecken PM, Binstead RA, Delgianni H, Andricacos PC (2005) The chemistry of additives in damascene copper plating. IBM J Res Dev 49:3–18
9. Moffat TP, Wheeler D, Edelstein MD, Josell D (2005) Superconformal film growth: mechanism and quantification. IBM J Res Dev 49:19–36
10. Myung NV, Park DY, Yoo BY, Sumodjo PTA (2003) Development of electroplated magnetic materials for MEMS. J Magn Magn Mater 265:189–198
11. Andricacos PC, Robertson N (1998) Future directions in electroplated materials for thin-film recording heads. IBM J Res Dev 42:671–680
12. Guan S, Nelson BJ (2005) Electrodeposition o f low residual stress CoNiMnP hard magnetic thin films for magnetic MEMs actuators. J Magn Magn Mater 292:49–58
13. Agarwala R, Agarwala V (2003) Electroles alloy/composite coatings: a review. Sadhana 28:475–493
14. Gawrilov GG (1979) Chemical (electroless) nickel plating, 1st edn. Portcullis Press Limited, Redhill
15. Khoperia TN, Tabatadze TJ, Zedgenidze TI (1997) Formation of microcircuits in microelectronics by electroless deposition. Electrochim Acta 42:3049–3055
16. Chen CJ, Lin KL (2000) Internal stress and adhesion of amorphous Ni–Cu–P alloy on aluminum. Thin Solid Films 370:106–113
17. Hsu JC, Lin KL (2003) Enhancement in the deposition behavior and deposit properties of electroless Ni–Cu–P. J Electrochem Soc 150:C653–C656
18. Osaka T (2004) Creation of highly functional thin films using electrochemical nanotechnology. Chem Rec 4:346–362
19. Shacham-Diamand Y, Inberg A, Sverdlov Y, Bogush V, Croitoru N, Moscovich H, Freeman A (2003) Electroless processes for micro- and nanoelectronics. Electrochim Acta 48:2987–2996
20. Sun L, Chien C-L, Searson PC (2004) Fabrication of nanoporous nickel by electrochemical dealloying. Chem Mater 16:3125–3129
21. Guell AG, Diez-Perez I, Gorostiza P, Sanz F (2004) Preparation of reliable probes for electrochemical tunneling spectroscopy. Anal Chem 76:5218–5222
22. West AC, Deligianni H, Andricacos PC (2005) Electrochemical planarization of interconnect metallization. IBM J Res Dev 49:37–48
23. Datta M (1995) Fabrication of an array of precision nozzles by through-mask electrochemical micromachining. J Electrochem Soc 142:3801–3805
24. Schuster R, Kirchner V, Allongue P, Ertl G (2000) Electrochemical micromachining. Science 289:98–101
25. Shaw KA, Zhang ZL, MacDonald NC (1994) SCREAM I: a single mask, single-crystal silicon, reactive ion etching process for microelectromechanical structures. Sens Actuators 40:63–70
26. Guckel H (1998) High-aspect-ratio micromachining via deep x-ray lithography. Proc IEEE 86:1586
27. Madou MJ (2002) Fundamentals of microfabrication, 2nd edn. CRC Press LLC, Boca Raton
28. Griffiths SK (2004) Fundamental limitations of LIGA x-ray lithography: sidewall onset, slope and minimum feature size. J Micromech Microeng 14:999–1011
29. Griffiths SK, Crowell JA, Kistler BL, Dryden AS (2004) Dimensional errors in LIGA-produced metal structures due to thermal expansion and swelling of PMMA. J Micromech Microeng 14:1548–1557
30. Goods SH, Kelly JJ, Yang NYC (2004) Electrodeposited nickel-manganese: an alloy for microsystem applications. Microsyst Technol Micro Nanosyst Inf Storage Process Syst 10:498–505

31. Kelly JJ, Goods SH, Yang NYC (2003) High performance nanostructured Ni–Mn alloy for microsystem applications. Electrochem Solid State Lett 6:C88–C91
32. Marquis EA, Talin AA, Kelly JJ, Goods SH, Michael JR (2006) Effects of current density on the structure of Ni and Ni–Mn electrodeposits. J Appl Electrochem 36:669–676
33. Yang JM, Zhu D, Qu NS, Lei WN (2004) Pulse electroforming of nanocrystalline Ni–Mn alloy. Adv Grinding Abrasive Process 259–2:596–601
34. Cohen A, Zhang G, Tseng F-G, Frodis U, Mansfeld F, Will F (1999) EFAB: rapid, low-cost desktop micromachining of high aspect ratio True 3-D MEMS. In: Twelfth IEEE international conference on micro electro mechanical systems, 1999. MEMS '99, pp 244–251
35. Chen R (2004) Micro-Fabrication Techniques. Wireless Des Develop 16–20
36. Malek CK, Saile V (2004) Applications of LIGA technology to precision manufacturing of high-aspect-ratio micro-components and -systems: a review. Microelectronics J 35:131–143
37. Alper SE, Ocak IE, Akin T (2006) Ultra-thick and high-aspect-ratio nickel microgyroscope using EFAB™ multi-layer additive electroforming. In: 19th IEEE international conference on micro electro mechanical systems, 2006. MEMS 2006 Istanbul, pp 670–673
38. Microfabrica (2004) Going beyond silicon MEMS with EFAB technology. pp 1–14
39. Yun M, Myung NV, Vasquez RP, Lee C, Menke E, Penner RM (2004) Electrochemically grown wires for individually addressable sensor arrays. Nano Lett 4:419–422
40. Ramanathan K, Bangar MA, Chen MYW, Mulchandani A, Myung NV (2004) Individually addressable conducting polymer nanowires array. Nano Lett 4:1237–1239
41. Im Y, Lee C, Vasquez RP, Bangar MA, Myung NV, Menke EJ, Penner RM, Yun M (2006) Investigation of a single Pd nanowire fo use as a hydrogen sensor. Small 2:356–358
42. Ramanathan K, Bangar MA, Yun M, Chen W, Myung NV, Mulchandani A (2005) Bioaffinity sensing using biologically functionalized conducting-polymer nanowire. J Am Chem Soc 127:496–497
43. Peng C-Y, Kalkan AK, Fonash SJ, Bu B, Sen A (2005) A "Grow-in-Place" architecture and methodology for electrochemical synthesis of conducting polymer nanoribbon device arrays. Nano Lett 5:439–444
44. Brumlik CJ, Martin CR (1991) Template synthesis of metal microtubules. J Am Chem Soc 113:3174–3175
45. Martin CR (1995) Nanomaterials: a membrane-based synthetic approach. Science 266:1961–1966
46. Martin CR (1995) Template synthesis of electronically conductive polymer nanostructures. Acc Chem Res 28:61–68
47. He H, Tao NJ (2003) Electrochemical fabrication of metal nanowires. In: Nalwa HS (ed) Encyclopedia of nanoscience and nanotechnology, vol X. American Scientific Publishers, pp 1–18
48. Kovtyukhova NI, Martin BR, Mbindyo JKN, Mallouk TE, Cabassi M, Mayer TS (2002) Layer-by-layer self-assembly strategy for template synthesis of nanoscale devices. Mater Sci Eng C 19:255–262
49. Kovtyukhova NI, Martin BR, Mbindyo JKN, Smith PA, Razavi B, Mayer TS, Mallouk TE (2001) Layer-by-layer assembly of rectifying junctions in and on metal nanowires. J Phys Chem B 105:8762–8769
50. Xu X, Chen L, Wang C, Yao Q, Feng C (2005) Template synthesis of heterostructured polyaniline/Bi_2Te_3 nanowires. J Solid State Chem 178:2163–2166
51. Li Y, Yang HH, You QH, Zhuang ZX, Wang XR (2006) Protein recognition via surface molecularly imprinted polymer nanowires. Anal Chem 78:317–320
52. Li Y, Yin X-F, Chen F-R, Yang H-H, Zhuang Z-X, Wang Z-R (2006) Synthesis of magnetic molecularly imprinted polymer nanowires using a nanoporous alumina template. Macromolecules 39:4497–4499
53. Park S, Chung SW, Mirkin CA (2004) Hybrid organic-inorganic, rod-shaped nanoresistors and diodes. J Am Chem Soc 126:11772–11773
54. Kovtyukhova NI, Kelley BK, Mallouk TE (2004) Coaxially gated in-wire thin-film transistors made by template assembly. J Am Chem Soc 126:12738–12739

55. Keller F, Hunter MS, Robinson DL (1953) Structural features of oxide coatings on aluminum. J Electrochem Soc 100:411–419
56. Huczko A (2000) Template-based synthesis of nanomaterials. Appl Phys A 70:365–376
57. Nielsch K, Choi J, Schwirn K, Wehrspohn RB, Gosele U (2002) Self-ordering regimes of porous alumina: the 10% porosity rule. Nano Lett 2:677–680
58. Xu T, Zangari G, Metzger RM (2002) Periodic holes with 10 nm diameter produced by grazing Ar^+ milling of the barrier layer in hexagonally ordered nanoporous alumina. Nano Lett 2:37–41
59. Masuda H, Hasegwa F, Ono S (1997) Self-ordering of cell arrangement of anodic porous alumina formed in sulfuric acid solution. J Electrochem Soc 144:L127–L130
60. Masuda H, Satoh M (1996) Fabrication of gold nanodot array using anodic porous alumina as an evaporation mask. Jpn J Appl Phys 35:L126–L129
61. Masuda H, Yamada J, Satoh M, Asoh J (1997) Highly ordered nanochannel-array architecture in anodic alumina. Appl Phys Lett 71:2770–2772
62. Yoo BY, Hendricks RK, Ozkan M, Myung NV (2006) Three-dimensional alumina nanotemplate. Electrochim Acta 51:3543–3550
63. Jee SE, Lee PS, Yoon B-J, Jeong S-H, Lee K-H (2005) Fabrication of microstructures by wet etching of anodic aluminum oxide substrates. Chem Mater 17:4049–4052
64. Cojocaru CS, Padovani JM, Wade T, Mandoli C, Jaskierowicz G, Wegrowe JE, AFi M, Pribat D (2005) Conformal anodic oxidation of aluminum thin films. Nano Lett 5:675–680
65. Chien CL, Sun L, Tanase M, Bauer LA, Hultgren A, Silevitch DM, Meyer GJ, Searson PC, Reich DH (2002) Electrodeposited magnetic nanowires: arrays, field-induced assembly, and surface functionalization. J Magn Magn Mater 249:146–155
66. Sun L, Searson PC, Chien CL (1999) Electrochemical deposition of nickel nanowire arrays in single-crystal mica films. Appl Phys Lett 74:2803–2805
67. Schonenberger C, vanderzande BMI, Fokkink LGJ, Henny M, Schmid C, Kruger M, Bachtold A, Huber R, Birk H, Staufer U (1997) Template synthesis of nanowires in porous polycarbonate membranes: electrochemistry and morphology. J Phys Chem B 101:5497–5505
68. Williams WD, Giordano N (1984) Fabrication of 80-a metal wires. Rev Sci Instrum 55:410–412
69. Brinker CJ, Lu Y, Sellinger A, Fan H (1999) Evaporation-induced self-assembly nanostructures made easy. Adv Mater 11:579–585
70. Lu Y, Fan H, Stump A, Ward TL, Rieker T, Brinker J (1999) Aerosol-assisted self-assembly of mesostructured sphereical nanoparticles. Nature 398:223–226
71. Lu Y, Gangull R, Drewien CA, Anderson MT, Brinker CJ, Gong W, Guo Y, Soyez H, Dunn B, Huang MH, Zink JI (1997) Continuous formation of supported cubic and hexagonal mesoporous films by sol-gel dip-coating. Nature 389:364–368
72. Lu Y, Yang H, Sellinger A, Lu M, Huang J, Fan H, Haddad R, Lopez G, Burns AR, Sasaki DY, Shelnutt J, Brinker CJ (2001) Self-assembly of mesoscopically ordered chromatic polydiacetylene/silica nanocomposites. Nature 410:913–917
73. Wang D, Luo H, Kou R, Gil MP, Xiao S, Golub VO, Yang Z, Brinker CJ, Lu Y (2004) A general route to macroscopic hierarchical 3D nanowire networks. Angew Chem Int Ed 43:6169–6173
74. Wang D, Zhou WL, McCaughy BF, Hampsey JE, Ji Z, Jiang Y-B, HuifangXu TJ, Schmehl RH, O'Connor C, Brinker CJ, Lu Y (2003) Electrodeposition of metallic nanowire thin films using mesoporous silica templates. Adv Mater 15:130–133
75. Morkved TL, Lu M, Urbas AM, Ehrichs EE, Jaeger HM, Mansky P, Russell TP (1996) Local control of microdomain orientation in diblock copolymer thin films with electric fields. Science 273:931–933
76. Xu T, Hawker CJ, Russell TP (2005) Interfacial interaction dependence of microdomain orientation in diblock copolymer thin films. Macromolecules 38:2802–2805
77. Kim SO, Solak HH, Stoykovich MP, Ferrier NJ, de Pablo JJ, Nealey PF (2003) Epitaxial self-assembly of block copolymers on lithographically defined nanopatterned substrates. Nature 424:411–414

78. Xu T, Zvelindovsky AV, Sevink GJA, Lyakhova KS, Jinnai H, Russell TP (2005) Electric field alignment of asymmetric diblock copolymer thin films. Macromolecules 38:10788–10798
79. Xu T, Goldbach JT, Leiston-Belanger J, Russell TP (2004) Effect of ionic impurities on the electric field alignment of diblock copolymer thin films. Colloid Polym Sci 282:927–931
80. Thurn-Albrecht T, Schotter J, Kastle GA, Emley N, Shibauchi T, Krusin-Elbaum L, Guarini K, Black CT, Tuominen MT, Russell TP (2000) Ultrahigh-density nanowire arrays grown in self-assembled diblock copolymer templates. Science 290:2126–2129
81. Lopes WA, Jaeger HM (2001) Hierarchical self-assembly of metal nanostructures on diblock copolymer scaffolds. Nature 414:735–738
82. Allred DB, Sarikaya M, Baneyx F, Schwartz DT (2005) Electrochemical nanofabrication using crystalline protein masks. Nano Lett 5:609–613
83. Dai HX, Choe WS, Thai CK, Sarikaya M, Traxler BA, Baneyx F, Schwartz DT (2005) Nonequilibrium synthesis and assembly of hybrid inorganic-protein nanostructures using an engineered DNA binding protein. J Am Chem Soc 127:15637–15643
84. Dai HX, Thai CK, Sarikaya M, Baneyx F, Schwartz DT (2004) Through-mask anodic patterning of copper surfaces and film stability in biological media. Langmuir 20:3483–3486
85. Thai CK, Dai HX, Sastry MSR, Sarikaya M, Schwartz DT, Baneyx F (2004) Identification and characterization of Cu2O- and ZnO-binding polypeptides by Escherichia coli cell surface display: toward an understanding of metal oxide binding. Biotechnol Bioeng 87:129–137
86. Li QG, Penner RM (2005) Photoconductive cadmium sulfide hemicylindrical shell nanowire ensembles. Nano Lett 5:1720–1725
87. Menke EJ, Li Q, Penner RM (2004) Bismuth telluride (Bi2Te3) nanowires synthesized by cyclic electrodeposition/stripping coupled with step edge decoration. Nano Lett 4:2009–2014
88. Walter EC, Murray BJ, Favier F, Kaltenpoth G, Grunze M, Penner RM (2002) Noble and coinage metal nanowires by electrochemical step edge decoration. J Phys Chem B 106: 11407–11411
89. Walter EC, Murray BJ, Favier F, Penner RM (2003) "Beaded" bimetallic nanowires: wiring nanoparticles of metal 1 using nanowires of metal 2. Adv Mater 15:396–399
90. Zach MP, Ng KH, Penner RM (2000) Molybdenum nanowires by electrodeposition. Science 290:2120–2123
91. Banks CE, Davies TJ, Wildgoose GG, Compton RG (2005) Electrocatalysis at graphite and carbon nanotube modified electrodes: edge-plane sites and tube ends are the reactive sites. Chem Commun 7:829–841
92. Fan YW, Goldsmith BR, Collins PG (2005) Identifying and counting point defects in carbon nanotubes. Nat Mater 4:906–911
93. Walter EC, Zach MP, Favier F, Murray BJ, Inazu K, Hemminger JC, Penner RM (2003) Metal nanowire arrays by electrodeposition. Chemphyschem 4:131–138
94. Zach MP, Inazu K, Ng KH, Hemminger JC, Penner RM (2002) Synthesis of molybdenum nanowires with millimeter-scale lengths using electrochemical step edge decoration. Chem Mater 14:3206–3216
95. Favier F, Walter EC, Zach MP, Benter T, Penner RM (2001) Hydrogen sensors and switches from electrodeposited palladium mesowire arrays. Science 293:2227–2231
96. Fennimore AM, Yuzvinsky TD, Han WQ, Fuhrer MS, Cumings J, Zettl A (2003) Rotational actuators based on carbon nanotubes. Nature 424:408–410
97. Chen XH, Chen CS, Xiao HN, Liu HB, Zhou LP, Li SL, Zhang G (2006) Dry friction and wear characteristics of nickel/carbon nanotube electroless composite deposits. Tribol Int 39:22–28
98. Kong J, Chapline MG, Dai HJ (2001) Functionalized carbon nanotubes for molecular hydrogen sensors. Adv Mater 13:1384–1386
99. Day TM, Unwin PR, Wilson NR, Macpherson JV (2005) Electrochemical templating of metal nanoparticles and nanowires on single-walled carbon nanotube networks. J Am Chem Soc 127:10639–10647

100. Huang JX, Kaner RB (2004) A general chemical route to polyaniline nanofibers. J Am Chem Soc 126:851–855
101. Huang JX, Virji S, Weiller BH, Kaner RB (2003) Polyaniline nanofibers: facile synthesis and chemical sensors. J Am Chem Soc 125:314–315
102. Liang L, Liu J, Windisch CF, Exarhos GJ, Lin YH (2002) Direct assembly of large arrays of oriented conducting polymer nanowires. Angew Chem Int Ed 41:3665–3668
103. Wang J, Chan S, Carlson RR, Luo Y, Ge GL, Ries RS, Heath JR, Tseng HR (2004) Electrochemically fabricated polyaniline nanoframework electrode junctions that function as resistive sensors. Nano Lett 4:1693–1697
104. Alam MM, Wang J, Guo YY, Lee SP, Tseng HR (2005) Electrolyte-gated transistors based on conducting polymer nanowire junction arrays. J Phys Chem B 109:12777–12784
105. Jin S, Whang DM, McAlpine MC, Friedman RS, Wu Y, Lieber CM (2004) Scalable interconnection and integration of nanowire devices without registration. Nano Lett 4:915–919
106. Tao A, Kim F, Hess C, Goldberger J, He RR, Sun YG, Xia YN, Yang PD (2003) Langmuir-Blodgett silver nanowire monolayers for molecular sensing using surface-enhanced Raman spectroscopy. Nano Lett 3:1229–1233
107. Whang D, Jin S, Lieber CM (2003) Nanolithography using hierarchically assembled nanowire masks. Nano Lett 3:951–954
108. Whang D, Jin S, Wu Y, Lieber CM (2003) Large-scale hierarchical organization of nanowire arrays for integrated nanosystems. Nano Lett 3:1255–1259
109. Yang PD, Kim F (2002) Langmuir-Blodgett assembly of one-dimensional nanostructures. Chemphyschem 3:503–506
110. Tanase M, Bauer LA, Hultgren A, Silevitch DM, Sun L, Reich DH, Searson PC, Meyer GJ (2001) Magnetic alignment of fluorescent nanowires. Nano Lett 1:155–158
111. Tanase M, Silevitch DM, Hultgren A, Bauer LA, Searson PC, Meyer GJ, Reich DH (2002) Magnetic trapping and self-assembly of multicomponent nanowires. J Appl Phys 91:8549–8551
112. Bentley AK, Trethewey JS, Ellis AB, Crone WC (2004) Magnetic manipulation of copper-tin nanowires capped with nickel ends. Nano Lett 4:487–490
113. Hangarter CM, Myung NV (2005) Magnetic alignment of nanowires. Chem Mater 17:1320–1324
114. Yoo B, Rheem Y, Beyermann WP, Myung NV (2006) Magnetically assembled 30 nm diameter nickel nanowire with ferromagnetic electrodes. Nanotechnology 17:2512–2517
115. Niyogi S, Hangarter C, Thamankar RM, Chiang YF, Kawakami R, Myung NV, Haddon RC (2004) Magnetically assembled multiwalled carbon nanotubes on ferromagnetic contacts. J Phys Chem B 108:19818–19824
116. Duan XF, Huang Y, Cui Y, Wang JF, Lieber CM (2001) Indium phosphide nanowires as building blocks for nanoscale electronic and optoelectronic devices. Nature 409:66–69
117. Smith PA, Nordquist CD, Jackson TN, Mayer TS, Martin BR, Mbindyo J, Mallouk TE (2000) Electric-field assisted assembly and alignment of metallic nanowires. Appl Phys Lett 77:1399–1401
118. Yamamoto K, Akita S, Nakayama Y (1998) Orientation and purification of carbon nanotubes using ac electrophoresis. J Phys D Appl Phys 31:L34–L36
119. Dong LF, Bush J, Chirayos V, Solanki R, Jiao J (2005) Dielectrophoretically controlled fabrication of single-crystal nickel silicide nanowire interconnects. Nano Lett 5:2112–2115
120. Edwards B, Mayer TS, Bhiladvala RB (2006) Synchronous electrorotation of nanowires in fluid. Nano Lett 6:626–632
121. Fan DL, Zhu FQ, Cammarata RC, Chien CL (2005) Controllable high-speed rotation of nanowires. Phys Rev Lett 94:247208
122. George T, Madsen L, Tang W, Chang-Chien A, Leskowitz G, Weitekamp D (2001) MEMS-based force-detected nuclear magnetic resonance spectrometer for in situ planetary exploration. In: IEEE, pp 273–278
123. Leskowitz GM, Madsen LA, Weitekamp DP (1998) Force-detected magnetic resonance without field gradients. Solid State Nucl Magn Reson 11:73–86

124. Xu BQ, Tao NJJ (2003) Measurement of single-molecule resistance by repeated formation of molecular junctions. Science 301:1221–1223
125. Chen F, Qing Q, Ren L, Wu ZY, Liu ZF (2005) Electrochemical approach for fabricating nanogap electrodes with well controllable separation. Appl Phys Lett 86:123105
126. Li CZ, He HX, Tao NJ (2000) Quantized tunneling current in the metallic nanogaps formed by electrodeposition and etching. Appl Phys Lett 77:3995–3997
127. Li CZ, Tao NJ (1998) Quantum transport in metallic nanowires fabricated by electrochemical deposition/dissolution. Appl Phys Lett 72:894–896
128. Liu B, Xiang J, Tian JH, Zhong C, Mao BW, Yang FZ, Chen ZB, Wu ST, Tian ZQ (2005) Controllable nanogap fabrication on microchip by chronopotentiometry. Electrochim Acta 50:3041–3047
129. Qing Q, Chen F, Li PG, Tang WH, Wu ZY, Liu ZF (2005) Finely tuning metallic nanogap size with electrodeposition by utilizing high-frequency impedance in feedback. Angew Chem Int Ed 44:7771–7775
130. Nicewarner-Pena SR, Freeman RG, Reiss BD, He L, Pena DJ, Walton ID, Cromer R, Keating CD, Natan MJ (2001) Submicrometer metallic barcodes. Science 294:137–141
131. Goolaup S, Singh N, Adeyeye AO, Ng V, Jalil MBA (2005) Transition from coherent rotation to curling mode reversal process in ferromagnetic nanowires. Eur Phys J B 44:259–264
132. Hao Z, Shaoguang Y, Gang N, Dongliang Y, Youwei D (2001) Study on magnetic property of Fe14Ni86 alloy nanowire array. J Magn Magn Mater 234:454–458
133. Lederman M, O'Barr R, Schultz S (1995) Experimental study of individual ferromagnetic sub-micron cylinders. IEEE Trans Magn 31:3793–3795
134. Leven B, Dumpich G (2005) Resistance behavior and magnetization reversal analysis of individual Co nanowires. Phys Rev B 71:064411(7 pages)
135. Martin JI, Velez M, Alameda JM, Briones F, Vicent JL (2002) Magnetotransport properties of patterned magnetic Ni wires of submicrometric dimensions. J Magn Magn Mater 240:14–16
136. Navas D, Asenjo A, Jaafar M, Pirota KR, Hernandez-Velez M, Sanz R, Lee W, Niesch K, Batallan F, Vazquez M (2005) Magnetic behavior of NixFe(100-x) (65 <= x <= 100) nanowire arrays. J Magn Magn Mater 290:191–194
137. OBarr R, Schultz S (1997) Switching field studies of individual single domain Ni columns. J Appl Phys 81:5458–5460
138. Ohgai T, Gravier L, Hoffer X, Lindeberg M, Hjort K, Spohr R, Ansermet J-P (2003) Template synthesis and magnetoresistance property of Ni and Co single nanowires electrodeposited into nanopores with a wide range of aspect ratios. J Phys D Appl Phys 36:3109–3114
139. Perez-Junquera A, Martin JI, Velez M, Alameda JM, Vicent JL (2003) Temperature dependence of the magnetization reversal process in patterned Ni nanowires. Nanotechnology 14:294–298
140. Pignard S, Goglio G, Radulescu A, Piraux L, Dubois S, Declemy A, Duvail JL (2000) Study of the magnetization reversal in individual nickel nanowires. J Appl Phys 87:824–829
141. Wegrowe JE, Gilbert SE, Kelly D, Doudin B, Ansermet JP (1998) Anisotropic magnetoresistance as a probe of magnetization reversal in individual nano-sized nickel wires. IEEE Trans Magn 34:903–905
142. Wegrowe JE, Kelly D, Franck A, Gilbert SE, Ansermet JP (1999) Magnetoresistance of ferromagnetic nanowires. Phys Rev Lett 82:3681–3684
143. Wernsdorfer W, Doudin B, Mailly D, Hasselbach K, Benoit A, Meier J, Ansermet JP, Barbara B (1996) Nucleation of magnetization reversal in individual nanosized nickel wires. Phys Rev Lett 77:1873–1876
144. Chen M, Searson PC, Chien CL (2003) Micromagnetic behavior of electrodeposited Ni/Cu multilayer nanowires. J Appl Phys 93:8253–8255
145. Dubois S, Marchal C, Beuken JM, Piraux L, Duvail JL, Fert A, George JM, Maurice JL (1997) Perpendicular giant magnetoresistance of NiFe/Cu multilayered nanowires. Appl Phys Lett 70:396–398

146. Evans PR, Yi G, Schwarzacher W (2000) Current perpendicular to plane giant magnetoresistance of multilayered nanowires electrodeposited in anodic aluminum oxide membranes. Appl Phys Lett 76:481–483
147. Murray BJ, Walter EC, Penner RM (2004) Amine vapor sensing with silver mesowires. Nano Lett 4:665–670
148. Kolmakov A, Zhang YX, Cheng GS, Moskovits M (2003) Detection of CO and O-2 using tin oxide nanowire sensors. Adv Mater 15:997–1000
149. Li C, Curreli M, Lin H, Lei B, Ishikawa FN, Datar R, Cote RJ, Thompson ME, Zhou CW (2005) Complementary detection of prostate-specific antigen using In(2)O(3) nanowires and carbon nanotubes. J Am Chem Soc 127:12484–12485
150. Wanekaya AK, Chen W, Myung NV, Mulchandani A (2006) Nanowire-based electrochemical biosensors. Electroanalysis 18:533–550
151. Snyder GJ, Lim JR, Huang CK, Fleurial JP (2003) Thermoelectric microdevice fabricated by a MEMS-like electrochemical process. Nat Mater 2:528–531
152. Hicks LD, Dresselhaus MS (1993) Thermoelectric figure of merit of a one-dimensional conductor. Phys Rev B 47:16631–16634
153. Lim JR, Whitacre JF, Fleurial JP, Huang CK, Ryan MA, Myung NV (2005) Fabrication method for thermoelectric nanodevices. Adv Mater 17:1488–1492
154. Wang W, Jia FL, Huang QH, Zhang JZ (2005) A new type of low power thermoelectric micro-generator fabricated by nanowire array thermoelectric material. Microelectron Eng 77:223–229
155. Kim SH, Misner MJ, Xu T, Kimura M, Russell TP (2004) Highly oriented and ordered arrays from block copolymers via solvent evaporation. Adv Mater 16:226–231
156. George T (2003) Overview of MEMS/NEMS technology development for space applications at NASA/JPL. SPIE 5116:136–148
157. Possin GE (1970) A method of forming very small diameter wires. Rev Sci Instrum 41:772–774
158. Luo, H, Wang D, He J, Lu Y (2005) Magnetic cobalt nanowire thin films. J Phys Chem B 109:1919–1922
159. Klein JD, Herrick RD, Palmer D, Sailor MJ, Brumlik CJ, Martin CR (1993) Electrochemical fabrication of cadmium chalcogenide microdiode arrays. Chem Mater 5:902–904
160. Sima M, Enculescu I, Visan T (2004) The electrodeposition of semiconductor nanowires with thermoelectric properties using "template" method. Revista De Chimie 55:743–746
161. Mukherjee PK, Chakravorty D (2004) Growth of CdS nanowires using Na-4 mica as template. J Appl Phys 95:3164–3169
162. Wang DH, Jakobson HP, Kou R, Tang J, Fineman RZ, Yu DH, Lu YF (2006) Metal and semiconductor nanowire network thin films with hierarchical pore structures. Chem Mater 18:4231–4237
163. Li Y, Meng GW, Zhang LD, Phillipp F (2000) Ordered semiconductor ZnO nanowire arrays and their photoluminescence properties. App Phy Lett 76:2011–2013

Chapter 14
Microelectronic Packaging Trends and the Role of Nanotechnology

Madhav Datta

14.1 Introduction

The microelectronic packaging industry is undergoing major changes to keep pace with the ever-increasing demands imposed by high performing chips and by end-use system applications. Solutions using advanced materials for microprocessor interconnect scaling and chip package interconnects, novel concepts in heat management systems, and improvements in package substrates continue to drive major packaging efforts. Advances in electrochemical technologies have played an important role in the evolution of such solutions for miniaturization of microelectronic devices and packages. Indeed, since the development of through-mask plating for thin film heads in the1960s and 1970s, an enormous amount of industrial and academic R&D effort has positioned electrochemical processing among the most sophisticated processing technologies employed in the microelectronics industry today [1–4]. Electrochemical processing is perhaps better understood than some of the dry processing technologies used in the microelectronics industry. Compared to other competing dry processing technologies, it has emerged as a more environmentally-friendly and cost-effective fabrication method. Electrochemical processing has, thus, become an integral part of advanced wafer processing fabs and an enabling technology for nanofabrication [5]. As the electronics industry faces the challenges of extending Moore's law, electrochemical processing is expected to continue to enable further miniaturization of high-performance chip interconnects, packages, and printed circuit boards. Evolving novel approaches to electrochemical processing using nano-materials and nano-fabrication techniques have started to make tremendous impact on further miniaturization of high performance devices and packages. A detailed discussion of different facets of technology advances in electronic packaging is difficult to present in the limited space of this chapter. The current chapter, therefore, makes an effort to capture some of the key

M. Datta (✉)
Cooligy Precision Cooling, Emerson Network Power, 800 Maude Avenue, Mountain View, CA, 94043, USA
e-mail: madhav.datta@emerson.com

T. Osaka et al. (eds.), *Electrochemical Nanotechnologies*,
Nanostructure Science and Technology, DOI 10.1007/978-1-4419-1424-8_14,

developments in microelectronic packaging while highlighting the impact of electrochemical processing. Also included is a brief discussion of some of the foreseeable applications of nano-materials and nano-structures in advanced packaging.

14.2 Microelectronic Packaging

Electronic packaging is the methodology for connecting and interfacing the chip technology with a system and the physical world. The objective of packaging is to ensure that the devices and interconnections are packaged efficiently and reliably. With continued miniaturization trends in integrated circuits, a steadily increasing percentage of wiring migrates into the chip, thus, making the semiconductor thin film wiring a very important aspect of microelectronic packaging [6, 7]. A typical microprocessor packaging hierarchy consists of several levels of packaging. The chip level packaging includes chip interconnect wiring (metallization), and provisions for chip-package interconnection such as flip-chip bumping, wire bonding, and tape automated bonding (TAB) [6–9]. The first level of packaging involves joining of chip(s) to a substrate, which may form a single chip module (SCM) or a multichip module (MCM). Depending on the thermal cycle environment, the substrates for SCM/MCM may be either organic or ceramic packages. In the second level of packaging, packaged SCMs, MCMs, and other components are assembled on a printed circuit board (PCB) or card. PCBs are generally copper-clad sheets of epoxy-glass laminates with plated through-hole interconnections. In some cases, the chips (without a substrate) are directly attached to boards (direct chip attach) known as chip-on-board (COB). The third level of packaging may vary depending on the system. In a desktop, several PCBs are plugged into a motherboard, while in a hand-held calculator the outer shell is the third level of packaging. On the other hand, a workstation or a mainframe uses several motherboards within an enclosed box.

Figure 14.1 is a schematic diagram showing the key components of a microprocessor assembly. The assembly consists of a chip, a package, and a printed circuit board that are joined together to form a second-level packaging. Chips with C4 solder bumps are flipped over and joined to the package substrate by reflow. An underfill is often used to improve reliability of the chip-package interconnect assembly. This first level assembly is further connected to the printed circuit board using ball grid arrays (BGA's) or pins. For efficient heat dissipation from the chip, an integrated heat spreader and a finned heat sink are attached to the back side of the chip using thermal interface materials (TIM1 and TIM2) Electrochemical processing technologies employed in the fabrication of all of these components, include low-end electroless nickel/gold coating of the copper heat sinks, wiring of packages and boards by through-hole plating, fabrication of plated C4 solder bumps and the fabrication of extremely precise nano-scale features of chip interconnect metallization. Accordingly, the degree of sophistication of tools and processes varies depending on whether they are applied in package/board fabrication on macro/micro scale or in semiconductor wafer processing on micro/nano scale [5].

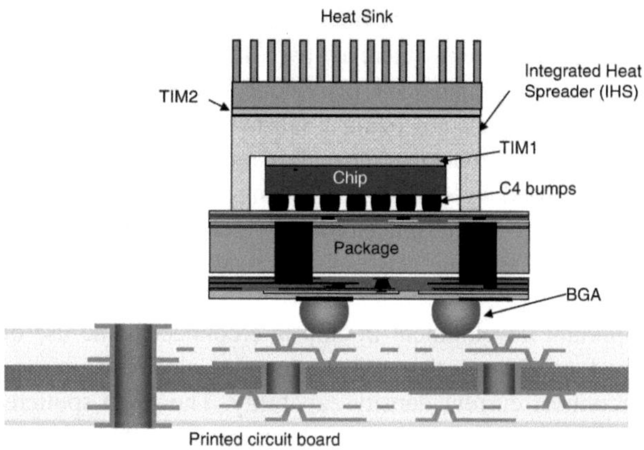

Fig. 14.1 Schematic diagram of a microprocessor assembly showing different levels of packaging and thermal management solution using a heat sink. *TIM* thermal interface material, *BGA* ball grid array

14.3 Semiconductor Packaging: Cu Metallization and Flip-Chip Technology

Introduction of electroplated copper metallization replacing aluminum brought about a paradigm shift in chip interconnect technology [10]. Advanced microprocessors now use electroplated Cu technology because of several advantages associated with copper wiring, and with the electroplating technology that has enabled the high volume manufacturing of Cu chips [10]. Development of a defect-free dual Damascene electroplating process, and the semiconductor equipment industry's drive to develop and market electroplating and CMP tools were the key enabling factors that made the implementation of copper interconnect technology possible [5]. Indeed, the development and availability of high volume manufacturing of electroplating and CMP tools that are compatible with ultra-clean fab standards represented a major shift in the semiconductor industry's strategy that was hitherto focused on vacuum processing. These developments together with an aggressive integration scheme have enabled the development and manufacturing of advanced interconnects. The defect-free fabrication of nano-scale, multilevel interconnect structures on a 300 mm wafer with high yield demonstrates the exceptional strength of electroplating technology. All these developments have placed electrochemical processing on a firm footing as an enabling technology for nano-processing.

As the semiconductor industry prepares for the 45 nm node and beyond, issues related to integration of advanced interlayer dielectric (ILD) material into finer Cu lines has become the key challenge. Accordingly, novel electrochemical processing methods that address issues related to planarization of the fragile Cu/ILD structure and electromigration of copper interconnects are evolving . A combination of electrochemical and mechanical means of Cu removal is being employed as a novel planarization technique. Following this direction, currently there are two different approaches being pursued by semiconductor equipment vendors [11,12]. In one approach, planarization is achieved during electroplating using electrochemical mechanical deposition (ECMD), which is then followed by electropolishing to remove the overburden, while in the other approach electroplating is followed by electrochemical mechanical polishing (ECMP). Both approaches emphasize electropolishing as the *key* metal removal method, thus positioning electropolishing at the center stage of planarization technologies for interconnect structure fabrication [11].

Another important development in chip interconnect is the use of electrolessly deposited capped nano-layer that reduces copper electromigration and provides improved interconnect current density capability. The PVD sputtered Ta(N) liner and PECVD Si(C)N cap technologies have been used as diffusion barriers for copper interconnects from the 0.25um to 65 nm process nodes [10, 13]. The PVD Ta(N) liner is a relatively high-resistance film and it accounts for ~15% of the metal area on lower metal layers. The key problem with the PECVD Si(C)N dielectric capping is that it forms a relatively weak chemical bond with copper that allows for excessive copper migration at the Cu-Si(C)N interface thus giving rise to electromigration problems, which limit the maximum current density in the underlying wire [13]. The PECVD Si(C)N cap film, also, has a high dielectric constant ($k \sim 7$).

Selective electroless deposition of cobalt alloys offers a novel approach for forming self-aligned metallic cap layers. Such films have better adhesion to copper than dielectric films.

An electroless metal cap deposition process is especially appealing because of low cost, intrinsic selectivity and superior film properties. Electroless deposition of Co and its alloys has been widely studied and interconnect capping processes using, Co, CoW, CoWP, and CoWB are available in literature [13–19]. Recent efforts have focused on developing Pd-free process for selective deposition of nano-layer of capped material [16, 17]. Long-term reliability data obtained in different research laboratories indicated that electroless CoW capping tremendously improved EM resistance [13, 17–19]. The electroless CoWP cap has also been shown to have excellent adhesion to copper, good corrosion and diffusion properties, as well as good selectivity. The semiconductor equipment industry is currently involved in the development and qualification of both Pd activation-based and Pd-free CoWP processes. Figure 14.2 shows SEM photographs of a CoWP-capped copper interconnect using a commercially developed Co alloy process and tool [19].

Chip-package interconnection technologies currently used in the semiconductor industry include wire bonding, TAB, and flip-chip solder connection. Flip-chip (C4) interconnection is an area array configuration in which the entire surface of the chip can be covered with bumps for the highest possible I/O counts. Various solder bumping technologies are used in high volume production, including evaporation, electroplating, and solder paste printing. IBM's original C4 technology involved evaporation of both seed layers and high melting temperature PbSn (90–97% Pb) solders and was mainly meant for high-end applications involving ceramic packages. However, with the increased demand of higher I/Os for consumer and mid-range products with

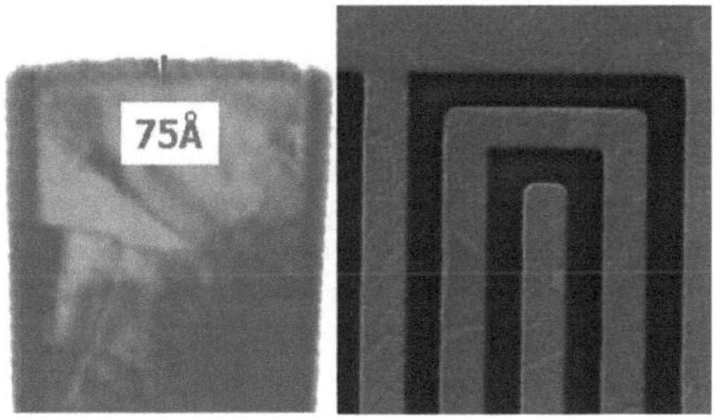

Fig. 14.2 SEM photographs of copper interconnects capped with 7.5 nm thick CoWP. *Left* cross section view, *Right* top view. Ref. [19]

plastic packages, which required a cost-effective C4 process with lower melting temperature eutectic solders (63Sn37Pb), the limitations of evaporation became apparent and paved the way for electroplating. Screening of solder paste is another means of fabricating solder bumps on wafers and packages. While this method is cost-effective for certain applications, inability to fabricate metal screens with fine dimensions limits the bump size/pitch and issues related to solder voids limit the process to low-end products. Electrochemical fabrication of C4s has now become the industry standard for advanced microprocessor assembly [20].

Electrochemical fabrication of C4s is an extremely selective and efficient process, which is extendible to finer pitch, larger wafers, and a variety of solder compositions including some lead-free alloys [20, 21]. These advantages, coupled with the advantages of area array interconnections, are making the plated C4 technology a preferable chip-package interconnection for a variety of products. Besides through-mask electroplating of solder alloys, the C4 fabrication involves careful etching of the underlying seed/ball-limiting metallurgy (BLM) layers. For some selected BLM layers such as phased CrCu, electroetching methods have been found to be the only means for their removal. This led to the development of various manufacturing processes and tools for electroetching.

Currently, two types of as-plated bump shapes are common in the microelectronics industry: mushroom bumps and column bumps. While mushroom bump technology has the advantage of using industry-standard-thin (up to 25 μm) photoresist, it may not be applicable to advanced products with narrower pitch bumps and high I/Os, due to possible bridging problems associated with these bumps. Furthermore, volume uniformity of mushroom bumps is difficult to control. On the other hand, the use of a thicker photo-resist mask to restrict the bump-plating process within the photo-resist feature permits fabrication of column bumps. Column bumps are extendible to finer pitch and higher I/Os, and provide better volume uniformity control. With the availability of cost-effective thick photo-resist technology, the semiconductor industry is rapidly moving from mushroom bumps to column bumps.

Another new method of solder bumping is based on injection-molded solder technique. This technology, which is known as C4NP (C4-new process) was developed

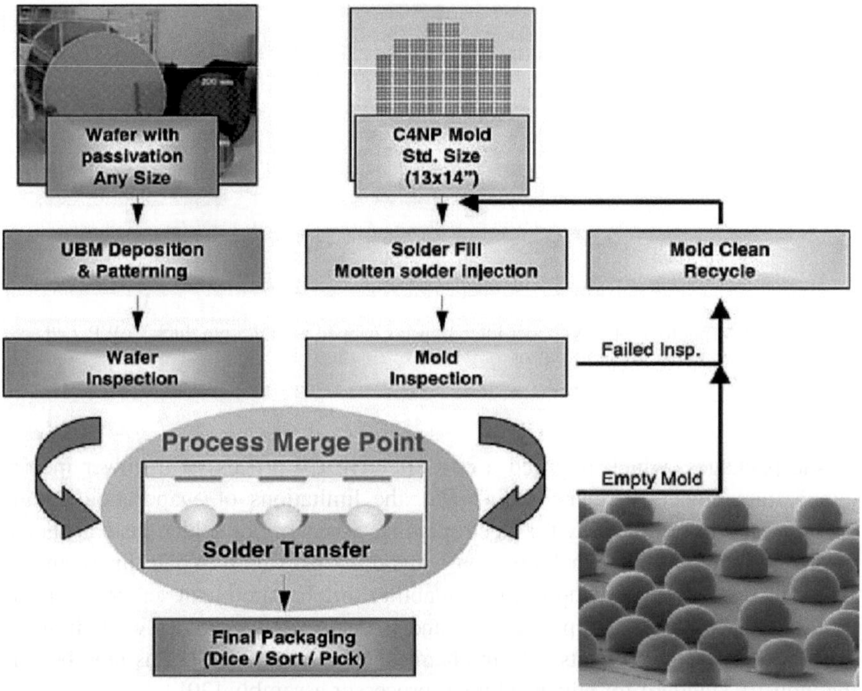

Fig. 14.3 Process steps involved in the C4NP technology. Wafers with BLM and solder-filled molds are processed in parallel and joined together. Solder bumps are thus transferred to the wafer. *Inset*: Solder balls formed by C4NP process. Ref. [23] reproduced with permission of the Electrochemical Society

at IBM and is now commercialized by Suss MicroTec [22, 23]. C4NP is a solder transfer technology where molten solder is injected into prefabricated and reusable glass templates (molds). The sequence of process steps involved is shown in Fig. 14.3. Borofloat glass plates, with the coefficient of thermal expansion matching that of silicon, are used to form the molds. Lithographic patterning and etching processes are used to form cavities with well-defined bump pitch and whose diameter and depth precisely determine the volume of the solder bump. The mold cavities are precisely filled with solder by injecting molten solder through a solder-injection head. After inspection, molds and wafer are brought into contact and solder bumps are transferred onto the entire wafer. C4NP technology is capable of fine pitch bumping while offering alloy selection flexibility. The technology is, therefore, particularly suited for exotic ternary or quaternary lead-free solder-bumping application.

Lead-tin alloys are the most commonly used solder materials for microelectronic packaging. Pb and Pb-rich alloys have one of the most desirable characteristics of C4 solders in that they are soft and compliant. The compliant nature of these alloys act a cushion for absorbing thermal and mechanical stresses and transfer minimum stresses to the die during microprocessor assembly processes. On the other hand, the melting temperature of these C4 materials is in the range of 310–328°C. The high joining temperatures required for these C4 solders make these materials incompatible with the advanced interconnect ILD materials and plastic packages. Indeed, integration of low-k/ULK dielectric in the chip necessitates a low temperature joining process, hence a low melting temperature solder. These integration issues coupled with increasing health and environmental concerns of lead-containing alloys, the microelectronic industry is now gearing up to the worldwide call for Pb-free solders.

In spite of a tremendous amount of effort for the search of lead-free C4s, no industry standard has evolved as yet. Commonly cited lead-free solders are Sn-rich alloys for which electroplating processes are available. The most popular lead-free solder among them is Sn3.9Ag0.6Cu with a melting temperature of 217°C. This solder is recommended by National Electronic Manufacturing Initiative (NEMI) and has been extensively studied and characterized by NIST. However, applicability of this solder as a C4 material in chips with advanced ILD is not known.

Indeed, Sn-rich solders present several issues, which need to be thoroughly examined for their applicability as advanced C4 materials. Sn-rich solders are normally 2.5–3.0 times harder than Pb-rich alloys (Table 14.1). Due to this reason, integration of Sn-rich C4s with fragile ULK will require significant modifications in the chip-package assembly process to avoid issues due to chip cracking. Another issue is high reactivity of Sn with copper. In some cases, Sn migration through BLM cracks may lead to chip failure [24, 25]. Sn-rich C4s, therefore, require selection of robust BLM layers and development of their etching processes. Finally, Sn-rich solders are prone to whisker growth, which may cause electrical shorts. Prevention of whisker growth, however, has been possible through alloying and through the use of additives in the plating bath.

Currently, there is no unique solution to lead-free C4 material that is available at this time. Based on the above discussions, it is anticipated that the selection of lead-free solder for C4s will be dictated by the interconnect materials in the chip and by the selected packaging solution. Therefore, the lead-free C4 material selection is expected to be application-specific.

The integrity of barrier layer metallurgy (BLM) is one of the key concerns for the reliability of microprocessor assemblies using Sn-rich solders. Commonly used BLMs consist of a combination of an adhesion layer and a solderable layer [20, 21].

Table 14.1 Thermo-mechanical properties of lead and tin rich solders [11, 20]

Solder type	Metal/alloy	Melting temperature, °C	Young's modulus, GPa
Lead-rich	Pb	328	21
	97Pb3Sn	310	22
	63Sn37Pb	183	36
Tin-rich	Sn0.75 Cu	227–229	50
	Sn3.5 Ag	221	51
	Sn3.9Ag0.6Cu	217	51
	Sn58Bi	139	28.5
	Sn52In	117	23.6

Due to fast kinetics of the Sn–Cu reaction, diffusion of Sn through the BLM must be prevented. The refractory metals, such as Ti, W, and TiW, are applicable as a barrier layer for Cu. However, as shown in Fig. 14.4, cracks and other defects present in the barrier layer or induced during thermal cycling, may act as migration path for Sn, which may react with the Cu conductors thus creating electrical shorts and eventually causing failure of the device. An example of such die-package interaction leading to microprocessor failure is shown in Fig. 14.4. The reliability tests of assembled microprocessor led to failure after baking at 170°C for 480 h [24–26]. Failure analysis included EDX analysis of cross-sectioned chip surface, which showed the presence of CuSn intermetallic around final copper lines leading to shorting of lines and eventual microprocessor failure. Further analysis of the BLM layer after selective etching of the C4 solder and the solderable layer showed cracks on the Ti layer. These data indicated that migration of Sn through BLM cracks and its reaction with copper is indeed a possible mechanism that is responsible for microprocessor failure. Grain boundary diffusion in crystalline material is another possible mode of compromising barrier properties. The use of amorphous layers, such as TiW, help alleviate such issues. For good bonding, the intermetallic reaction between Cu in the BLM and the Sn in the solder produces Cu6Sn5 that adheres well both to the solder and to the Cu in the BLM. To obtain good adhesion, it is imperative that all Cu is not consumed in intermetallic reaction because the Cu6Sn5 intermetallic does not adhere well to the underlying Cr, Ti or TiW layer. With high tin-containing solders, complete consumption of the thin Cu in the BLM causes loss of adhesion and a weak interface. Therefore, one of the accepted processes used for eutectic 63Sn/37Pb solder bumping is the use of a Cu stud as the BLM. The copper stud is electroplated through a photo-resist mask on top of a sputtered TiW/Cu BLM. The function of the Cu stud (or "minibump") is to provide an increased solder wettable area and to provide a solder diffusion barrier. However, these functions of the Cu stud have to be evaluated vis-à-vis its impact on the solder fatigue life. The Cu stud process is also becoming widely popular for Sn-rich alloys that are being implemented as Pb-free solders for C4 bumping.

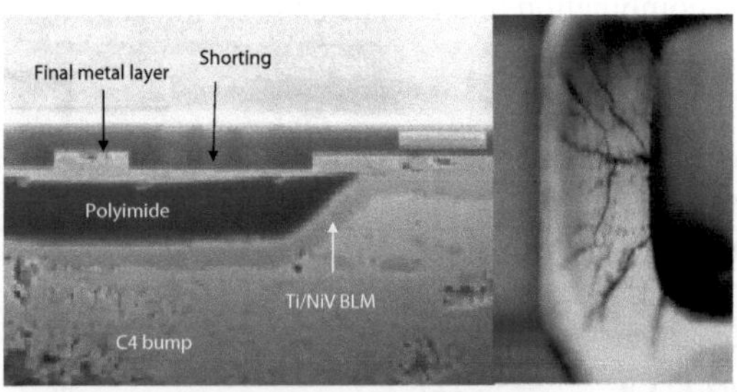

Fig. 14.4 FIB cross section of a failed chip showing the two last layers of copper metallization and C4 bump (*left*). Migration of Sn through cracks in the BLM layer and its reaction with cooper metallization led to shorting and microprocessor failure. Cracks in the Ti layer (after selective etching of solder and solderable layer) is clearly visible (*right*)

14.4 First-Level Packages and Printed Circuit Boards

First-level packages provide interconnection between the printed circuit board and the chip. These packages must have the desired number of wiring layers, provide thermal expansion compatibility with the chip, provide a thermal path for heat dissipation from the chip, and keep electrical noise and transmission delay to the minimum. The two types of packages used are plastic packages and ceramic packages.

Plastic or organic packages offer many advantages over ceramic packages in terms of size, weight, performance, cost, and availability. Tremendous improvements in hermeticity, and reliability of organic packages have increased use of plastic packages in a wide range of products spanning from consumer to low-end to high-end microelectronic assemblies and account for 97% of the first-level package market [27]. The substrate technology is generally based on a single-level metal frame, which is the foundation of the molded plastic package. The two most common lead frame metals employed by the industry are nickel–iron alloy (alloy 42, 42%Ni/58%Fe), and copper alloy [27, 28]. The chip is bonded with a die attach adhesive onto the die paddle. Gold wire bonding is used to make the connections between the chip and the lead frame. Injection molding process is used to encapsulate the IC and bond wires. The outer leads are then solder-screened or plated for board mount. Plastic ball grid array (PBGA) package with an organic substrate technology has now become the technology of choice for IC packaging. It is based on the concept of printed circuit board manufacturing, which uses plated through-hole technology used to form the copper interconnections. Development of these denser but lighter packages were made possible through advances in electrochemical processing technologies including the ability to obtain micro-via plating coverage, fine wiring, and high aspect ratio through hole plating.

Ceramic packages use co-fired multilayer ceramic substrates and provide the highest wiring density of all packaging technologies. Co-firing of as many as 63 layers in full production and 100 layers in development have been reported [6]. Indeed, hermiticity and exceptional dimensional stability of ceramic substrates make these packages superior to organic packages. Since high density of wiring is possible in thin film layers, the addition of thin film layers on ceramic packages reduces the total number of layers required in the package. Electrochemical processing technologies played a significant role in the development of these packages. Some of these process steps include: electroplating and etching to form vias and conductor wires, electroless plating of Co(P) as diffusion barrier layer, electroless deposition of Ni/Au on sintered molybdenum pads, and chemical mechanical polishing for planar structures.

For PCB, plated through-hole technology enables interconnection of various layers through formation of vias and holes. The sequential lamination involves lamination of subassemblies with drilled and plated-through holes interconnection. Evolution of PCB industry owes much to the development in electrochemical processing technologies, which enabled efficient fabrication of circuits on non-metallic surfaces through the availability of high volume plating and etching technologies.

PCB wiring requires high ductility copper, for which both electroless and electrolytic baths have been developed [29, 30]. Electroless plating provides good throwing power, hence is suitable for high aspect ratio multilayered boards with densely packed inhomogeneous circuitry.

The on-going materials development efforts in the packaging industry are aimed at developing nano-engineered packages with polymers and metals having improved thermo-mechanical properties. Some of the challenges include nano-material synthesis, placement at nano-scale, material property control at molecular level, and new metrologies for characterizing morphology and properties at nano-scale.

14.5 Thermal Management

Microprocessor scaling for increased performance and reduced cost places significant challenges on power delivery and heat removal. Heat removal is essential for an electronic package to offer optimal performance without failure. The substrate and circuit often possess different rates of thermal expansion. Thermal cycling and other thermo-mechanical stresses can lead to breaks in interconnections and package failure. Solutions using advanced materials and thermal management systems such as heat spreaders and efficient cooling systems are needed to facilitate heat removal. Key thermal management challenges include increasing power dissipation and the need to cool regions of local power concentrations on the die known as hot spots. A typical microprocessor assembly shown in Fig. 14.1 includes thermal management features such as integrated heat spreader (IHS) to spread heat while transporting heat from the die to the heat sink, which in turn dissipates heat to the environment through a fan. Successful thermal management requires the use of a thermal interface material (TIM) that makes good thermal contact between the die and the integrated heat spreader and the heat spreader and the heat sink. TIMs are generally thermal greases and gels that are made up of a polymeric material loaded with thermally conducting metallic or ceramic fillers [31, 32]. Heat dissipation through these materials occurs through percolation. Metallic TIMs such as solders materials provide heat dissipation entirely through conduction. However, the integration of metallic TIMs in microprocessor assembly is an enormous challenging task involving issues related to stresses and expensive processing steps.

14.5.1 Advanced Cooling Systems

Solutions using advanced materials and thermal management systems, such as heat spreaders and cooling systems, are currently employed to facilitate heat removal. Heat sinks and heat pipes are among the commonly used solutions for cooling microprocessor assemblies. In each case, successful thermal management requires the use of a TIMthat makes good thermal contact between the die and the heat sink and the integrated heat spreader. TIMs are generally thermal greases or gels that are made up of a polymeric material loaded with thermally conducting metallic or ceramic fillers. Advanced thermal management solutions are now evolving, which include the use of cooling systems such as micro-channel (also known as micro-structure), liquid cooling, and attention to enhanced thermal conductivity TIMs. The use of a liquid-cooling system is attractive because of higher heat transfer coefficients or lower thermal resistance as compared to traditional heat pipe or heat sink solutions.

A liquid-cooling system (LCS) for cooling microprocessors is shown in Fig. 14.5, which shows the schematic of a closed-loop LCS for a typical application that requires cooling of two microprocessors [33]. Two microheat exchangers are attached to the microprocessors with a TIM in between the heat exchanger and the microprocessor. Cold liquid, driven by a pump, enters the microheat exchangers where the flowing liquid extracts the heat from the microprocessors. The warm liquid flows into a fan-cooled radiator where it rejects the heat to the air thus cooling the liquid in preparation for repeating the cycle.

A microheat exchanger is one of the key components of LCS. Figure 14.6a shows a schematic diagram of a typical microheat exchanger. The microheat exchanger base (also known as the cold plate) is placed on top of the heat source with a thermal interface material in between. Effective heat transfer in a cooling system requires the cooling fluid to be in contact with as much surface area as possible of the material that is designed to extract the heat. In microheat exchangers, where the non-dimensional heat transport rate, as expressed by its Nusselt number, is a constant quantity – the heat-transfer coefficient is inversely proportional to the

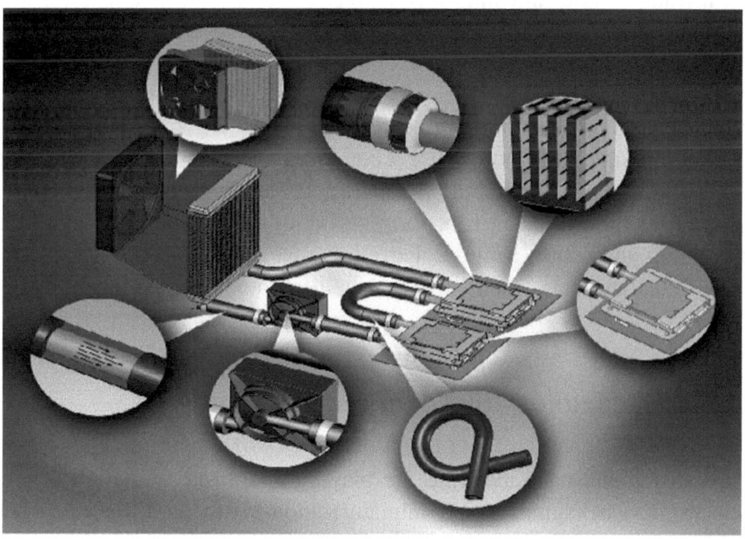

Fig. 14.5 A commercially available Custom Designed Liquid Cooling System [33]

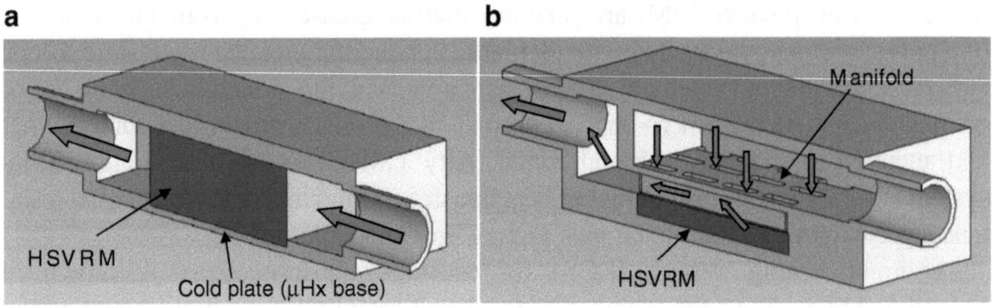

Fig. 14.6 Schematic diagram showing cross section of microheat exchangers (**a**) without and (**b**) with manifold [33, 34]

hydraulic diameter. Heat exchangers with *high surface-to-volume-ratio microstructures* (HSVRMs) through which the liquid flows should, therefore, provide significantly enhanced heat transfer. Fabrication of heat exchangers with reliable and efficient HSVRMs is, therefore, extremely critical.

HSVRMs with very fine openings provide significantly enhanced heat transfer. However, the problem with fine openings is that they also increase the pressure drop, imposing additional requirements on the pumping system. This problem has been circumvented by incorporating a manifold into the heat exchanger, while retaining the microstructure adjacent to the surface to be cooled [34]. Use of a properly designed manifold helps to uniformly distribute the cooling liquid flow

through the HSVRM, thus significantly minimizing pressure drop. A typical microheat exchanger with a manifold over HSVRM is shown in Fig. 14.6b. It is apparent from the above discussion that developing a microheat exchanger with a high heat-transfer rate and minimized pressure drop is a challenging task that requires proper design of the HSVRM structure.

In addition to efficient microheat exchangers, high thermal performance of LCS require high efficiency radiator designs and the use of a high thermal conductivity cooling liquid that provide minimum resistance to heat transfer. Highly reliable and robust LCS marketed by Cooligy use a compact, low-noise, and reliable pumping system; optimized cooling liquid with particular attention to the highest level of corrosion protection by using film-forming inhibitors; and ultra-low permeability flexible tubing and robust triple-sealed joints that ensure a completely leak-proof system with insignificant water loss over the lifetime of microprocessors [33].

14.6 Nanotechnology in Advanced Packaging

Nano-processing and nano-materials are playing a key role in the advancement of microelectronics packaging. Steady miniaturization of silicon devices and interconnects represents an evolutionary nano-technology where nano-processing has been integrated with conventional microscale processing. An enormous amount of exciting research is currently underway in the area of nano-processing that is expected to shape the next generations of electronic packages. Such topics include: the use of nano-particles to influence thermo-mechanical properties of solders for flip-chip, BGAs, and packaging materials; the use of carbon nano-tubes, nano-fibers and their metallic composites for advanced chip interconnects and thermal interface materials; fabrication of porous nano-structures and development of nano-fluids for advanced cooling systems; ink-jet printing for low-cost printed electronics; and precise fabrication of n- or p-type nano-films and nano-wires for application in thermoelectric coolers. Some of these topics are briefly discussed below.

14.6.1 CNT/CNF and Their Metallic Composites

Carbon nanotube (CNT) is one of the most-researched topics of nanotechnology. A CNT is a long, thin, cylindrical, hexagonal lattice (graphene) of carbon molecules, one molecule thick, only 10 to 20 atoms around, and up to 100 micrometer long. CNTs display extraordinary mechanical, electrical, and thermal properties [35, 36]. They have much higher current carrying capability than copper, they conduct heat as well as a diamond, and are about 100 times stronger than steel at 1/6th of the weight per volume. Their characteristics vary depending on how they are rolled, and how thick they are (single or multi-walled). Depending on the direction in which CNTs are rolled up they demonstrate either metallic or semiconducting properties, which is one reason why they can carry very high currents while emitting little heat. Because of their extremely desirable properties of high mechanical and thermal stability, high thermal conductivity and large current-carrying capacity, CNTs have aroused a lot of research interest in their applicability as ULSI interconnects of the future [37, 38]. Mixing together nano-tubes with different electrical properties could simplify the design of future chips. Le et al. used PECVD to grow whisker-like CNTs on the surface of a silicon wafer [38]. A layer of silicon dioxide was then deposited to encapsulate the CNTs and the chip surface. The top part of the CNTs and part of the silicon dioxide layer are then polished by CMP planarization. Using this method processing of multilayer interconnects using vertical carbon nano-tubes was demonstrated. However, the establishment of a method to separate and sort CNTs of different conductivity types remains one of the most important topics in nano-tube research.

Carbon nano-tubes and carbon nano-wires (CNF) have unique thermal and thermoelectric properties that give rise to new opportunities in thermal management of electronic devices. However, long and free-standing tubes and fibers may not be able to withstand the rigorous thermo-mechanical stresses in packaging process flows. Gap-filling copper between vertically aligned CNFs provides a suitable mechanical anchor for the nano-fibers to the substrate while also serving as a lateral heat spreader. The robust physical characteristics of the CNF–Cu composite also allow one to take advantage of increased contact surface area to the target material. Carbon nano-fiber composites are a strong candidate material to provide thermal solutions for advanced cooling systems. Ngo et al. used DC-powered PECVD to fabricate vertically aligned, free-standing CNF arrays on silicon wafers of ~500 um thickness [39]. Copper electrodeposition was used for gap-filling high aspect ratio trenches, thereby creating CNF–Cu composite array. Obtained data demonstrated the mechanical strength and efficient interfacial heat conduction of CNF–Cu composite arrays suitable for next-generation heat-sink devices. A layer of titanium (300 Å) was used as both an adhesion layer for a thin layer of nickel catalyst used for the CNF array growth and as a seed layer for the subsequent copper electrodeposition. As grown vertically aligned carbon nano-fiber arrays are shown in Fig. 14.7. A superfilling copper electrodeposition bath was employed that consisted of 100 ppm chloride ions (Cl^-), 400 ppm polyethylene glycol (PEG), 10 ppm bis(2-sulfopropyl) disulfide (SPS), 10 ppm Janus Green B (JGB), 0.6 mol/l copper sulfate, and 1.85 mol/l sulfuric acid. An etching/electropolishing step in 85% ortho-phosphoric acid solution was used to improve surface finish and to remove excess copper thereby revealing the nano-fiber ends. Ngo et al. compared the thermal resistance measurements for as grown CNF array versus CNF-Cu composite with the same length and diameter distribution of fibers. CNF-Cu composites showed lower thermal resistance indicating their potential applicability as efficient thermal interface materials in cooling devices.

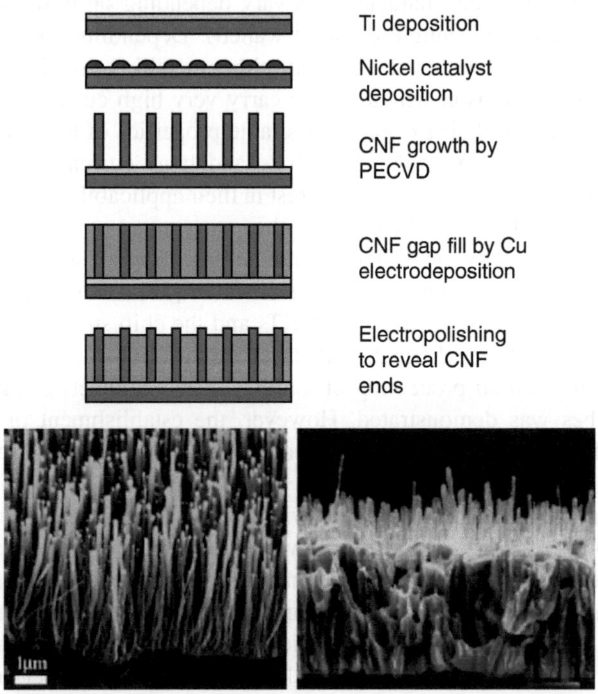

Fig. 14.7 Top: Process steps involved in the fabrication of CNF-Cu composite. *Bottom left* SEM photographs of as grown CNFs, *Bottom right* CNF-Cu composite after copper electropolishing. Reprinted with permission from ref. [39]. Copyright (2004) American Chemical Society

Fabrication of carbon nanofiber-copper composite by electroplating was studied by Arai and Endo [40]. They used commercially available vapor-grown carbon nano-fibers fabricated by catalyst-assisted chemical vapor deposition as a filler material in a copper sulfate/sulfuric acid plating bath. The 20 mm long and 100–200 nm diameter carbon nano-fibres were fabricated by catalyst-assisted chemical vapor deposition. Two forms of polyacrylic acid, PA5000 and PA25000, were added to the bath to aid dispersion of CNFs with the aid of stirrer agitation. A relatively uniform dispersion of CNFs within copper grains was observed on the film. Cross section of the films indicated that CNFs were not only deposited on the surface but were distributed within the deposited thickness. These composites were easily separated from the substrate by ultrasonic waves in acetone, and by changing electrodeposition parameters the composites could be obtained in film or powder forms.

14.6.2 Nanoparticles in Advanced Packaging

Nano-particles are expected to play a major impact in electronic packaging. Compared to their bulk counterparts, nano-scale materials exhibit large surface area and size-dependent distinct chemical, electrical, optical, thermal and magnetic properties. The thermo-mechanical properties of interconnect materials, packaging materials, and thermal interface materials can be significantly modified by the inclusion of nano-particles. Nano-particles are also finding increasing applications in nano-printing, and in nano-fluids for liquid-cooling devices. These aspects are briefly described in the following section.

14.6.2.1 Nanoparticle Seeding for Interconnects

Electrochemical deposition is an effective method of seeding the inert barrier layer due to its inherent advantages in filling high aspect ratio features, as well as low processing cost. While palladium seeding is commonly used to initiate the autocatalytic reaction of subsequent electroless copper deposition, Pd reduces the stability of the electroless Cu plating bath and the formation of CuPd alloys increases the resistivity of electroless Cu deposits [41]. Li et al. studied the deposition of copper nano-particles to act as metal activation seed layers on TaSiN barrier films for subsequent electroless or electrolytic copper deposition applicable in nano-scale interconnect metallization [42]. The process is based on an electrochemical displacement mechanism in which the more noble metal ions in the organic solution are reduced to metal nano-particle crystals at the less noble solid metal substrate. The organic deposition solution consisted of conventional solvent extractants that are very poor electrolytic conductors but can sustain short-range spontaneous reactions. Additives consisting of low formula weight organics were used to enhance the copper nano-particle deposition. An aqueous15 g/L CuSO4 solution was mixed with an equal volume of organic extraction solution consisting of 30 vol%

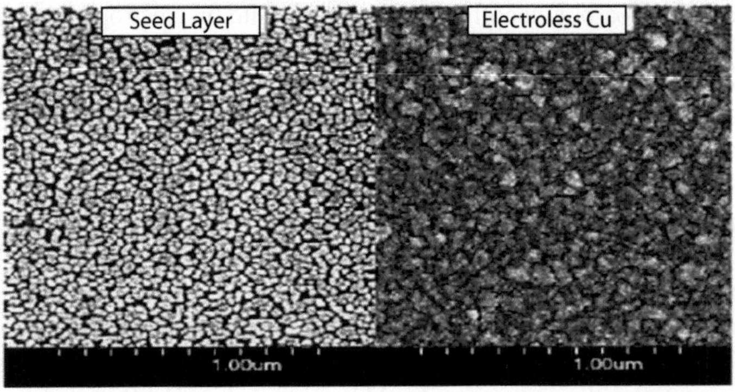

Fig. 14.8 SEM photographs of Cu seeding on TaSiN (*left*) and subsequent electroless copper coating (*right*). Ref. [42] reproduced with permission for the Electrochemical Society

di-2-ethylhexy phosphoric acid, 20 vol% tri-*n* -butyl phosphate, and 50 vol% kerosene. The Cu concentration in the organic phase ranged from 10 to 300 ppm. A 2 vol% HF aqueous solution (50% concentration) was added into the deposition solution for in situ activation of the substrate surface. A proprietary additive was used, the concentration of which influenced the seed morphology. Without the additive, very few Cu nano-particles were present while the seed density increased with increasing additive concentration. A very dense, uniform Cu seed crystal layer was obtained with 10 vol% of the additive. Increasing the concentration of the additive to 20 vol% resulted in a bi-modal size distribution. The study of Li et al. demonstrated that the organic composition, degree of agitation, the seeding time, the additive type, and concentration all effect the morphology, density, and distribution of the Cu nano-particles [42]. Using optimized solution and deposition conditions, highly dense and uniform Cu nano-particles were successfully deposited on the TaSiN barrier at near ambient temperature and pressure. Electroless deposition of copper was performed on the Cu-seeded TaSiN. The morphology and quality of the electroless Cu plating was correlated with the morphology of the Cu seed layer. SEM photographs of Fig. 14.8 show that by using the optimized conditions for nano-particle seed layer deposition and electroless copper deposition, a very uniform and smooth cooper deposit could be obtained [42].

14.6.2.2 Lead-Free Solders

The lead-free solders are mostly based on Sn-containing binary and ternary alloys. Among them, the Sn–Ag system is one of the earliest commercially available lead-free solders and has been recommended for general-purpose use as a substitute for Sn–Pb eutectic solder. Addition of nano-particles of second phase helps in improving thermo-mechanical properties such as melting temperature, mechanical strength, mechanical fatigue resistance, creep resistance and solder-joint reliability.

Lee et al. studies the influence of nickel concentration and inclusion of Ni3Sn4 nano-particles on the morphology of SnAgNi solders [43]. While inclusion of Ni_3Sn_4 nano-particles did not influence the melting temperature of the solder, the wettability of the solder to copper substrate increased significantly as evidenced by smaller contact angle of the Ni_3Sn_4 doped solders. Shen et al. studied the strengthening effects of ZrO_2 nano-particles on the microstructure and microhardness of Sn3.5Ag solder [44]. The addition of ZrO_2 nano-particles in Sn-3.5Ag solder resulted in a finely dispersed Ag_3Sn phase. The refinement of microstructure is due to the adsorption of ZrO_2 nano-particles on the surface of the nucleating Ag_3Sn phase during solidification. ZrO_2 nano-particles suppress growth of Ag_3Sn intermetallic precipitates, and thus reduce the particle size of the formed Ag_3Sn phase and lead to uniform dispersion of the phase. The finely dispersed Ag_3Sn phase provides dispersion strengthening and therefore, enhances the microhardness of Sn-3.5Ag-ZrO_2 solder.

14.6.2.3 Thermal Interface Materials

As shown in Fig. 14.1, in a microprocessor assembly, TIMs are used for conducting heat from chip to heat spreader and from heat spreader to heat sink. The TIMs are subject to challenging requirements, including the ability to reduce thermal stress between regions with vastly differing thermal expansion coefficients, the ability to be reworked, and high thermal conductivity [31, 32]. Commonly used TIMs include a variety of polymer-based materials with high thermal conductivity particle inclusions, typically with diameters of 2–25 μm. The effective thermal conductivities of particle-filled polymer interface materials are typically about an order of magnitude higher than the polymer matrix alone, i.e. of the order of 2 W/mK. Inclusion of nano-particles, in particular bundles of carbon nano-tubes or nano-fibres, is the subject of current research for improving the conductivity of interface materials. CNT/NF inclusions in soft materials promise near-ideal thermal and mechanical properties of TIMs. CNTs/CNFs offer very high directional thermal conductivities along with flexible geometry. This unique combination provides an opportunity for combining low-effective values of resistance and elastic modulus. In a recent study, thermal conductivities of silicone-based interface materials containing varying concentrations of nano-tubes and Ni particles were measured [32, 45]. The relative impact of an increase in CNT volume fraction was found to be amplified by the presence of nickel particles for CNT volume fraction above ~0.02%. This transition is consistent with a percolation model including thermal interconnections formed by CNTs between nickel particles.

14.6.2.4 Nanofluids

Fluids containing suspensions of nanometer-sized solid particles are popularly known as nano-fluids. The solid nano-particles' or nano-fibers' size vary typically in the range of 1–100 nms. Nano-fluids have attracted great interest recently, particularly in

Table 14.2 Thermal conductivities of selected solids and liquids [48]

Material type	Material	Thermal conductivity (W/m – K)
Solids	Carbon nanotubes	1,800–2,000
	Diamond	2,300
	Graphite	110–190
	Fullerenes (film)	0.4
	Silver	429
	Copper	401
	Aluminum	237
	Nickel	158
	Silicon	148
	Alumina	40
Liquids	Water	0.613
	Ethylene glycol	0.253
	Engine oil	0.145

liquid cooling systems, because of their greatly enhanced thermal properties. Solid materials have typically orders of magnitude larger thermal conductivity than liquids that are commonly used in heat transfer systems (Table 14.2). Because of the extremely HSVR of nano-particles, a dramatic improvement in the effective thermal conductivity is expected by decreasing the particle size in a solution compared to the incremental improvement that can be obtained by altering the shape of large particles. As an example, the surface-area-to volume ratio is 1,000 times larger for particles with a 10 nm diameter than that of a 10 um diameter particle. Nano-fluids are, therefore, expected to have superior heat-transfer properties compared to conventional fluids and fluids containing micrometer-sized particles. Indeed, a small amount (<1% volume fraction) of Cu nano-particles or carbon nano-tubes dispersed in ethylene glycol or oil is reported to increase the inherently poor thermal conductivity of the liquid by 40% and 150%, respectively [46, 47]. Nano-fluids with carbon nano-tubes are expected to possess even better heat-transfer properties due to the non-spherical shape of the carbon nano-tubes [48]. The optimization of nano-fluid thermal properties requires successful synthesis procedures for creating stable suspensions of nano-particles in liquids. Depending on the requirements of a particular application, many combinations of particle materials and fluids are of potential interest. For example, nano-particles of oxides, nitrides, metals, metal carbides, and non-metals with or without surfactant molecules can be dispersed into fluids such as water, ethylene glycol, or oils [46].

14.6.2.5 Ink-jet Printing

Nanotechnology is bringing a paradigm shift in packaging through the development of low-cost printed electronics based on novel inks and printing fluids [49]. Ink-jet is noncontact digital printing that produces consistent drop volume with accurate drop displacement. It is an additive process that does not waste expensive material.

The deposition system consists of ink-jet, fluid, printer and software. Ink-jet printing allows the ability to print directly on package and substrates. Development of printable fluids containing nano-particles of conducting, semiconducting, dielectric, and insulating materials is central to the ink-jet technology for electronics. The sintering of nano-particles is significantly lower than the bulk materials. This property enables the printing of high-conducting metallic structures on packaging materials such as ceramics and polymers [49]. The ink-jet fluids can be manipulated and blended to build up layer-by-layer structures of electronic components. Linking computer-aided-design to the printer thus makes it possible to create functional electronic structures.

14.6.3 *Active Micro and Nanostructures for Thermal Management*

As noted earlier, design and development of a high surface area active microstructure is an essential consideration of an efficient liquid cooling system. Indeed, effective heat transfer in a cooling system requires the cooling fluid to be in contact with as much surface area as possible of the material that is designed to extract the heat.

Several different types of High Surface to Volume Ratio Microstructures (HSVRM) and their use in microheat exchangers have been mentioned in the literature [50–56]. For brevity, we classify the HSVRMs into two types: (1) Microchannel type, and (2) Microporous type. Figure 14.9 shows typical microchannel type of HSVRMs. Silicon micro channels are one of the commonly cited heat collector structures used in liquid cooling systems [50]. High aspect ratio channels are easily fabricated by anisotropic etching of silicon, which has found widespread use in micromachining and MEMS. However, the low thermal conductivity of silicon makes it a less-effective heat exchanger material relative to metallic materials. Metallic microchannels are formed by different mechanical or electrochemical means including EDM, dicing, LIGA and plating through thick photo resist masks. Folded fins are another class of microchannel HSVRMs that are used as heat exchanger structures and they can be straight or louvered fin types [52].

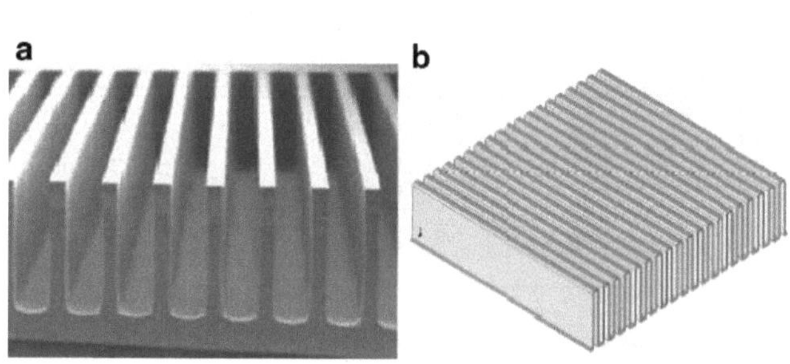

Fig. 14.9 (**a**) Micro/minichannel, and (**b**) folded fin HSVRMs [50, 51]

Microprorous types of HSVRM structures provide higher values of surface area per unit volume than commonly used micro/minichannels. Porous structures can be ordered or un-ordered. Mesh and woven mesh structures shown in Fig. 14.10 are ordered porous structures. Woven mesh structures are made of metallic wires of defined diameter to produce parts of defined mesh numbers. These commercially available structures have been studied and reported in the heat transfer literature [53]. Mesh structures shown in Fig. 14.10 can be fabricated by mechanical, or wet-etching methods [54].

Metallic foams are the most common form of un-ordered microporous structures that have undergone thermal transport studies in the literature [55]. Open cell metal foams shown in Fig. 14.10 are created by a variety of methods including sintering, metal deposition by evaporation, CVD or electrodeposition. Metal sintering is the most cost-effective means of manufacturing metal foams. In this process, a polymeric foam substrate is coated with a slurry of metallic particles and is heated in a furnace. The polymer foam skeleton vaporizes and the metal particles sinter together leaving behind a metallic foam structure. The open-cell foams are generally compressed to a desired size for their application in heat exchangers.

Metal foams can also be created by electrodeposition. Indeed, the surface of area of electrochemically formed porous foam structure is several orders of magnitude higher than the other types of foams. Electrodeposition can easily create a 3-D dendritic structure. However, it is difficult to control the microstructure since trunks and branches in the porous dendritic structure are often unable to support the weight of numerous sub-branches. Shin et al. developed a technique of making self-supported nano-ramified deposits that involves electrodeposition accompanied by hydrogen evolution [56]. In this process, hydrogen bubbles function as a dynamic negative template. During the deposition, the growth of the dendritic copper structure was blocked by the hydrogen bubbles and they functioned as a

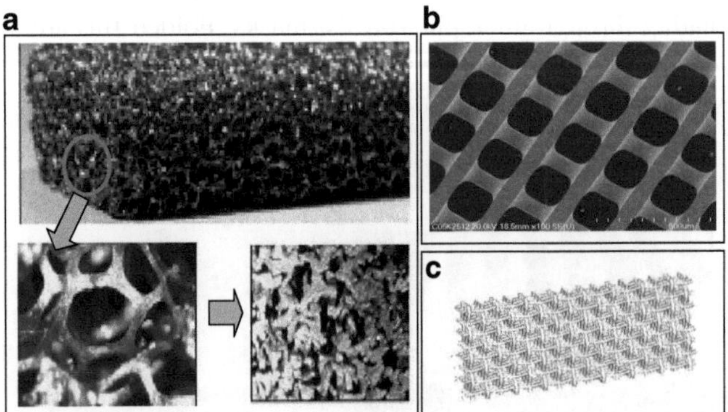

Fig. 14.10 Microphotographs of porous HSVRMs [53–55]. (**a**) Metallic foam, (**b**) mesh structure, and (**c**) woven mesh structure

dynamic masking template during the deposition. The hydrogen bubbles depart from the surface, rise and merge into larger bubbles, and as a result, the pore size of the deposited copper structure increases with distance from the surface. The deposition process can be described as a competition between hydrogen evolution and coalescence away from the surface and metal deposition on to the surface. Using this technique, Shin et al. fabricated 3-D free-standing copper and tin foams with highly porous ramified (dendritic) walls. Figure 14.11 shows SEM photographs of porous copper deposits created by electrodeposition at different deposition times. The data indicted that the surface pore size increased with deposition time. Enhanced boiling heat transfer from nano-dendritic microporous copper structures was investigated by Furberg [57]. The study involved a detailed investigation of different fabrication parameters including current density, time, electrolyte concentration, additives, temperature and annealing. At low current density (<2 A/cm^2) the frequency and nucleation density of the hydrogen bubbles were low, resulting in a dense dendritic structure without any pores. At increasing current density (≥3 A/cm^2), the bubble population, frequency and coalescence increased to such an extent that the bubbles created permanent voids above the cathode and thereby functioned as a masking template, producing the desired structure. Longer deposition time and higher copper concentration in the electrolyte resulted in a thicker structure with larger pores.

Pore size and wall structure could be varied by using different additives. Hydrochloric acid, as an additive, considerably reduced the elementary branches in the structure thus influencing the pore shape. The influence of temperature on pore structure is shown in Fig. 14.12. The pore size increased considerably by increasing the electrolyte temperature from 20°C to 65°C. The dendritic branches comprised nano-sized grains between 10 and 20 nm. The electrical resistance decreased from 0.23 to 0.01 ohm after annealing at 500°C for 5 h [57].

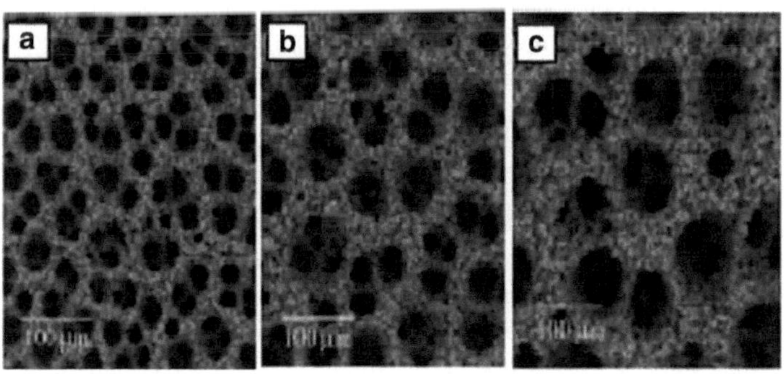

Fig. 14.11 SEM photographs porous copper deposits at different deposition times of 5 s (**a**), 10 s (**b**), and 20 s (**c**). Ref. [56] Copyright Wiley-VCH Verlag GmbH & Co.KGaA. Reproduced with permission

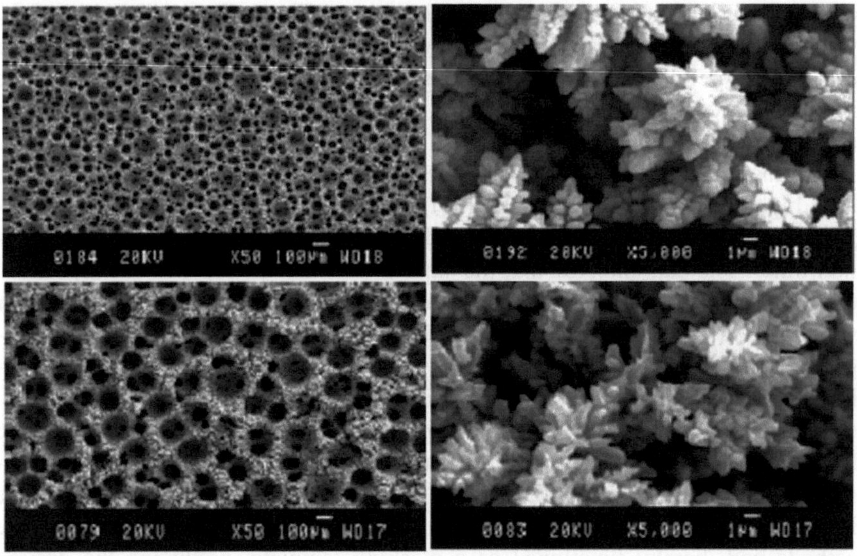

Fig. 14.12 Electrodeposited micro-porous structures fabricated at 20°C (*top*) and at 65°C (*bottom*) electrolyte temperatures. Reproduced from ref. [57]

14.6.4 Nanowires for Thermoelectric Probes

Thermoelectric temperature control systems are being used in an increasing number of applications. Typical applications of these systems are in thermal regulation of electronic enclosures, recirculating chillers for laser cooling, semiconductor process control and management of patient hypo- and hyperthermia. Thermoelectric coolers (TECs) are essentially solid state devices that pump heat against a temperature differential utilizing electrical energy. Based on the Peltier effect, each cooler consists of a matrix of thermoelectric p–n thermocouples connected electrically in series and thermally in parallel. Thermoelectric microprobes or arrays of such probes suitable for highly localized cooling and heating, in situ temperature sensing, and temperature control have potentially significant applications where thermal manipulation at micro/nano levels is required. A thermoelectric microprobe for temperature manipulation consists of an n-branch and a p-branch forming p–n junctions. When a current is supplied to the microprobe, the temperature will increase at one junction and decrease at the other. In order to be useful for the targeted applications, the thermoelectric microprobe or arrays must be from 500 to 750 µm tall.

Efficient thermoelectric devices require the use of high figure of merit thermoelectric materials. The thermoelectric figure of merit, ZT, can be expressed as $\alpha 2\sigma T/\kappa$, where α is the Seebeck coefficient, σ, the electrical conductivity, T, the temperature, and κ, the thermal conductivity. Among the various thermoelectric materials, bismuth telluride (Bi2Te3) has been the main focus of research because of its superior ZT near room temperature [58].

Bismuth telluride with the stoichiometric composition ratio (Bi_2Te_3) is electronically neutral. However, the common carrier of the bismuth telluride alloy can be modified by adjusting the percentage of tellurium or bismuth incorporated. If the alloy is telluride-rich, it can be n-type. In order to be p-type, the composition of the bismuth telluride alloy should be bismuth-rich instead of telluride-rich [59]. This is advantageous since the material deposition development can focus on a single material system which will yield both the n- and p-branches of the thermoelectric probes.

Yoo et al. studied the electrodeposition of n-type BiTe alloy thin films [60]. Bi_xTe_y films were electrodeposited on a Bi_2Te_3 seed layer from an aqueous nitric acid bath at 23°C. The bath composed of 0 to 0.008 M Bi^{3+} from $Bi(NO_3)_3$ and 0.01 M $HTeO_2^+$ from TeO_2 in 1 M HNO_3. The pH was maintained at pH 0.1, and a 1 μm thick Bi_2Te_3 seed layer was deposited on a SiO_2/Si wafer (lightly doped p-type) by rf sputtering. The film compositions, which varied from 57 to 63 at.% Te were strongly dependent on the deposition conditions. Surface morphologies varied from needle-like to granular structures depending on the deposited Te content. Electrical and thermoelectric properties of these electrodeposited Bi_xTe_y thin films were measured before and after annealing and compared to those of bulk Bi_2Te_3. Annealing at 250°C in reduced hydrogen atmosphere enhanced thermoelectric properties by reducing film defects. In-plane electrical resistivity was highly dependent on composition and microstructure. In-plane Hall mobility decreased with increasing carrier concentration, while the magnitude of the Seebeck coefficient increased with increasing electrical conductivity to a maximum of −188.5 μV/K. The thermoelectric properties of electrodeposited n-type BiTe thin films after annealing were comparable to those of bulk BiTe films.

Theoretical analysis by Hicks and Dresselhaus indicated that ZT can be significantly enhanced by reducing the structural dimensions to a single dimension [61]. The analysis further showed that ZT for nano-wires will be significantly improved by decreasing the diameter. As a result, research into the fabrication and characterization of nano-wire structures of thermoelectric materials has received much attention in recent years [62–66].

Huang et al. used the LIGA(Lithographie Galvano Formung Abformung) process to fabricate bismuth telluride microposts of 150 μm diameter and heights up to 500 μm [62]. Bismuth telluride alloys were electrodeposited galvanostatically into microfeatures on a titanium substrate. The electrolyte was prepared by dissolving BiN and tellurium in nitric acid solution of zero pH. The monophasic BiTe alloy microposts had a polycrystalline structure with excellent adhesion to the substrate and good mechanical strength. By controlling the electrolyte composition and the current density either p- or n-type BiTe alloy microposts could be produced.

Synthesis of arrays of BiTe nano-wires by electrodeposition has been studied by different authors [63–66]. Li et al. employed pulsed electrodeposition into the nanochannels of porous anodized alumina membranes to fabricate nano-wire arrays of Be_2Te_3 [63]. The anodized alumina membrane with the pore sizes of about 40 and 60 nm were used. A 200 nm thick layer of Au film was sputtered onto one side of the anodized alumina membrane to serve as the working electrode in a two-electrode plating cell, and a graphite plate was used as the counter-electrode.

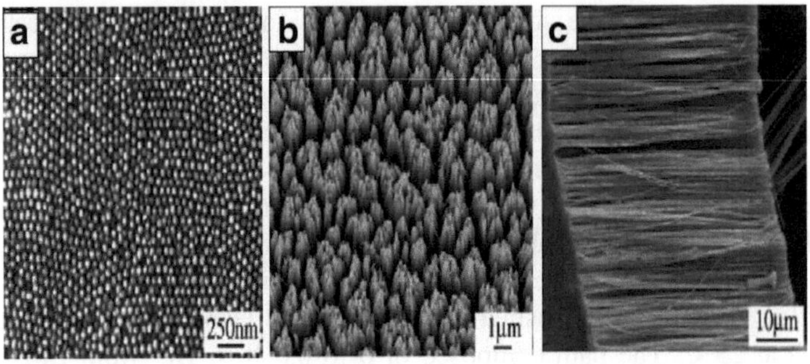

Fig. 14.13 SEM photographs of the bismuth telluride array showing (**a**) bottom view, (**b**) surface view, and (**c**) cross sectional view respectively. Ref. [63] reproduced with permission of the IOP

Bi_2Te_3 nano-wire arrays were deposited under modulated voltage control and a potential of −1.3 V was applied between the two electrodes. During the pulse time, 3 ms, metal species were electrochemically reduced on the pore ground of the membrane. The delayed time, 10 ms, provided time for the recovery of the ion concentration. The electrolyte was prepared by dissolving 0.01 M $Bi(NO_3)_3{\cdot}5H_2O$ and 0.015 M TeO_2 in HNO_3 solution. The pH of the final electrolyte was adjusted to about 1 with HNO_3. High-filling-rate, highly oriented, and ordered Bi_2Te_3 nano-wire arrays could be thus fabricated. The Bi_2Te_3 nano-wires were single crystalline with a preferential orientation along the [015] direction. Optimum growth condition could be achieved by the proper choice of the pulsed parameters and the deposition potentials and ion concentrations in the electrolyte in Fig. 14.13. The electrical transport measurements showed that the Bi_2Te_3 nano-wires with the diameters between 40 and 60 nm exhibited typical semiconductor behavior, and the 40 nm nano-wires have lower resistance sensitivity to temperature than the 60 nm nano-wires. Similar studies were conducted by Sanders et al., who employed DC electrodeposition into porous anodic alumina template to fabricate nano-wire arrays of Bi_2Te_3 [64]. Purkayastha et al. developed a template-free method to synthesize rod-shaped rohmbohedral nanocrystals of bismuth telluride [65]. Menke et al. prepared polycrystalline bismuth telluride using step edge selective cyclic electrodeposition on highly oriented pyrolytic graphite surfaces [66]. Control of the number of cathodic/anodic scans allowed the diameter of bismuth telluride to be varied between 100 and 300 nm.

14.7 Concluding Remarks

This chapter provided a brief description of the current trends in microelectronic packaging technologies and some examples of the role of nanotechnology in the advancement of packaging materials and microelectronic packages. The discussion was mostly limited to topics that highlighted the impact of electrochemical processing in microelectronic packaging. Dual Damascene plating for Cu chip metallization and electroplated C4 technology have enabled a paradigm shift in semiconductor packaging. As the industry moves to 45 node technology, implementation of electropolishing-based planarization and electroless-capped layer in manufacturing are the key challenges for the next generation of chip interconnects. Future planarization efforts are expected to focus on the development of integrated electropolishing method for barrier layer removal and elimination of final CMP process. Several different approaches are expected to evolve to address electromigration issues in chip and chip-package interconnects. In the chip-package interconnection arena, the current focus is on the development of low melting temperature, Pb-free flip-chip technology. Some other packaging trends include the use of novel materials for packages and boards with improved compliance and CTE mismatch and advanced approaches to heat dissipation.

While micro-scale technologies are expected to continue to thrive, nano-processing and nano-materials are already playing a key role in the advancement of microelectronic packaging. Steady miniaturization of silicon devices and interconnects has been possible due to smooth integration of nano-processing with conventional micro-scale processing. An enormous amount of exciting research is currently underway in the area of nano-technology that is expected to shape the next generations of electronic packages. Indeed, due to their high surface to volume ratio, nano-materials have the ability to revolutionize the packaging materials. A brief description of such applications has been presented, examples of which include: the use of nano-particles to influence thermo-mechanical properties of solders, adhesives, and packaging materials; the use of carbon nano-tubes, nano-fibers and their metallic composites for advanced chip interconnects and thermal interface materials; fabrication of porous nano-structures and development of nano-fluids for advanced cooling systems; ink-jet printing for low-cost printed electronics, and precise fabrication of n- or p-type nano-films and nano-wires for application in thermo-electric coolers. In summary, nano-materials and nano-processing are expected to address the key issues related to continued miniaturization and increased performance of next generation of microelectronic packages.

References

1. Romankiw LT, Croll I, Hatzakis M (1970) IEEE Trans Magn 6:729
2. Romankiw LT (1997) Electrochim Acta 42:2985
3. Osaka T (1997) Electrochim Acta 42:3015
4. Datta M, Landolt D (2000) Electrochim Acta 45:2535
5. Datta M (2003) Electrochim Acta 48:2975
6. Rymaszewski EJ, Tummala RR, Watari T (1997). In: Tummala RR, Rymaszewski EJ, Klopfenstein AG (eds) Microelectronic packaging handbook, part I, 2nd edn. Chapman and Hall, New York
7. Tummala RR, Garrou P, Gupta T, Kuramoto N, Niwa K, Shimda Y, Terasawa M (1999). In: Tummala RR, Rymaszewski EJ, Klopfenstein AG (eds) Microelectronic packaging handbook, part II, 2nd edn. Kluwer Academic Publishers, Boston

8. D.P. Seraphim, D.E. Barr, W.T. Chen, G.P. Schmitt, and R.R. Tummala (1997). In: Tummala RR, Rymaszewski EJ, Klopfenstein AG (eds) Microelectronic packaging handbook, part III, 2nd edn. Chapman and Hall, New York
9. Datta M (2005). In: Datta M, Osaka T, Schultze WJ (eds) Microelectronic packaging, CRC Press, pp 3–27
10. Edelstein DC (1997). Tech Dig IEEE Intl electron devices conference, 773, 1997; IBM Res Mag, No. 4, 16
11. Datta M. In: Krongleb S, Bonhote C, Osaka T, Kitamoto Y (eds) Proceedings, 8th Intl. symposium on magnetic materials processes and devices, Electrochem Soc, NJ, PV2004-23, pp 126–143
12. Basol BM (2004) J Electrochem Soc 151:C765–C771
13. Hu C, Gignac L, Rosenberg R, Liniger E, Rubino J, Sambucetti C, Domenicucci A, Chen X, Stamper AK (2002) Appl Phys Lett 81:1782–1784
14. Dubin VM, Lopatin S, Kohn A, Petrov N, Eizenberg M, Shacham-Diamand Y (2004). In: Datta M, Osaka T, Schultze WJ (eds) Microelectronic packaging, CRC Press, pp 65–110
15. Kohn A, Eizenberg M, Shacham-Diamand Y, Israel B, Sverdlor Y (2001) Microelectronic Eng 155:297–303
16. Nakano H, Itabashi T, Akahoshi H (2005) J Electrochem Soc 152(3):C163–C166
17. Moon P, Dubin V, Johnston S, Leu J, Raol K, Wu C (2003). Proc IEDM, IEEE Intl, pp 35.1.1–35.1.4
18. Hu C, Gignac L, Liniger E, Herst B, Rath DL, Chen ST, Kaldor S, Simon A, Wang W-T (2003) Appl Phys Lett 83:869
19. Lee B, Ivanov I (2009). In: Shacham-Diamand Y, Osaka T, Datta M, Ohba T (eds) (2009) Advanced nanoscale ULSI interconnects: fundamentals and applications, Springer
20. Datta M (2004). In: Datta M, Osaka T, Schultze WJ (eds) Microelectronic packaging, CRC Press, pp 167–200
21. Datta M, Shenoy RV, Jahnes C, Andricacos PC, Horkans J, Dukovic JO, Romankiw LT, Roeder J, Deligianni H, Nye H, Agarwala B, Tong HM, Totta PA (1995) J Electrochem Soc 142:3779
22. Gruber PA, Belanger L, Brouillete GP, Danovitch DH, Landreville JL, Naugle DT, Oberson VA, Shi DY, Tessler CL, Turgeon MR (2005) IBM J Res Dev 49(4/5):621
23. Gruber PA, Budd RA, Buchwalter SL, Shi DY, Busby JA, Grant JJ, Giri AP, Knickerbocker SH, Longworth HP, Naugle DT. Abstract #1634, 120th ECS meeting, Oct. 29–Nov. 3, 2006, Cancun, Mexico
24. Datta M, Emory D, Huang T-L, Joshi SM, King CA, Ma Z, Marieb T, McKeag M, Suh D, Yang S. US Patent No. 6,740,427, May 25, 2004
25. Datta M, Emory D, Joshi S, Menezes S, Suh D. US patent no. 6,853,076, February 8, 2005
26. Moon P, Zhiyong Ma, Datta M. US patent no. 6,703,069, March 9, 2004
27. Pecht MG, Nguyen LT (1999). In: Tummala RR, Rymaszewski EJ, Klopfenstein AG (eds) Microelectronic packaging handbook, part II, 2nd edn. Kluwer Academic Publishers, Boston
28. Breedis JT (1986). J Metals AIME 48
29. Van Tiburg GC (1984) Plat Surf Finish 71(6):78
30. Houma H, Mizushima S (1984) Met Finish 82(1):47
31. Schelling PK, Shi L, Goodson KE (2005). Materials Today, 30–35
32. Parasher RS, Chang J-Y, Sauciuc I, Narasimhan S, Chou D, Chrysler G, Myers A, Prstic S, Hu C (2005). Intel Technol J 9(04)
33. Datta, M, Lin E, Choi H, McMaster M, Brewer R, Werner D, Hom J, Upadhya G, Gopalakrishnan S, Rebarber F (2007). Transactions of the Electrochemical Society, 6(8): 13–31
34. Brewer R, Upadhaya G, Zhou P, McMaster M, Tsao P. US patent # 7,188,662, March 13, 2007
35. Wei BQ, Vajtai R, Ajayan PM (2001) Appl Phys Lett 79(8):1172–1174
36. Collins PG, Hersam M, Arnold M, Martel R, Avouris Ph (2001) Phys Review Lett 86(14):3128–3131

37. Kreupl F, Graham AP, Duesberg GS, Steinhogl W, Lieban M, Unger E, Honlein W (2002) Microelectronic Eng 64:399–408
38. Li J, Ye Q, Cassell A, Ng HT, Stevens R, Han J, Meyyappan M (2003) Appl Phys Lett 82(15):2491–2493
39. Ngo Q, Cruden BA, Casselle AM, Sims G, Meyyappan M, Li J, Yang CY (2004) Nano Lett 4(12):2403–2407
40. Arai S, Endo M (2004) Electrochem Solid State Lett 7(3):C25–C26
41. Dubin VM (1992) J Electrochem Soc 139:633
42. Li J, You S, O'Keefe MJ, O'Keefe TJ (2006) J Elecrochem Soc 153(10):C722–C727
43. Lee H-Y, Duh J-G (2006) J Electronic Met 35(3):494–503
44. Shen J, Liu YC, Han YJ, Gao HX (2006) J Electronic Metals 33(8):1672–1679
45. Xu J, Fisher TS (2006) Int J Heat Mass Transf 49:1658–1666
46. Eastman JA, Choi SVS, Li S, Yu W, Thompson LJ (2001) Appl Phys Lett 78(6):718–720
47. Xuan Y, Li Q (2000) Int J Heat Fluid Flow 21:58–64
48. Marquis FDS, Chibante LPF (2005). JOM 57(12):32–43
49. Butler P (2006) The packaging professional. 6–7
50. Tuckerman DB, Pease RFW (1981) IEEE Electron Dev Lett 2(5):126–129
51. Kandilkar SG, Grande WJ (2003) Heat Transf Eng 24(1):3–17
52. Marthinuss J, Hall G (2004). Electronics Cooling
53. Park JW, Ruch D, Wirtz RA. American Association of Aeronautics and Astronautics, AIAA, 2002-0208, 1–9
54. Datta M, McMaster M, Brewer R, Zhou P, Tsao P, Upadhaya G, Munch M. Patent pending
55. Boomsa K, Poulikakos D, Zwick F (2003) Mech Mater 35:1161–1176
56. Shin H-C, Dong J, Liu M (2003) Adv Mater 15(19):1610–1614
57. Furberg R (2006) Enhanced boiling heat transfer from a novel nano-dendritic microporous copper structure, licentiate thesis, KTH School of Industrial Engineering & Management, Department of Energy Technology, Stockholm
58. Rowe DM (1995) CRC handbook of thermoelectrics. CRC Press, London
59. Yim WM, Rosi FD (1972) J Solid State Electron 15:1131–1140
60. Yoo BY, Huang C-K, Lim JR, Herman J, Ryan MA, Fleurial J-P, Myung NV (2005) Electrochim Acta 50:4371–4377
61. Hicks LD, Drwsselhaus MS (1997) Phys Rev B47:631
62. Huang L, Wang W, Murphy MC (1999) Microsystem Technol 6:1–5
63. Li L, Yang Y, Huang X, Li G, Zhang L (2006) Nanotechnology 17:1706–1712
64. Sander MS, Prieto AL, Gronsky R, Sands T, Stacy AM (2002) Adv Mater 14(9):665–667
65. Purkyastha A, Lupo F, Kim S, Borca-Tasciuc T, Ramnath G (2006) Adv Mater Des 18:496–500
66. Menke EJ, Li Q, Penner RM (2004) Nano Lett 4(10):2009–2014

Chapter 15
Electrochemical Fabrication Process for ULSI Interconnects

Tetsuya Osaka and Masahiro Yoshino

15.1 Introduction

In the field of ultra-large-scale integration (ULSI) technologies, electrodeposited copper (Cu) has been used since 1997 as the interconnection material [1, 2]. Recently, most manufacturers have switched over to electrodeposited Cu interconnect technology. Before the introduction of electrodeposited Cu to interconnects, Al and/or Al–Cu alloys were used. The Al interconnects are easily fabricated by the subtractive etching process. In this process, the Al layers are deposited by physical vapor deposition (PVD), followed by reactive ion etching (RIE). Al is preferable for interconnects because this material does not diffuse into the SiO_2 substrate and the layers adhere well to the substrate (Fig. 15.1a). However, resistivity of Al is relatively high (2.65 μΩ cm), and the layers suffer from the disadvantage of its poor electromigration resistance. With an increase in the density of interconnects and a decrease in the dimensions of interconnects, the problem of an increase in latency, or RC delay, and electromigraion become much more critical. Therefore, the search for new interconnect materials enabling a further miniaturization of semiconductor devices has mainly focused on minimizing RC delay and electromigration [3–10]. Cu has been regarded as a potential candidate for an interconnection material because of its lower resistivity (1.68 μΩ cm) compared with that of Al. The other important advantage of Cu interconnects is that Cu offers much better electromigration resistance than Al does. At a comparable dimension, the time-to-failure of a Cu interconnect was about a 100 times as long as that of an Al one [11–13]. Therefore, Cu interconnection is able to support higher current density, and this makes it possible to accelerate further miniaturization of interconnects. The Cu interconnection technologies were originally developed through the implementation of the "Damascene process" introduced by IBM (Fig. 15.1b) [1, 2].

T. Osaka (✉) and M. Yoshino
Faculty of Science and Engineering, Waseda University,
3-4-1 Okubo, Shinjuku-ku, Tokyo, 169-8555, Japan
e-mail: osakatets@waseda.jp

T. Osaka et al. (eds.), *Electrochemical Nanotechnologies*,
Nanostructure Science and Technology, DOI 10.1007/978-1-4419-1424-8_15,

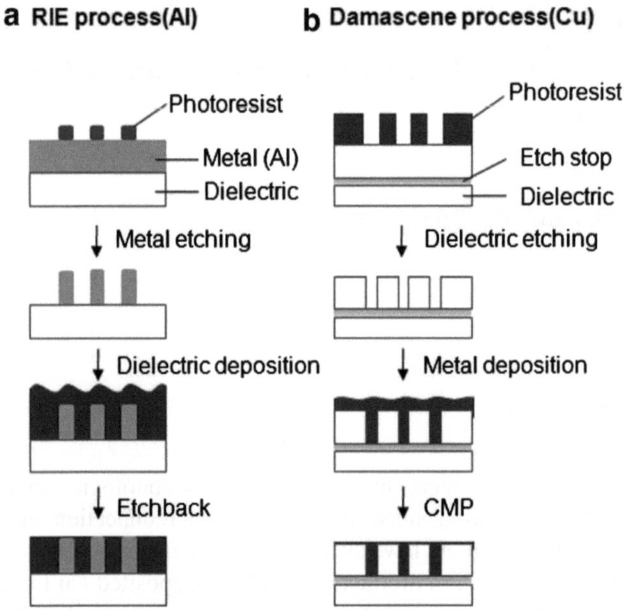

Fig. 15.1 Comparison of subtractive etching process and damascene process

The name of the damascene process originates in an ancient inlayer technique developed in the city of Damascus as early as the twelfth century. In this ancient metal work, decorative metals such as Au, Ag, or Cu were hammered into grooves made in metal armaments after which the excess material was removed by polishing. The damascene process has made clear to us the usefulness of wet processes, in ULSI technologies. Although only Cu wiring structures are fabricated by the "wet fabrication process" of electro deposition in the practical ULSI process, it is expected that several steps in conventional dry processes will be also accomplished by wet processes for advanced interconnect technology. For the successful achievement of Cu interconnections in ULSI, it is essential to introduce an effective barrier layer as well as capping layer because Cu diffuses readily into the inter-level dielectric such as SiO_2, and is oxidized easily, and both of these occurrences cause the degradation of electrical performance in ULSI devices. In the current ULSI structure, sputtered SiN is mainly used as the capping layer that protects the Cu surface from corrosion and prevents Cu diffusion. As device integration has advanced, however, it has become inconvenient to use SiN as a capping material, because the dielectric constant of SiN is rather high, which should cause the rise of wiring capacitances. For advanced wiring technology, hence, the application of electroless metal film to a capping layer is an attractive and practical process. Introducing the electroless metal cap is expected to decrease the wiring capacitances. Besides, electroless metal deposits onto Cu wires, selectively. As for the barrier layer, several materials, such as TaN, TiN, etc., are employed in practical use, and these barrier layers are deposited by a dry process such as sputtering or chemical vapor deposition (CVD) and atomic layer deposition (ALD) [14–17]. With an increasing degree of integration, difficulties in uniform formation of thin films have appeared in fine-patterned structures with high aspect ratios, due to the intrinsic limitation of the sputtering technique itself. Recently, a process of fabricating a barrier layer by electroless deposition has been proposed [18–21]. Electroless deposition has several advantages, compared with the dry method, such as its simplicity of operation and its low cost. In particular, it is attractive for producing uniform deposits independent of the size and geometry of the structure, which could help us overcome the above-mentioned problem on coverage. However, the electroless deposition of barrier layers usually requires the sputtering process of seeds to initiate its reaction [18–21], resulting in a lessening in the advantage of the electroless deposition method. We have proposed a concept for all-wet Cu wiring fabrication technology, as shown in Fig. 15.2b. Figure 15.2a shows the schematic illustration of conventional dry damascene process. In this proposal, the capping and barrier layers are fabricated by electroless deposition; in particular, we propose a novel process for fabricating a barrier layer without sputtered seeds by using a layer of organic molecules as an adhesive and catalyst immobilizing layer for the electroless Ni-alloy deposition on SiO_2 [22–28]. The layer of organic molecules is bonded to the SiO_2 surface by silane-coupling reaction, leading to the formation of a layer. Amino groups exposed at the top surface of the layer immobilize Pd catalysts, which is necessary for activating electroless plating.

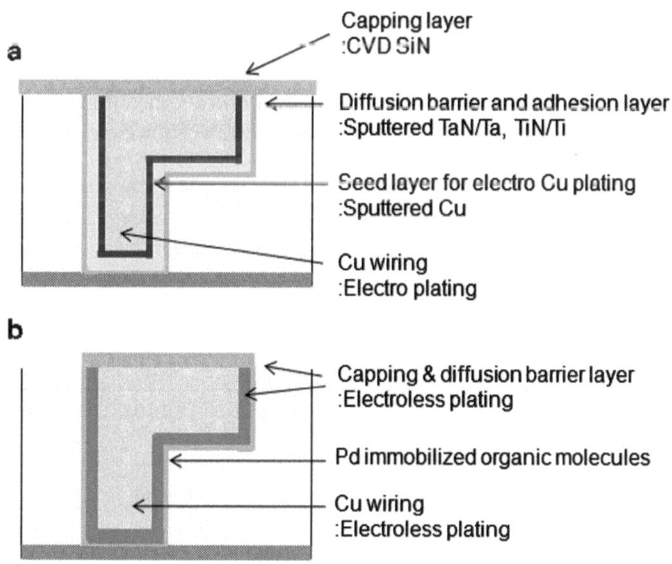

Fig. 15.2 Schematic illustrations of fabrication process for Cu wiring. (**a**) Current dry process, (**b**) our proposal wet process

15.2 Electroless Process for ULSI Interconnects Fabrication Process

Electroless deposition processes have distinct advantages, as shown in the first section for ULSI fabrication process. We have proposed one candidate for the electroless process, and we have found Ni-alloy films show superior properties for the capping and barrier layer. Moreover, we introduce our new process called the "all-electroless process" for ULSI interconnect fabrication.

15.2.1 Thermal Stability of Electroless Ni-Alloy Films

The sheet resistance of the Ni-alloy films was measured to evaluate the thermal stability of the electroless deposited Ni-alloy films for preventing Cu diffusion. The specimens discussed in this section were prepared on a Cu/Ta/SiO_2/Si substrate. Details of the film composition depending on the bath composition and plating condition are listed in the Table 15.1 [22–24]. The sheet resistance of each film varied with annealing temperature, as shown in Fig. 15.3. One candidate for cap and barrier material is Ni–P alloy films obtained from a solution containing sodium hypophosphite. In the case of NiP film, the sheet resistance was significantly increased at 300°C, which suggested that interdiffusion occurred at this temperature. On the other hand, all the NiReP films were stable up to 400°C, irrespective of the Re content of the film. The XRD pattern of NiReP included small peaks attributable to a Ni_3P phase, and their intensity did not change by heat treatment up to 450°C. The thermal stability of NiP and NiReP films was also confirmed by cross-sectional FE-SEM. Figures 15.4 and 15.5 show the NiP/Cu and NiReP/Cu interfacial regions, respectively. Figure 15.4a shows the specimen before annealing; the interface of NiP/Cu was clearly observed. After annealing to 300°C, however, it is obvious that

Table 15.1 Formulation and operating conditions of electroless Ni-alloy deposition solutions

Chemicals	mol dm^3			
	NiWP	NiReP	NiB	NiWB
$NiSO_4\ 6H_2O$	0.027	0.075	0.10	0.10
$NaH_2PO_2\ H_2O$	0.066	0.10	–	–
$(NH_4)_2\ SO_4$	0.23	–	–	–
Sodium citrate	0.12	0.10	0.20	0.05
$Na_2WO_4\ 2H_2O$	0.11	–	–	0.01,0.20
$(NH_4)_2ReO_4$		0.03	–	–
Dimethylamine-borane (DMAB)			0.05	0.05

pH. 9.0 (adjusted with NaOH) temperature, 70°C

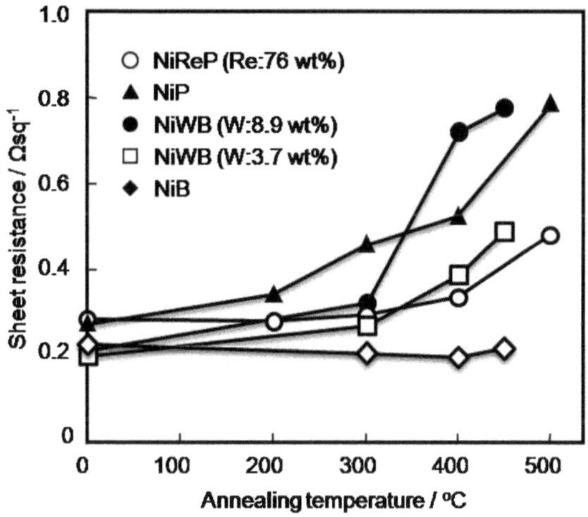

Fig. 15.3 Variation of sheet resistance of Ni-alloy films with annealing temperature. Films were formed on Cu (100 nm)/Ta (30 nm)/SiO_2 (500 nm)/Si substrate

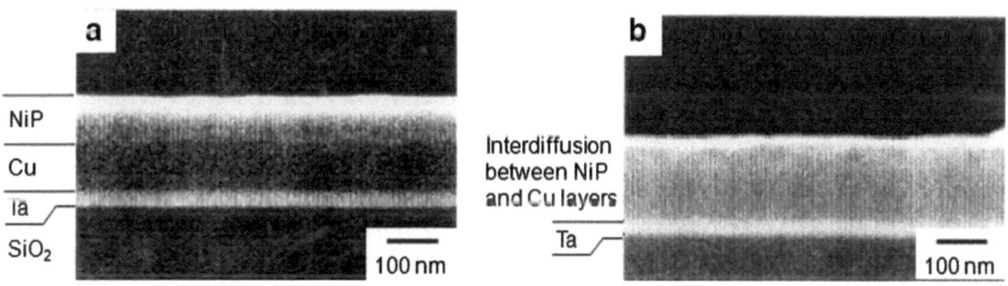

Fig. 15.4 Cross-sectional FE-SEM images of NiP/Cu interfacial region; (**a**) before and (**b**) after annealing to 300°C [22]

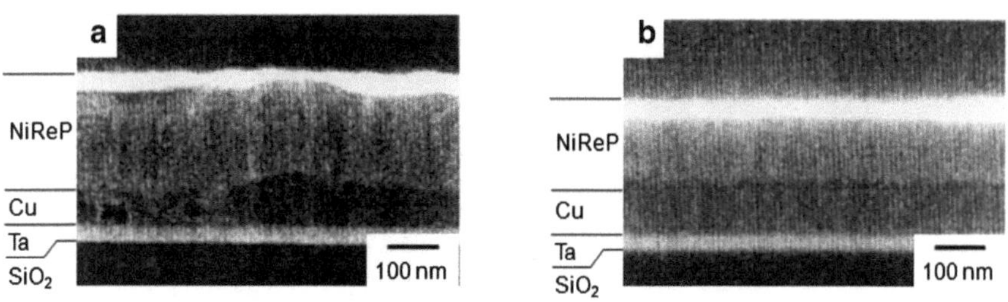

Fig. 15.5 Cross-sectional FE-SEM images of NiReP(Re: 65 wt%)/Cu interfacial region; (**a**) before and (**b**) after annealing to 400°C [22]

interdiffusion occurred in the interfacial region as observed in Fig. 15.4b. In contrast, the NiReP/Cu interfacial region was still retained after annealing to 400°C, as shown in Fig. 15.5. From these results, it is apparent that the electroless deposited NiReP films have sufficient thermal stability against interdiffusion between the Cu and barrier layer. Although we also investigated NiWP films [23], the improvement of thermal stability was difficult to achieve, maybe due to insufficient W content (~5 wt% at most) independent of plating conditions. Electroless Ni–B alloy films obtained from a solution containing dimethylamine-borane (DMAB) as a reducing agent are also candidates for cap and barrier material. These types of films are able to deposit directly onto Cu surfaces without a catalyzing process.

The sheet resistance of the NiWB films began to increase at 300°C, while NiB film was stable up to 450°C, suggesting that the electroless deposited NiB film has an excellent thermal stability against interdiffusion in the interfacial region of the Cu/barrier layer. The XRD patterns of NiB and NiWB showed differences in the annealing temperature. Before annealing, only the peaks attributable to the substrate were seen in both the films. In accordance with rising temperature, the peaks corresponding to the Ni_3B phase appeared and became stronger for the NiB film, whereas the diffraction pattern of the NiWB film hardly changed, showing an amorphous structure. The NiB film with excellent thermal stability shows a crystal growth when annealing, while the NiWB film showing the thermal degradation of its barrier property maintains an amorphous structure even after annealing. Figure 15.6 shows a cross-sectional SEM image of the capping layer of electeoless NiB film deposited on a Cu wiring pattern substrate. Electroless deposition was carried out only on the Cu wiring.

15.2.2 All-Wet Barrier-Layer Fabrication Process on SiO_2 Substrate

As mentioned in the introduction, exclusion of sputtered seeds is considered to be highly advantageous for future ULSI technology. In general, however, a seed layer fabricated by sputtering is essential to initiate the electroless deposition reaction of

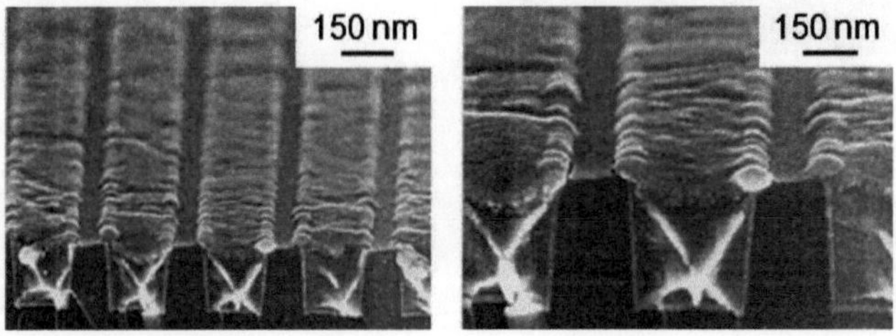

Fig. 15.6 Surface and cross-sectional SEM images of capping layer on Cu wiring pattern on Ta/SiO_2/Si substrates [24]

a barrier layer on the inter-level dielectric, in this case, SiO_2. At the beginning, the fabrication of electroless NiP films on SiO_2 without sputtered seeds was investigated to obtain fundamental information for the formation of barriers, and then this method was applied to other Ni-alloys. A layer of organic molecules with a functional group such as an amino or pyridyl group was utilized as a catalyst-immobilizing layer because this layer of organic molecules is known to be highly effective for electroless metallization of a SiO_2 surface. After optimization to shorten the treatment time, we decided to use a toluene solution of 3-aminopropyltriethoxysilane (APTES) [26]. Figure 15.7 shows the schematic illustrations of the formation process for a layer of organic molecules (APTES), following the Pd catalyst process. The formation of the layer of organic molecules was found to yield uniform and glossy NiP films on the surface (Fig. 15.8b).

On SiO_2 without a layer of organic molecules, however, only a small area was coated with NiP deposit (Fig. 15.8a). This deposit seemed to have been formed on a few catalytic Pd particles, which remained barely attached to the surface. In the

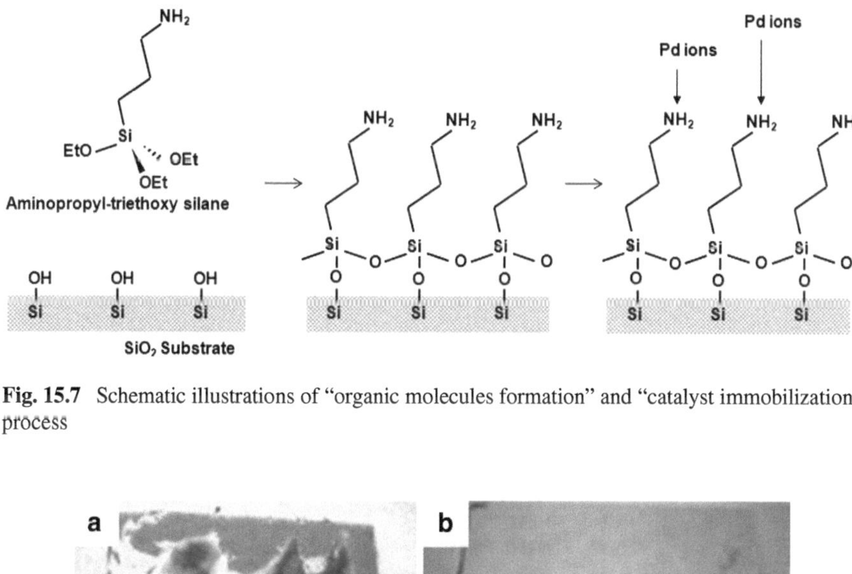

Fig. 15.7 Schematic illustrations of "organic molecules formation" and "catalyst immobilization" process

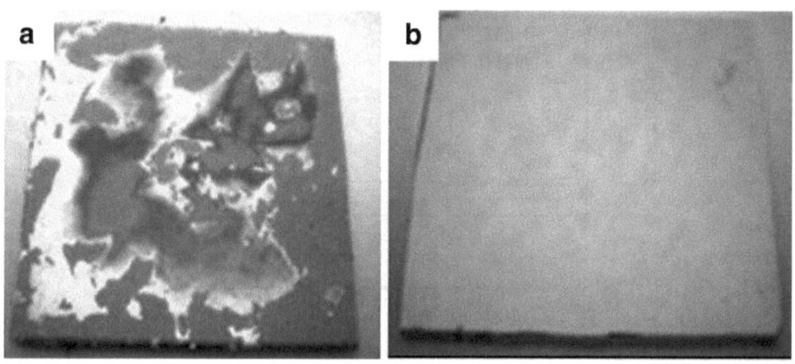

Fig. 15.8 Electroless NiP film formed on SiO_2 substrate, (a) without layer of organic molecules, (b) with layer of organic molecules

formation of NiReP films on the organic molecules-coated SiO_2 substrate, strike plating in the acid NiP bath must be applied before immersion in the NiReP bath [22, 23]. The layer of organic molecules is probably degraded by immersing in NiReP solution, because the deposition rate of electroless NiReP is extremely slow. The pre-deposited NiP nuclei protect the organic molecules from the potential damage triggered by the alkaline solution. By adopting this two-step process, nearly perfect and uniform NiReP films with a bright appearance could be produced. As shown in Fig. 15.9, the NiReP films formed on an SiO_2 substrate with a layer of organic molecules were confirmed to be stable up to 400°C, which was also proved by the AES depth profile results. Although the sudden rise of sheet resistance at over 500°C is suggestive of less stability of the NiReP barrier layer on organic molecules/SiO_2/Si than that on Cu/Ta/SiO_2/Si, thermal stability up to 400°C is *sufficient* for practical use. The electroless deposited NiB film is another superior candidate for barrier materials. Electroless NiB film can deposit on SiO_2 substrate with a Pd-immobilized organic layer directly. Continuous films, which have smooth surfaces with Ra of 1.842 nm were obtained. The result of a peel test showed an excellent adhesion of the NiB film with a thickness of 20–60 nm. This film also has good thermal stability (Fig. 15.9). Compared to other Ni-alloys, the increase rate of the sheet resistance of this film is very low. This indicates the viability of the NiB layer as a diffusion barrier layer for the Cu wiring. Figure 15.10 shows a cross-sectional SEM image of an SiO_2 trenches pattern substrate with an electroless NiB barrier layer. The width and the aspect ratio of the trench were 130 nm and 2.5, respectively. It is observed that NiB films covered homogeneously on a whole side of the substrate surface, and that the thickness increased with the increase in the deposition time. In addition, this film on a patterned substrate shows adhesion as good as that on a non-patterned substrate.

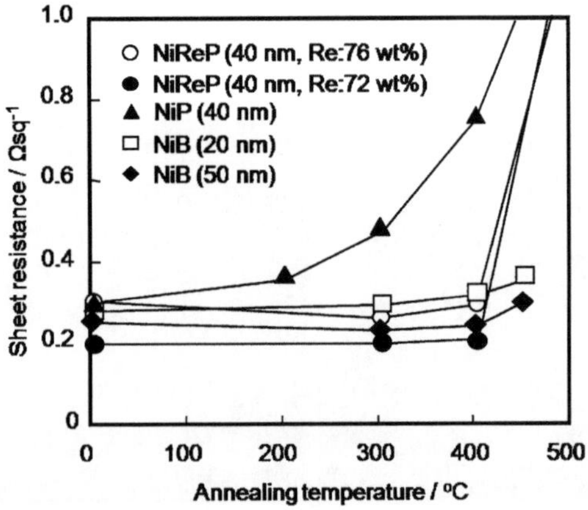

Fig. 15.9 Variation of sheet resistance of Ni-alloy films with annealing temperature. Films were formed on SiO_2 (500 nm)/Si substrate with layer of organic molecules

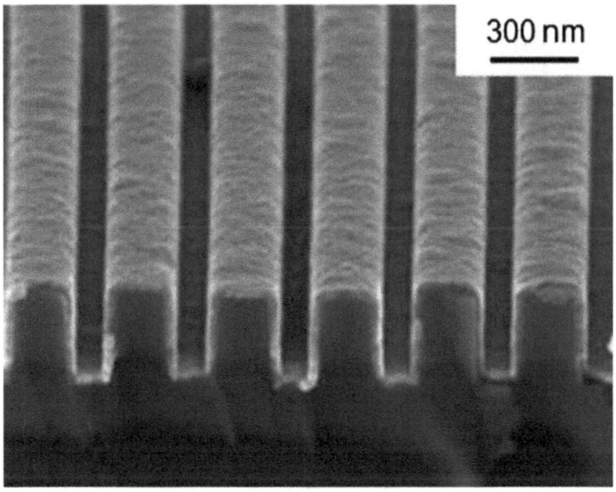

Fig. 15.10 Cross-sectional SEM image of NiB film as barrier layer deposited on the trench patterned SiO_2 substrate

15.2.3 Evaluation of Catalyst Process

The catalyzed surface was confirmed by the appearance of a Pd 3d 5/2 peak in the XPS spectrum. Figure 15.11a shows the XPS spectrum obtained from the SiO_2 substrate without a catalyst process. Figure 15.11b shows the XPS spectrum obtained from a Pd-treated specimen. A peak at 337.3 eV was observed in the Pd, which means that the Pd is present in an oxidized state. In addition, it means that the Pd ions exist in the form of hydroxo-Pd oligomers. This result is consistent with the finding reported by H. Kind et al. [29]. In the subsequent process, immobilized hydroxo-Pd oligomers will be reduced to the metallic state by a reducing agent (Fig. 15.11c). The surface roughness of the modified SiO_2 substrate was observed by AFM. The average roughness values of the surface *before* and *after* the formation of organic molecules, and *after immobilization* of the Pd ions, were 0.27, 0.16, 0.36 and 0.31 nm, respectively, showing that the smoothness of the surfaces remained unchanged after each step. The catalyzed surface was also observed by in-plane TEM observation. All of the specimens for TEM observation were treated with DMAB solution before observation to reduce the Pd ions to the metal state. Figure 15.12a shows the TEM image of a C grid treated with APTES and a standard Pd solution containing 5.5×10^{-5} mol dm^{-3} $PdCl_2$ and 0.01 mol dm^{-3} HCl. Significant small grains, present as dark spots, are observed. The dark spots indicate the Pd particles, whose existence is confirmed by the Pd lattice length. The Pd particles were discrete, and the coverage in the Pd area was almost 12.5%. Several researchers reported that the discrete catalysts show high catalytic activity. The grain size and coverage of the Pd catalyst are controllable by controlling the condition of the Pd solution. For example, by adding Cl ions to the Pd solution, the equilibrium state of Pd solution was drastically changed. Figure 15.13 shows normalized compositions of the Pd solution at pH = 2 as a function of log[Cl^-]. In the case of Cl ions,

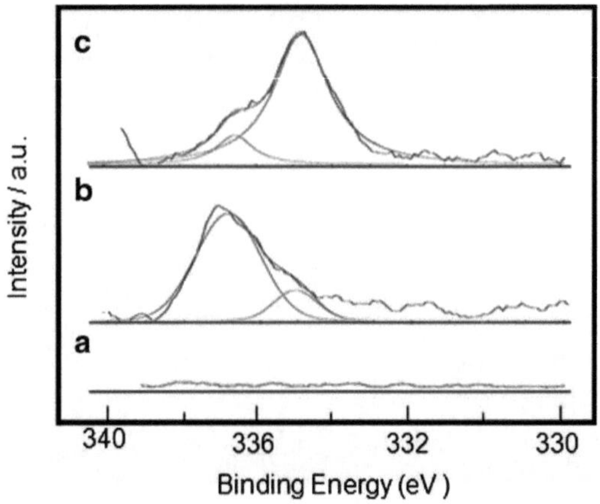

Fig. 15.11 (**a**) Pd 3d5/2 region of the XPS spectrum of an APTES coated oxide Si wafer; (**b**) an APTES coated oxide Si wafer treated with $PdCl_2$ solution; (**c**) wafer from part (**b**) after treatment with DMAB solution

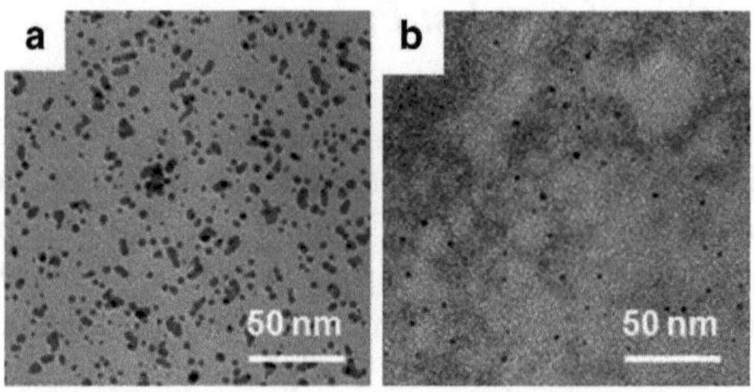

Fig. 15.12 (**a**) TEM image of APTES coated C grid treated with standard $PdCl_2$ solution following reduction process; (**b**) APTES coated C grid treated with $PdCl_2$ solution containing 1.0 mol dm^{-3} of Cl ions following reduction process

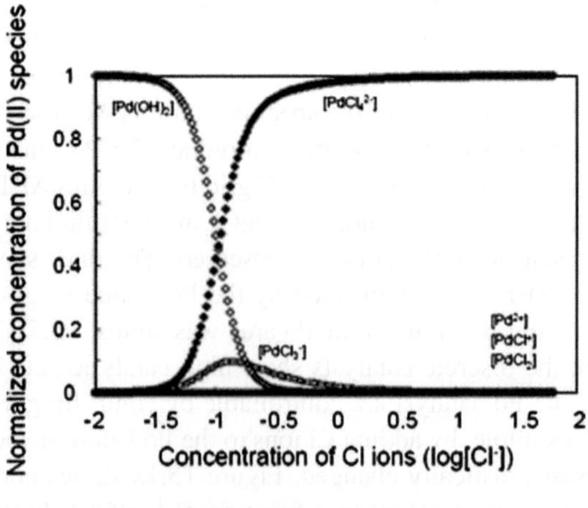

Fig. 15.13 Normalized compositions of the catalyst solution at pH = 2 as a function of log[Cl^-]. The aqueous solution contains NaCl, $PdCl_2$, HCl

concentration was low; mainly, Pd $(OH)_2$ aq. is present. On the other hand, in solution containing a high concentration of Cl ions, Pd $(OH)_2$ aq. are changed to $PdCl_4^{2-}$. Figure 15.12b shows the TEM image of a specimen treated with Pd solution containing 1.0 mol dm^{-3} of Cl ions. The grain size was extremely small, with average grain diameters under 1 nm.

15.2.4 Void-Free Filling by Electroless Cu Deposition

In this section, we introduce the Cu filling process by electroless deposition using a bath containing a very small amount of polyethylene glycol (PEG) as an inhibiting additive [30]. Electroless Cu deposition is expected to be useful for the fabrication of further miniaturized interconnection structures because it is, in principle, capable of forming a conformal thin film on large substrates.

The electroless plating bath contained 0.04 mol dm^{-3} $CuSO_4$, 0.08 mol dm^{-3} glyoxilic acid, and 0.08 mol dm^{-3} tetrasodium ethylenediamine tetraacetic acid. We employed polyethylene glycol (PEG) with an average molecular weight of 4,000 as an inhibiting additive. The bath was operated at 70°C without forced agitation. To investigate the effect of PEG on electroless deposition, the deposition rate and the deposition potential were measured with a non-patterned substrate (Fig. 15.14). The deposition rate measured on the non-patterned substrate (Fig. 15.14, open diamond) decreased with an increase in the concentration of PEG. The extent of the decrease in the rate of deposition was significant only at PEG concentrations between 0 and 1 ppm. The deposition rates were 11.0 μm/h for the bath without PEG (additive-free bath), and 1.95 μm/h for the bath with 1 ppm of PEG (PEG (1 ppm) bath). The deposition potential shifted in the negative direction with the decrease in deposition rate (Fig. 15.14, solid diamond). The deposition potential for the additive-free bath was –0.68 V, while that for the PEG (1 ppm) bath was –0.75 V. Electroless

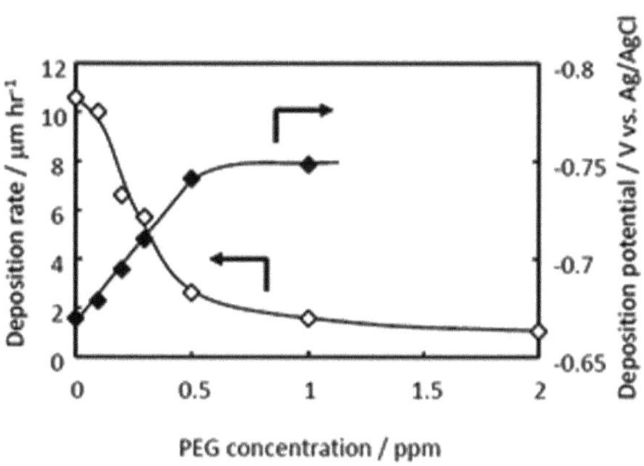

Fig. 15.14 Effects of PEG concentration on the rate of Cu deposition (*open diamond*) and the deposition potential (*solid diamond*) on an unpatterned substrate

Cu deposition was carried out on the trench-patterned substrate with the PEG (1 ppm) bath for various deposition times (Fig. 15.15). Figure 15.15a shows the image of trenches before the deposition. The thickness of Cu at all positions on the trench walls after 30 s of deposition (Fig. 15.15b) was similar to that observed before the deposition (Fig. 15.15a). This indicates that the Cu deposition scarcely occurred during the first 30 s (Fig. 15.15a, b). Then, a significant bottom-up growth of Cu took place during the period between 30 s (Fig. 15.15b) and 1 min (Fig. 15.15c), and Cu filling was completed after 5 min of deposition (Fig. 15.15f). Furthermore, as shown in Fig. 15.15g, Cu filled the trenches without the formation of overfill bumps. The absence of the overfill phenomenon observed in this study agrees with the finding reported by Shingubara et al [31]. Figure 15.16a shows the thickness of electroless Cu at the bottom (T_{bottom}) and that at the top (T_{top}) of trenches, which were measured from cross-sections of Cu deposits shown in Fig. 15.15. Figure 15.16b shows the plot of the deposition rate at trench bottom against time. The data were

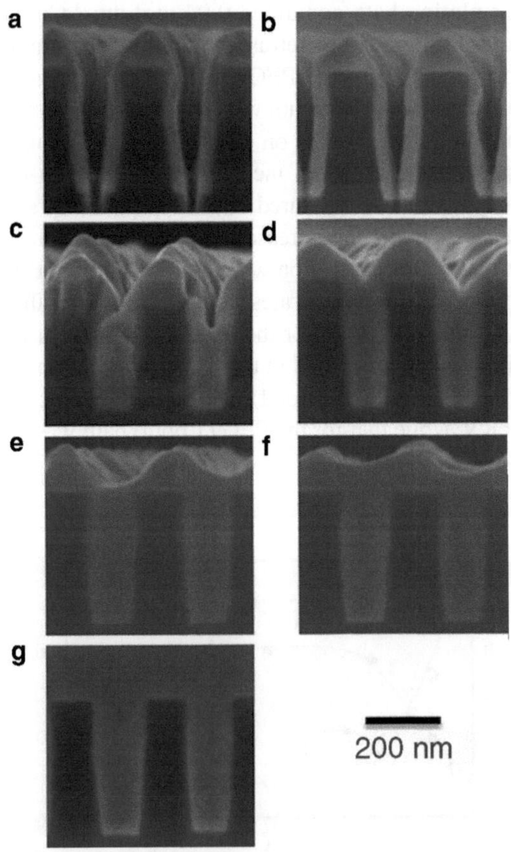

Fig. 15.15 SEM images showing the change with time of cross-sectional features of copper deposited in trenches from PEG (1 ppm) bath. Deposition time, (**a**) 0 min, (**b**) 0.5 min, (**c**) 1 min, (**d**) 2 min, (**e**) 3 min, (**f**) 5 min, and (**g**) 10 min

obtained from the slope of the deposit thickness–time curve shown in Fig. 15.16a. As noted above, a *significant* change in thickness of Cu deposit was found to occur at trench bottoms during the period between 30 s and 1 min, while the extent of increase in the thickness of Cu was very small during the first 30 s. As shown in Fig. 15.16b, the deposition rate at 1 min was 16.9 μm/h. It is interesting to note that this deposition rate was much *higher* than that for the additive-free bath measured on a non-patterned substrate (11.0 μm/h). The rate of deposition at the bottom decreased gradually after 1 min and continued to decrease until the trenches were completely filled with Cu (5 min). After that, the deposition rate remained constant at least for 5 min (5–10 min), and the value of the deposition rate during this period was nearly the same as that of the PEG (1 ppm) bath measured on the unpatterned substrate (1.95 μm/h). Contrary to the deposition behavior at the bottom, the deposition rate at the trench opening was essentially nil during the filling process (0–3 min). After the trenches were almost filled with Cu, the thickness of Cu at the opening started to increase. During the period between 5 and 10 min, the deposition rate at the opening was similar to that at the bottom. Figure 15.17 shows the potential–time curves measured with a trench-patterned substrate immersed into the additive-free

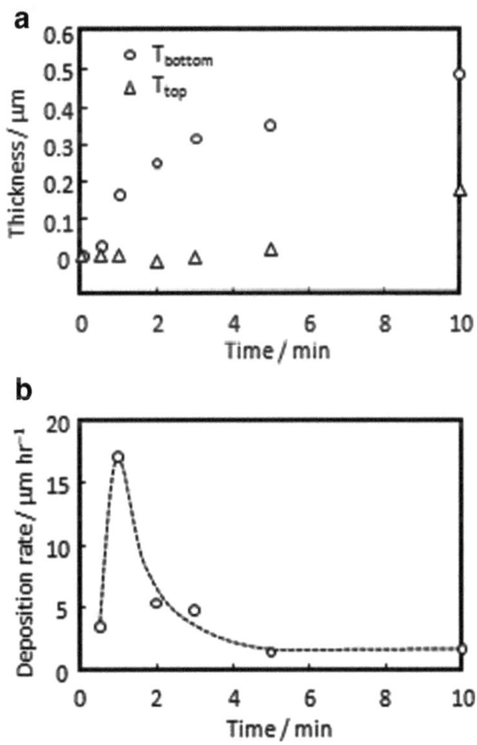

Fig. 15.16 (**a**) Change with time of the copper deposit thicknesses at bottom (*Tbottom*) and top (*Ttop*) of trenches, and (**b**) change with time of deposition rate at the bottom of trenches

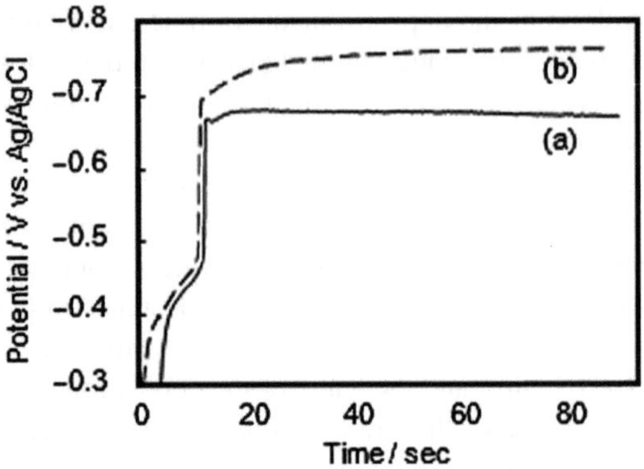

Fig. 15.17 Change with time of the potential of a patterned substrate during electroless copper deposition. Measurements were made in (**a**) additive-free bath, and (**b**) PEG (1 ppm) bath

(curve *a*) and the PEG (1 ppm) (curve *b*) baths. Before the measurements, the patterned substrate was activated with H_2SO_4, and rinsed with UPW. We should emphasize here that the substrate pretreatment procedure immediately before the electroless deposition was the rinse in UPW. In this trench size of 130 nm in width and 350 nm in depth, the initial process that takes place after the UPW rinse must be to fill the trenches with UPW, and this process is followed by the replacement of the UPW by the electroless plating solution just before the electroless plating. For the first 10 s after the substrate was immersed into the bath, the potentials measured in both the additive-free and PEG (1 ppm) baths were less negative than –0.5 V, where Cu deposition did not proceed. This initial period of 10 s was regarded as the *induction period* involved in the overall electroless Cu deposition. The induction period measured with non-patterned substrate was very short, and was less than 1 s. It was also found that, when the patterned substrate was immersed, as an extra intermediate step, into the additive-free bath at 20°C between the UPW rinse and the immersion in the additive-free bath, the incubation period became much shorter than that observed in the measurement carried out without the extra intermediate step (Fig. 15.17b).

The longer induction period observed in the potential measurement with the patterned substrate (Fig. 15.17), as compared with that for the unpatterned substrate, is attributed to the presence of *rinse water* remaining in the trenches of the patterned substrate after the pretreatment process. Because the trenches were very small in diameter, the mass-transfer from the bulk of the plating solution into the trenches must have been greatly restricted. Consequently, the concentrations of the bath constituents, such as Cu(II)-EDTA complex and glyoxylic acid, were considered to be essentially nil inside the trenches at the initial stages of the filling process. Therefore, it is assumed that the diffusion of bath constituents from the bulk of the solution into the trenches began to proceed when the substrate was immersed into the bath, and that the Cu deposition began when the concentrations of bath constituents in trenches became sufficiently high. After the induction period was over, the deposition potential for the PEG (1 ppm) bath shifted to −0.7 V (Fig. 15.17b), which was similar to the deposition potential of an unpatterned substrate in the additive-free (PEG-free) bath. The deposition potential in the PEG (1 ppm) bath continued to shift gradually in the negative direction during the period between 10 and 70 s (Fig. 15.17b), while that in the additive-free bath was almost constant (Fig. 15.17a). The potential measured with a patterned substrate after 70 s was −0.76 V. This potential was essentially identical to that measured on a unpatterned substrate in the PEG (1 ppm) bath. The observed change in the potential is attributed to the change in PEG concentration in trenches during the initial diffusion process of PEG. The increase in the concentration of PEG is assumed to be slower than that for the other bath constituents, because this additive is a large polymer molecule. According to the literature [32, 33], the diffusion coefficient of PEG (Mw 4,000) is less than 1/10th of that of Cu(II)-EDTA complex at the same temperature. Thus, the concentration of PEG at the trench bottom must have been lower than that at the opening during trench filling. We believe that the observed difference in the deposition behavior at the bottom and at the opening (Fig. 15.16) is due to the *lower* concentration of PEG inside the trenches, where the mass transfer of bath constituent species is greatly restricted.

15.3 Electroless Barrier Formation Process Application to the Low Dielectric Materials

15.3.1 Surface Modification of Low Dielectric Materials

Two types of low dielectric films were used in this study. One was an inorganic hydrogen silsesquioxane, HSQ (dielectric constant, 2.9), while the other was an organic methyl silsesquioxane, MSQ, (dielectric constant, 2.7). These materials were spin-coated on Si (100) substrates. The modification of the low dielectric film surface by UV irradiation in the O_2 atmosphere was investigated. Note that the HSQ and MSQ materials do not contain -OH groups while they do contain a large number of Si–H and Si–CH_3 groups, respectively. Therefore, surface modification was required because -OH groups were necessary for the silane-coupling reaction to form the organic molecules. The surface modification by UV irradiation was carried out, and the change in surface condition was followed by measuring the change in contact angle with DI water and also by FT-IR measurements. Wettability was found to depend on the temperature during the UV irradiation. The low dielectric film surface was found to change from hydrophobic to hydrophilic. The surface of the *inorganic* dielectric material was more easily modified than that of the *organic* material. Significant changes took place even at temperatures lower than 30°C. The surfaces of the UV-irradiated low dielectric films were analyzed by FT-IR. A significant peak for the -OH group was observed for the organic material treated at 250°C. With the inorganic material, a peak for the -OH group was observed after the treatment at lower temperatures. Such an

-OH group peak in the FT-IR spectra (the peak near 3,000–4,000 cm^{-1}) was not observed before the UV treatment. It is thus confirmed that the -OH group was influenced by the UV treatment on the low dielectric film surface. The formation of an APS on a low dielectric film, MSQ, without UV treatment was investigated by XPS. No NH_2 peak (398.7 eV) attributable to APS was detected. However, with UV treatment, a small peak for NH_2 and a more intense C 1s peak originating from APS were detected. The surface morphology of the modified film was evaluated by AFM. A smooth and defect-free surface was obtained with *R*a value lower than 1.0 nm even after organic molecules formed. These results confirmed the successful formation of the organic molecules by immersion into the toluene solution of APS. It was revealed that UV treatment was *critical* for high-quality surface modification.

15.3.2 Effect of Low-k Dielectric Film Composition on Barrier Layer Formation

The formation of an electroless NiB layer as a barrier on catalyzed low dielectric film was carried out. The film thickness was controlled by immersion time. The deposited layer had a metallic luster and uniform surface, and exhibited an excellent adhesion on both HSQ and MSQ films. Figure 15.18 shows FE-SEM images of the electroless NiB layer. The thickness of these samples was 20 nm. The layer consisted of small grains, and no pits were observed. On the low dielectrics films without a layer of organic molecules, the electroless deposition reaction did not proceed. The surface morphology of the barrier layer was evaluated by AFM. The surface was very smooth, i.e., surface roughness (*R*a) values of the as-deposited films were 0.704 nm for the organic film and 1.20 nm for the inorganic film, as compared to an initial surface roughness of 0.334 nm for the organic film and 0.628 nm for the inorganic substrate. The formation of a NiB layer on the surface of deep sub-micron trenches was carried out. Figure 15.19 shows a cross-sectional TEM image of trench-patterned substrate coated with the electroless barrier layer. In this figure, electroless barrier layer was deposited on a SiO_2/p-SiOC/SiO_2 substrate with a trench pattern. The thickness of the NiB layer was 30 nm. The NiB layer was found to be very uniform and conformal over a wide range of trench width from about 80 nm to 3 μm.

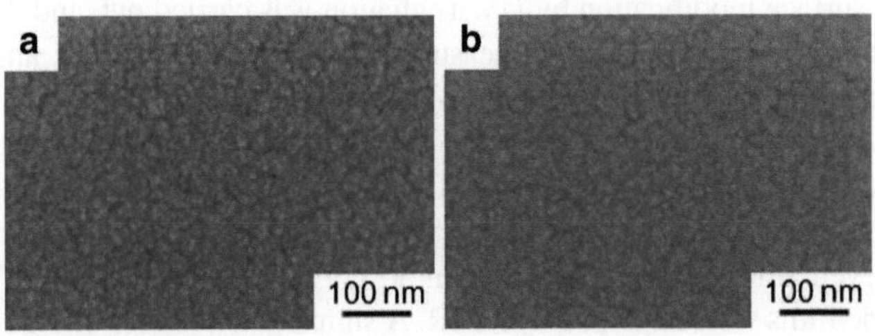

Fig. 15.18 FE-SEM images of electroless NiB film deposited on (**a**) the organic substrate, (**b**) the inorganic substrate

15.3.3 Barrier Property and Low-k Dielectric Property

Thermal stability of the electroless NiB barrier layer was evaluated by measuring the change in sheet resistance with time and temperature. The samples were prepared on bare low dielectric films of MSQ. A 100 nm thick layer of Cu was deposited on the barrier layer by evaporation as a wiring material. Figure 15.20 shows the sheet resistance of the Cu layer *versus* annealing temperature. The increase in sheet resistance indicates the occurrence of Cu diffusion into the barrier layer. Plots of sheet resistance *versus* annealing temperature for barrier layers of electroless NiReP, NiP, and NiWP 50 nm thick are also included in Fig. 15.20 for comparison. It was found that the sheet resistance of specimens containing either HSQ or MSQ film remained unchanged up to the annealing temperature of 400°C, thereby indicating that no

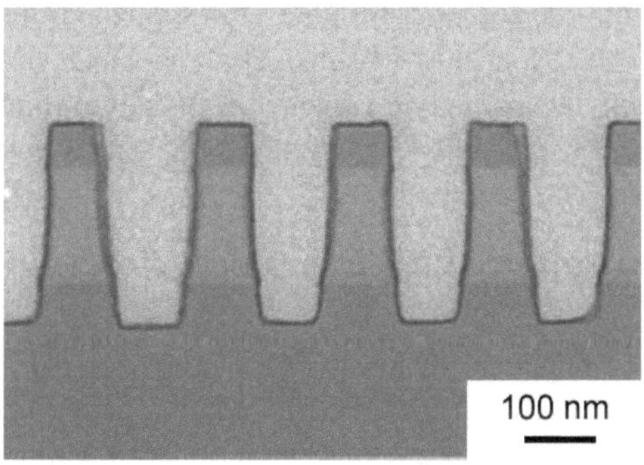

Fig. 15.19 Cross-sectional TEM image of trench patterned substrate coated with the electroless NiB film. The thickness of the film was 6 nm. This TEM observation was performed by Semiconductor Leading Edge Technologies Inc.

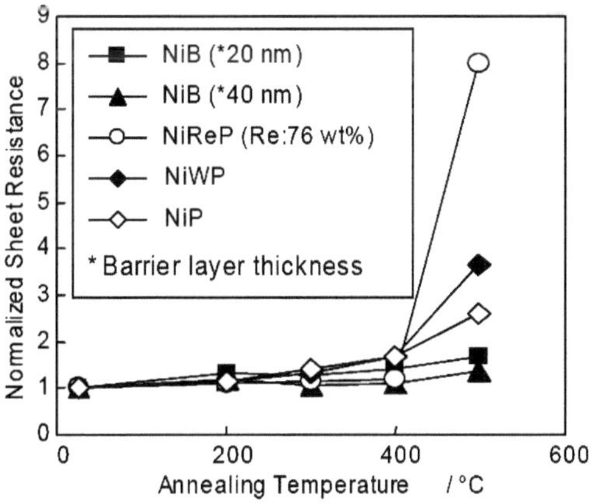

Fig. 15.20 Sheet resistances of electroless Ni-alloy films as a function of annealing temperature. Cu layer was deposited by evaporation on electroless Ni-alloy film

interdiffusion occurred between the Cu wiring film and the low dielectric film. This result suggests that the electrolessly deposited NiB layer prevented Cu atoms from diffusing into the substrate. Additionally, the barrier property of the layer on HSQ film seemed better than that on MSQ film, perhaps because the inorganic film surface is more easily modified by UV irradiation. A large number of OH groups exist on the HSQ surface. A surface with a high density of OH groups promotes the formation of a high-quality layer of organic molecules. The surface condition is an important factor in this process. We also studied the degradation of the dielectric constant of low dielectric films during this wet processing. The dielectric constant was measured after each processing step, i.e., after organic molecules formation, after Pd activation, and after electroless plating. The result showed that the dielectric constant remained *unchanged* during the entire wet formation process. Additionally, we attempted to use a porous, low dielectric film with pore diameter smaller than 1 nm. This film consisted of an inorganic material, and its dielectric constant was equal to 2.8. After the organic molecules formation process, the dielectric constant increased by only 5%. From these results, it is concluded that the proposed wet process *could possibly be* applied on various low dielectric films.

15.3.4 Demonstration of All-Electroless Process

Cu filling on the trench pattern by electroless plating was performed with an electroless barrier layer of NiB. Electroless Cu solution containing PEG was used for this experiment. Figure 15.21 shows a cross-sectional SEM image of the specimen. Cu is filled in the trenches without any defects. We demonstrated the establishment of "all-electroless" Cu wiring process. By combining with our on-going investiga-

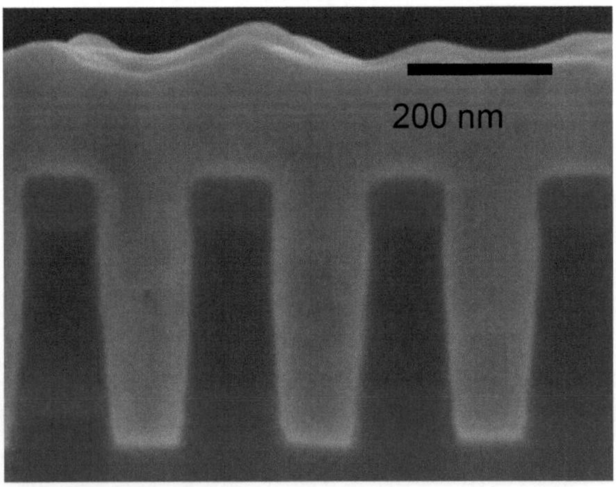

Fig. 15.21 Cross-sectional SEM image of Cu filled trench patterned substrate fabricated by "all-electroless process"

tions concerning fundamental aspects of Cu superfilling, this technique can be more sophisticated. We believe that our proposal is a promising technology in advanced ULSI applications.

References

1. Andricacos PC, Uzoh C et al (1998) Damascene copper electroplating for chip interconnections. IBM J Res Dev 42:567–574
2. Andricacos PC (1999) Copper on-chip interconnections a breakthrough in electrodeposition to make better chips. Electrochem Soc Interface 8:32–37
3. Ismail YI, Friedman EG et al (2001) Exploiting the on-chip inductance in high-speed clock distribution networks. IEEE Trans Very Large Scale Integr (VLSI) Syst 9:963–973
4. Deutsch A, Kopcsay GV et al (2001) Frequency-dependent losses on high-performance interconnections. IEEE Trans Electromagn Compat 43:446–465
5. Ismail YI et al (2001) Repeater insertion in tree structured inductive interconnect. IEEE Trans Circuits Syst II – Analog Digit Signal Process 48:471–481
6. Deutsch A et al (2001) On-chip wiring design challenges for gigahertz operation. Proc IEEE 89:529–555
7. Ismail YI, Friedman EG et al (2000) Effects of inductance on the propagation delay and repeater insertion in VLSI circuits. IEEE Trans Very Large Scale Integr (VLSI) Syst 8:195–206
8. Ismail YI et al (2000) Equivalent Elmore delay for RLC trees. IEEE Trans CAD Integr Circuits Syst 19:83–97
9. Sakurai T (1983) Approximation of wiring delay in MOSFET LSI. IEEE J Solid-State Circuit 18:418–426
10. Edelstein DC et al (1995) VLSI on-chip interconnection performance simulations and measurements. IBM J Res Dev 39:383–402
11. Edelstein DC et al (1997) Full copper wiring in a sub-0.25 μm CMOS ULSI technology. In: Technical digest IEEE international electron devices meeting, 773–776
12. Venkatesan S et al (1997) A high performance 1.8 V, 0.20 μm CMOS technology with copper metallization. In: Technological digest, IEEE international electron devices meeting, 769–772
13. Rosenberg R et al (2000) Copper metallization for high performance silicon technology. Annu Rev Mat Sci 30:229–262
14. Suntola T (1984) Atomic Layer Epitaxy. 16th international conference on solid state devices and materials, 647–650
15. Suntola T, Antson J US patent No. 4058430
16. Ritala M, Leskelä M (2002) Deposition and processing of thin films. In: Nalwa HS (ed) Handbook of thin film materials, volume 1. Academic, San Diego
17. Suntola T (1992) Atomic layer epitazy. Thin Solid Films 216:84–89
18. Shacham Diamand Y, Lopatin S (1997) High aspect ratio quarter-micron electroless copper integrated technology. Microelectron Eng 37(38):77–88
19. Shacham Diamand Y, Lopatin S (1999) Integrated electroless metallization for ULSI. Electrochim Acta 44:3639–3649
20. Shacham Diamand Y, Sverdlov Y (2000) Electrochemically deposited thin film alloys for ULSI and MEMS applications. Microelectron Eng 50:525–531
21. O'Sullivan EJ et al (1998) Electrolessly deposited diffusion barriers for microelectronics. IBM J Res Dev 42:607–620
22. Osaka T et al (2002) Fabrication of electroless NiReP barrier layer on SiO_2 without sputtered seed layer. J Electrochem Solid-State Lett 5:C7–C10
23. Osaka T et al (2002) Electroless nickel ternary alloy deposition on SiO_2 for application to diffusion barrier layer in copper interconnect technology. J Electrochem Soc 149:C573–C578

24. Osaka T et al (2003) Characterization of chemically-deposited NiB and NiWB thin films as a capping layer for ULSI application. Surf Coat Technol 169(170):124–127
25. Osaka T et al (2004) Formation of high functional thin films in electronics. J Surf Fin Soc Jpn 55:753–757
26. Yoshino M et al (2005) All-wet fabrication process for ULSI interconnect technologies. Electrochim Acta 51:916–920
27. Yoshino M et al (2006) Fabrication of the electroless NiMoB films as a diffusion barrier layer on the low-k substrate. ECS Trans 1:57–67
28. Osaka T et al (2003) Microfabrication of electro- and electroless-deposition and it application in the electronic field. Surf Coat Technol 169(170):1–7
29. Kind H et al (1998) Electroless deposition of metal nanoislands on aminothiolate-functionalized Au(111) electrodes. J Phys Chem B 102:7582–7589
30. Hasegawa M et al (2007) Evidence for "superfilling" of submicrometer trenches with electroless copper deposit. Appl Phys Lett 90:101916
31. Shingubara S et al (2004) Bottom-up fill of copper in deep submicrometer holes by electroless plating. Electrochem Solid-State Lett 7:C78–C80
32. Shimada K et al (2005) Precise measurement of the self-diffusion coefficient for poly(ethylene glycol) in aqueous solution using uniform oligomers. J Chem Phys 122:244914
33. Norkus E, Vaskelis A (1987) Russ J Inorg Chem 32:130

Index

MIX
Papier aus verantwortungsvollen Quellen
Paper from responsible sources
FSC® C105338

If you have any concerns about our products,
you can contact us on
ProductSafety@springernature.com

In case Publisher is established outside the EU,
the EU authorized representative is:
Springer Nature Customer Service Center GmbH
Europaplatz 3, 69115 Heidelberg, Germany

Printed by Libri Plureos GmbH
in Hamburg, Germany